For Phyllis + Jay;
In memory of Maxine + Frank Falenger
who provided our first home on top
of a hill near Coalinga. Fondly
Martha Tracy Frey

PORTRAIT

OF A WESTERN FAMILY

MARTHA TRACY FREY

OLIVE PRESS PUBLICATIONS

Printed in the United States of America by

Thomson-Shore, Inc.

Editing and electronic file by

Jane Suzanne King

THE OLIVE PRESS
PUBLICATIONS

Olive Press Publications
P.O. Box 99
Los Olivos CA 93441
805/ 688-2445

Acknowledgements

It takes more than one!

Yes, it takes more than one to tell the story. I would like to dedicate this one to my entire family, who has assisted me in its formation.

I am indebted to my sister Frances Tracy Selvidge for her numerous contributions of childhood recollections. To her son JB Selvidge I am indebted for his accurate accounting of past events and family history. It is he who is the true family historian and guardian to the site of the original Tracy home.

Thank you Bill, John and David Tracy, sons of my brother Tilton and my sister-in-law Kathryn, for sharing precious memories of your beloved parents. Thank you to Bill's wife, Susie Tracy, and all of the others who have so generously provided photographs for this book, and to John's wife, Donna Tracy, for recording insight into the pleasures and perils of horsemanship.

Thanks to Victor Brown, close friend and childhood companion to Tilton, for his detailed account of their pack trip when, at the tender age of 13, they undertook the treacherous journey to the top of Mt. Whitney, and thanks to Robert Montgomery, another of Tilton's boyhood friends, who shared his colorful reminiscences of summers spent of the ranch as a youth.

A special thank-you to my niece Marilyn Tracy, daughter of my brother Darrel, for her poignant memories of a devoted father.

Thank you to the many members of the community, too numerous to list here, who have supplied many of the details that have added so much to my efforts.

Thank you to Jack Frey, my husband of 64 years, who has always allowed me the freedom to pursue my own interests. To him, and to all of his partners who have put this old ranch back on its feet, I owe my eternal gratitude. Without their efforts there would be no story to tell.

I wish to express my gratitude to my daughter Nancy for having the courage to transcribe and edit this book, and for having the ability to deal with the new technologies that have played a part in accomplishing it. My gratitude, also, to her husband Randy Trahan who spent many hours scanning and formatting images.

Most of all I want to thank the publishers, Jim and Lynne Norris, for their patience and careful guidance in making *Portrait of a Western Family* a reality.

Sincerely,

Martha Tracy Frey

THE INCEPTION OF A BOOK

Five years ago I would have laughed at the idea of publishing a book. After all, I had seen how much work it entailed when my mother, Fannie Tracy, published her book in 1962 at the age of 83. Now, in 2003 at the age of 88, guess who is doing the same!

It is quite by accident that *Portrait of a Western Family* was born. In 1998, our brother-in-law Lawrence Parker was terminally ill at a hospital in Paso Robles. This is where he and my husband's sister had made their home. I packed my suitcase and prepared to help her make it through this difficult time.

Once in Paso Robles, I found myself with a great deal of time on my hands. To fill the void I took pen in hand and began writing about the colorful character that was my eldest brother: "Uncle Cecil." Cecil had enjoyed the title of uncle because, although he had been married four times, he had not fathered any children. Since he now had a number of nieces and nephews, he liked to think of himself as the benevolent uncle.

After their father passed away, Lawrence's two sons, John and Robert Parker, came to their mother's assistance and looked after her until she too passed away in 2002. At that time, we had to say farewell to a sister-in-law who had been an integral part of our family.

By the time I returned home from Paso Robles I had finished writing about Cecil and just kept right on writing. We could have wallpapered the house with the rejects that flooded the wastebaskets or were burned in the fireplace. After a couple of years, I had the ingredients for a book. But like Uncle Cecil's junk pile, it was simply one great big pile of disorganization. Not unlike the chef that had the ingredients but no recipe to follow, I pondered my situation.

It was then that my daughter Nancy came to my rescue. She volunteered to edit the material for me! This gave me the courage to go on. Nancy is a certified gemologist and her office is equipped with all the new technology vital to her business. This would be a necessity in editing a book. With the magic of her computer, one section of the story could be shifted to another position with the touch of a key. Best of all, my poor spelling would automatically be corrected! For more than a year, Nancy sacrificed time from her profession to work on my project. Even her husband Randy spent many hours on photographs for the book. Finally the story of a western family was beginning to emerge. But there was still a long way to go.

It was about this time that my nephew John Tracy handed me a copy of the book, *Don't Get Me Started*, by Henry Twisselman, a pioneer cattle rancher in our area. I was impressed with the book, and when I contacted the publisher, Olive Press Publications, I found it to be a husband and wife team, namely Jim and Lynne Norris. They lived in the coastal town of Los Olivos ("the olive trees"). Since this was not too far away, I invited them to visit our ranch in Buttonwillow.

The Brushfield property lies right next to the large Frito Lay plant on Highway 58, east of Buttonwillow. It is here that our family grows cotton and carrots. They also grow potatoes for Frito Lay and use the waste products for the cattle feedlot next door.

It was a memorable day when Lynne and Jim Norris arrived. As we toured the feedlot together, I snapped pictures of the two of them standing in front of a huge mound of carrots (see page 326). They marveled at the great stacks of potato chips located in the open fields by the cattle pens. The Tracy Ranch feedlot is John Tracy's domain; here he runs a boarding house for other ranchers' cattle as well as our own.

Herds of many colors enjoy the hospitality of John's "restaurant," where they dine on carrots, chopped alfalfa, hay, and whatever else is on the menu of the day. Herds of Holstein cows with their patterns of black and white form a sharp contrast to the knee-deep muck of mud and manure in the pens. White-faced Herefords, Black Angus, and Jerseys are all part of the heterogeneous mix.

The most spectacular of all are the enormous rodeo bulls that dine in the lot between rodeos. These magnificent monsters will reveal their might when they have a cowboy on their backs! No less awesome is the stately Brahma steer, easily distinguished by the pronounced hump on his shoulder. This fellow does not look as if he would back away from a fight either.

Long rows of bright orange carrots stand out against the cement troughs as we pass by in the lemon yellow pickup, but the cattle pay little attention to us; they are munching on carrots, their favorite meal and the special of the day.

After we finished our tour of the Brushfield property, we stopped by the site of the original Tracy home, now the Tracy Ranch headquarters. Although a modern dwelling has risen upon the ashes of the home that burned in 1974, we wanted the Norrises to see the ancient structures that my father, Will Tracy, had built when he was a young man.

We visited the homestead cabin where he had lived with his older brother. We passed by the "old" barn built in 1898 and restored in 1978. It was originally used to house the great draft horses that were raised for market. Nearby is the old blacksmith shop where the horses were shod. We then visited the "new" barn. Will Tracy was proud of this barn that he built in 1915, the very year I was born.

The house belonging to my nephew JB now stands on the exact spot where our father built the first home for his bride, the spirited Fannie Rowlee. She had come to teach at Wildwood. Fannie was 27 and Will was 38. Here she planted the eucalyptus trees that tower over the home site today, and here she gave birth to the six Tracy children.

Still standing is the three-story tank house that provided water as well as a playhouse for the Tracy children. Near the tank house rests Cecil's accumulation of antique farm equipment, now miraculously restored and mounted for display!

Last of all we showed our guests the dinner bell that was used to summon family and ranch hands alike to the table. It has been almost sixty years since my brothers, Cecil and Darrel, erected this giant tuning fork as a memorial to the father we loved so well.

Now we bid goodbye to our new friends, Jim and Lynne Norris. It had been a memorable day and I knew that they were the kind of people I would wish to publish my book.

THEY DID!

CONTENTS

Childhood Memories

The Dinner Bell

One of the fondest memories of my childhood is that of the old dinner bell. It must have been installed on our father's ranch before he was married, for not one of his children could remember it not being in existence. The dinner bell was shaped like a giant iron tuning fork. It was about five feet high and weighed about 150 pounds. It was a standing family joke that whenever a stray cat or dog disappeared from the ranch, the dinner bell must have fallen on it.

Our father had secured this fork from the oil fields, and he suspended it from a tree. When a meal was ready to be served, a "bell ringer" was sent to sound the call. It was rung by alternately striking each side of the interior with a stout iron rod. It was a joyful sound that summoned hungry children and field hands alike to the warmth of the family kitchen. The sound of the bell ringing at odd hours of the day signaled an emergency and a call for help.

A typical breakfast would consist of fried eggs and bacon (from which the rinds had not been removed), perhaps some jam or jelly, and mother's wonderful baking powder biscuits. These biscuits were so light and tender that it was said they could almost float. It was always a wonder to me that she could accomplish this without even measuring the ingredients. The breakfast was served at a very early hour, since the horses must be harnessed and ready for work by sunrise.

Other meals often consisted of chicken in one form or another, the necks usually wrung by mother's own hands. This could not have been a pleasant task, but it was the lot of the countrywoman, and mother was never one to shirk a duty. Chicken with noodles was one of her children's favorites. For this dish she selected an old "stewing hen," a hen that had outlived her ability to contribute eggs to the family larder. The bird was made tender by several hours of slow simmering. In the meantime, dough was made of flour, water, eggs and salt. It was rolled out into a thin sheet like piecrust, then rolled up like a jellyroll and cut into one-quarter-inch slices. The slices were allowed to dry until some child was designated to unroll them. They were then added to the pot with the chicken. The antiquity of the hen, combined with the process of slow cooking, resulted in a dish that was rich and flavorful.

Occasionally our father would slaughter a pig in order to add some variety to our diet. After the pig was slaughtered it was shaved and butchered, then placed in a large iron cauldron over an outside fire. A by-product of the pork was mother's "cracklin's" which were made by scoring the rinds of the pork and baking them for several hours. They were delicious, and not to be compared with the ones we now buy in packages.

Beef was not a staple in our diet, for electricity had not yet come our way, and there were no horse-drawn ice delivery wagons in the country. Consequently, there was no way to refrigerate beef. When our father did butcher, however, he always shared the meat with his neighbors; he was a generous man, and it gave him pleasure to do so. Butchering was usually done in the summer, for a large part of our meat was made into jerky. And how we loved jerky! It did not take six children long to demolish a sack of jerky. It was made by cutting slender strips of meat across the grain, seasoning them with salt and pepper, and then hanging them on a line to dry. We children anxiously awaited the curing of the meat and carefully examined the tiniest pieces, as these would be the first ready for consumption. Jerky remains one of the most pleasant of our childhood memories and bears no resemblance to that now served in bars or purchased at the supermarket.

Since our market was several miles away by horse and wagon, it was a necessity for mother to make her own bread, and she took great pride in doing so. Although meals were not always bountiful, one could usually ward off hunger by partaking of a bowl of the ever-present pink beans and mother's delicious bread. Trips to the Miller and Lux Headquarters for groceries were infrequent, and it was an embarrassment to mother to have to send someone for staples more often than once every two or three months.

Dairy products were usually available, since there were nearly always a few cows to milk. At one time we had a considerable dairy of about fifty cows. My brothers Cecil and Darrel were in charge of the dairy, and they employed one hired hand. One day the hired hand did something that displeased Cecil, and he fired him. That evening, when Cecil was faced with fifty cows to milk, he suddenly realized the error of his ways and hired him back. Cecil always enjoyed telling this story on himself. Cecil and Darrel alleviated the boredom of milking by reciting a long litany of names describing the cows. It would start out with, "You cockeyed, lump-jawed, lopsided, hammer-hoofed, crooked-horned…" and so on. The list of names had to be recited in the exact same order each time; then, as they thought of a new name, it was added to the list. Another form of amusement was to squirt milk from a cow's teat directly into the mouth of a cat waiting patiently a few feet away.

There were times during the Depression when the menu became pretty monotonous. I remember one year when mother "put down" a crock of sauerkraut and another of corned beef. Meal after meal, the menu remained the same until, eventually when I smelled it cooking, my stomach would almost rebel. However, when I sat down to eat, it tasted pretty good; I guess it was better than the alternative of going hungry. Another time, when wealthy duck hunters from Los Angeles sat at our table, mother was challenged to put forth her very best culinary efforts. At times like these, delicacies such as wild duck or apple and lemon pies might be added to the menu; these were real treats, and mother did them very well.

So this was the fare that nourished our bodies when we responded to the melodious sound of the old dinner bell. By today's nutritional standards I fear it would not have been considered a very healthful diet, but we did survive and, strangely enough, we were rarely sick. In fact, nothing short of a broken bone would send us to the doctor's office.

Early Childhood

Because I was the next to the last of six children and the second girl, my childhood was not always sheltered. My mother's older sister, Dolly, had no children, and our home was overflowing with them; consequently, I spent considerable time, as I was growing up, with "Aunt Dee" and her kind and gentle husband, George Taylor. In fact, our brood sometimes seemed to spill over into their home, and we were always welcome there. Quite a belle in her day, Aunt Dee was a little odd in some ways, but she had a kind heart and she was good to me.

In fact, on May 7, 1915, the *Lusitania* was sunk, the United States was drawn closer to the war in Europe, and I was delivered into the front bedroom of Aunt Dee's home at 2323 Chester Lane in Bakersfield. Mother had gone to Bakersfield to visit her sister when she suddenly went into labor, thus depriving my grandmother Martha Rowlee the pleasure of delivering Fannie's fifth child.

Recollections of my childhood include gathering eggs with Teddy, my older brother (he was lost to us when he died of scarlet fever at the age of eight). Chickens were allowed to run free on the ranch, and made their nests in the haystacks. It was very exciting to find a nest of a dozen or more eggs! Later, gathering eggs with my youngest brother, Tilton, was a little different, since this included bird eggs. Bird eggs were a fascination to me—they still are. Our favorite pastime was to search beneath the bridges near our home for swallows' eggs. These nests were made of daubs of mud and lined with softer materials. Orioles made their nests of horsehair, but they hung from high tree branches and were not accessible to us.

I remember one time when I was about eight or nine and Tilton was four or five, and we spotted a nest in an umbrella tree. Since the tree was not very tall, I conceived the idea of pushing the nest out of the tree with a broom, while Tilton was designated to catch it as it fell to the ground. I was able to push the nest free, but Tilton was such a poor 'catcher' that it provoked me into taking after ***him*** with the broom!

About this time there was such a profusion of chicken eggs that I decided to package them and trade them in to our friendly grocer in Buttonwillow. Mother agreed to this plan. However, the eggs had been gathered over some period of time, and although some were fresh, many were not. Nevertheless, I cleaned them all, and they were delivered to the store. I've often wondered if mother informed the grocer as to the vintage of the eggs. If not, some customers were in for an unpleasant surprise.

Ranch life could be pretty lonesome and monotonous for a little girl. I recall the "dog days" of summer, when the heat was so oppressive that even the energies of a small child were very low. When electricity came our way, we acquired a wonderful new fan. We placed it at the end of our long kitchen table, and how we each waited for it to turn in our direction and bring a welcome breath of air. I remember how it would sometimes blow a little bit of sugar out of its bowl and onto the table.

Along with the electricity came the first electric refrigerator and ice cubes! I recall freezing fresh peaches, and what a gourmet treat that was!

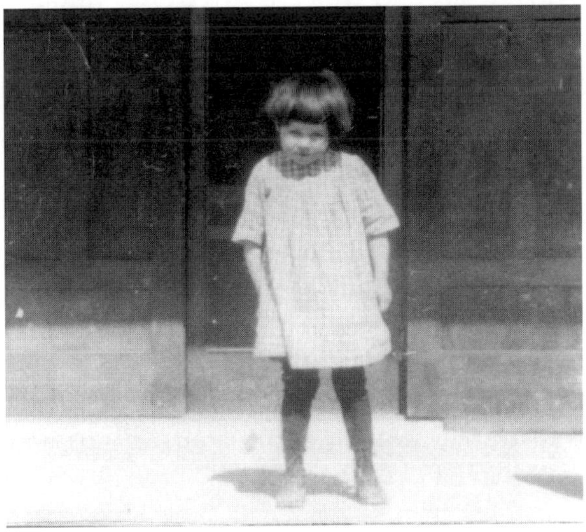

Above left: Camera-shy preschooler, taken at Bowerbank School near Buttonwillow. Right: Asleep in the alfalfa field with Uncle George Taylor, Aunt Dee's husband.

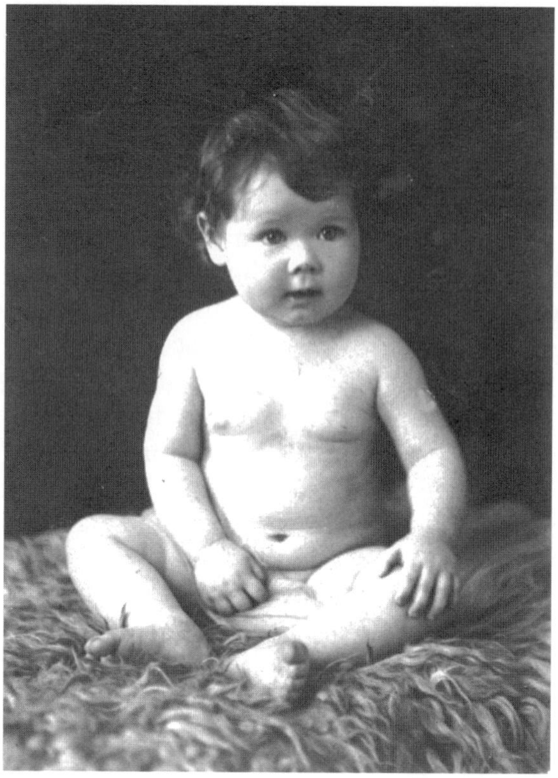

Above: Martha May Tracy, born to Fannie and William Tracy on May 7, 1915, the same day as the sinking of the ship *Lusitania*

Martha Did It

One day when I was about five years of age, we children were all playing in Aunt Dee's back yard. I found a ball of gray yarn and was having a nice time playing with it. I first attached it to a water hydrant, and then I proceeded to wind it around a nearby tree. I continued attaching it to trees and shrubs until I had enclosed a large area of the yard. (The gray yarn had been secured to knit helmet liners for our soldiers in W.W.I. Women had been encouraged to knit these to keep our boys warm as they fought in Germany. I don't recall having seen a finished helmet liner, but I'm sure the intentions had been good. At any rate, I had fun playing with the yarn!)

I was quite pleased with my "pen," until mother appeared on the back porch and demanded to know who had strung the yarn around in such a manner. Without hesitation, Cecil exclaimed, "Martha did it!" Then Darrel said, "Martha did it!" Frances chirped in with the same accusation; even my little brother Teddy told Mother, "Martha did it!" The family reports that with hands on hips I replied, "Yes, Martha did it, and Martha don't care!!!"

Now, mother was not one to take much sass, and I fully expected some retribution. Therefore, I was surprised when she spoke not a word but turned with an amused look and went back into the house, no doubt to share the episode with her sister.

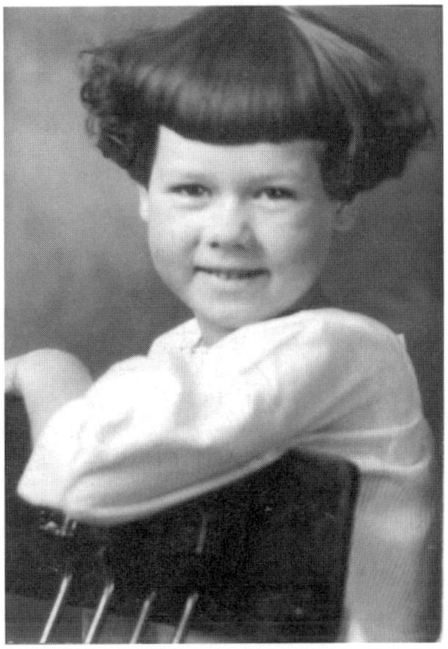

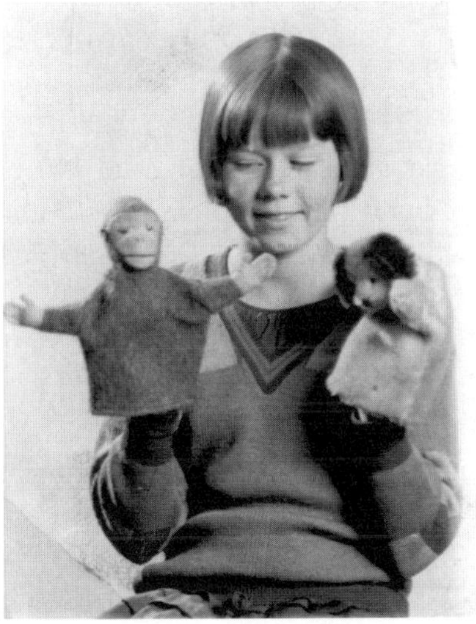

Growing up, a large part of my childhood was spent at the home of my mother's sister, "Aunt Dee," who had no children of her own.

"Baching" in Bakersfield

Most of my grammar school education took place at the one-teacher Bowerbank School where my mother was the teacher. When I reached the eighth grade, however, I was sent to live in Aunt Dee and Uncle George's home, and I attended Emerson Junior High in Bakersfield. This school had an enrollment of about 500 children, and it was quite a culture shock for me. But parents didn't worry about such things as "culture shock" in those days. Since Aunt Dee was then teaching in the nearby town of Delano and came home only on weekends, Uncle George and I "bached." We ate fairly well; Uncle George would bring home some meat and a vegetable or two, and I would make a salad. I rode to school on my bicycle and became so adroit at cycling that I would sometimes ride without touching the handlebars. Since I had no lunch money, I would ride the several blocks home at noon to forage what I could from the icebox.

During this time I was plagued by a chronic sore throat, and mother decided I should have my tonsils removed. One day Mother happened to encounter Cecil's former wife, Florence, who was then working for the county health doctor. Florence told mother that she could arrange for me to have my tonsils removed through the county for $15. An appointment was set up, and when the day arrived I rode my bicycle to the doctor's office. Florence told me that they had decided to remove my tonsils with a "local anesthetic." She said that, since this was a new procedure, they planned to take moving pictures while the operation was being performed. This proved to be one of the most unpleasant experiences of my life.

First they anesthetized my throat, a terrifying experience in itself. While your throat is paralyzed, it is impossible to swallow, and it feels as if you will surely suffocate. As the camera rolled, the doctor first took an instrument that looked like a pair of scissors and clipped my adenoids. He then applied a sharp knife to each side of my throat. As he bore down with what seemed like all of his might, I could hear the sound of the knife cutting through my flesh as I gagged over and over again. The operation complete, I was allowed to lie down in an adjoining room for an hour or so, after which I rode home on my bicycle. My throat was unbelievably sore, and I lay in agony for the next several days. With Uncle George at work there was no one to care for me, but I could not have eaten anyway. I remember watching ants crawl on the soiled newspaper that lay beside my bed. I never saw the movie version of the operation, but when I saw Florence a year or two later she told me how "funny" it was.

Top: In 1925 Fannie Tracy was teacher at the Bowerbank School, located about three miles east of Buttonwillow on what is now Highway 58. The class was composed of some 20 children whose families worked for the San Joaquin Light and Power Company, plus the Antongiovanni family. Martha, at the extreme right, wears the "boyish bob" styled by sister Frances. Above: Later, Aunt Dee taught at Bowerbank. Martha is again at the extreme right.

The Tank House

Our father built the tank house in about 1900. Its purpose, of course, was to supply water for domestic use. It was a three-story building with a windmill at the top. The third story contained a very large tank holding about 500 gallons of water. Although its primary function was to provide water, the tank house provided a great deal more as a fascinating playhouse for the Tracy children. There was only one way to reach the third story, and that was by climbing a ladder made from slats nailed between the studs on an inside wall of each story. By the time I came along the older children were scaling these walls like monkeys. But the exploration of this territory was denied to me, for I was unable to reach from the first rung of the ladder to the second.

I shall never forget the day when I had finally grown big enough to barely work my fingers around that second rung. I had come of age! It was a miracle, and a whole new world was now open to me. What a thrill it was to reach the top and survey the tiny world so far below!

Horned owls were free to rest on the rim of the open tank, and occasionally the skeleton of a frog or two were found at the bottom. It is doubtful that our water would have passed present-day standards of sanitation, but we survived in spite of these conditions and, in fact, we were proud of our "pure" soft water.

Memories abound surrounding the tank house. There was the time that Cecil and Darrel conceived the idea of jumping to the ground from a second-floor window with the aid of an umbrella. This experiment did not prove successful, but fortunately there were no broken limbs. Since the tank house was sometimes used for butchering, meat hooks hung from a first-story rafter. One day Darrell was found hanging helplessly in midair, suspended from a meat hook through his foot!

Another time, an eighth grade party was held at the ranch after dark, and it was at the very top of the tank house roof that I received my first kiss from a boy!

Although a stairway was later added to the exterior, the tank house remains very much as it did in our youth, and is a happy reminder of our childhood.

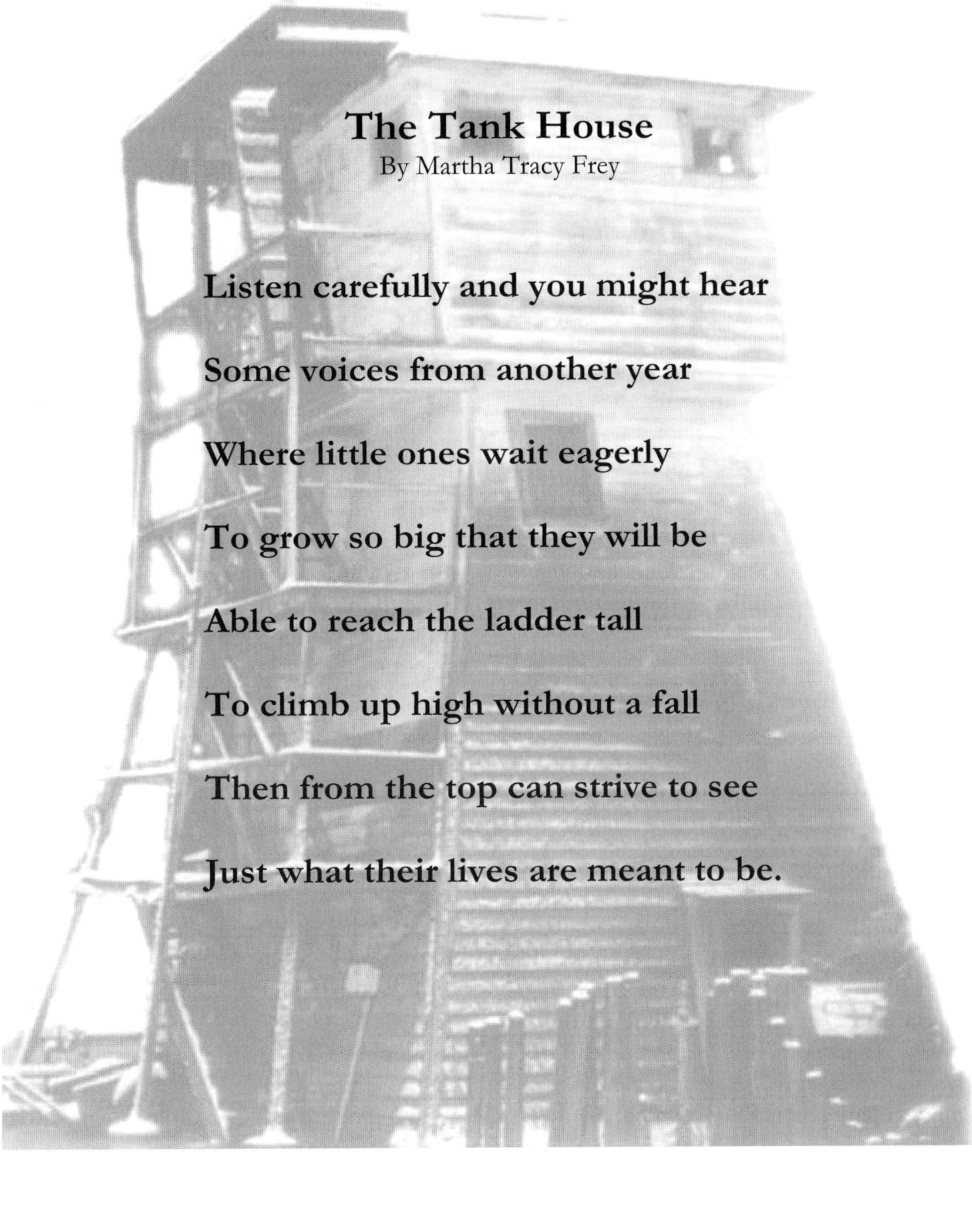

The Tank House

By Martha Tracy Frey

Listen carefully and you might hear

Some voices from another year

Where little ones wait eagerly

To grow so big that they will be

Able to reach the ladder tall

To climb up high without a fall

Then from the top can strive to see

Just what their lives are meant to be.

TRACY RANCH HISTORY

By JB Selvidge

Overview

Most of today's family-held ranches raise cotton, wheat, potatoes, tomatoes, carrots, almonds and pistachios; beef-raising rounds out the operations of the Buttonwillow Land and Cattle Company, a partnership made up of three families: the Freys, Selvedges and Tracys.

Tracy Ranch Incorporated is a family-held corporation owned by widespread stockholders, also Freys, Selvidges and Tracys, and includes female descendants who, through marriage, have assumed new last names.

The family legacy in Kern County began with Ferdinand Tracy (1829-1908) in the late 1850s in the gold mines near Kernville. However, the family uses the official date of 1862 when Ferdinand and Wellington Canfield formed the partnership Canfield and Tracy, a rangeland cattle operation. Their herds roamed the lower San Joaquin Valley, grazing on wild grass in an untamed dominion ruled by the likes of rattlesnakes, jackrabbits, coyotes and waterfowl and, at times, subjugate to the ruthless whims of the mighty Kern River.

In 1898, Ferdinand's nephew, William Tracy, established what would be today's headquarters/historical park on Wildwood Road, five miles northeast of Buttonwillow. This era, marked by the raising of Belgian draft horses and later ostriches, abounds with entertaining tales. His marriage in 1904 to the daughter of another pioneer family, Fannie C. Rowlee, would yield six children, including one son who succumbed to scarlet fever at age eight.

With the death of William Tracy on July 4, 1941, his widow Fannie rallied her children and their spouses back to the ranch. This was a turning point that saved a ranch ravaged by bad luck of the '20s and the Great Depression of the '30s. Fueled by the talents and resources of the Freys, Selvidges and Tracys and made urgent by World War II, the ranch was transformed from an equine epoch into mechanized farming.

Today in the fifth generation, diversity is their strength, the achievement of their unity and of all hearts being in one place—the Ranch. Reaching its 140th year in 2002, the Tracy Ranch is joined today by only a handful of other surviving Kern County operations with roots going back to early California. All still alive with the romance of our Western Culture, they share one common beginning… rangeland cattle.

Tejon Ranch	1843	Tracy Ranch	1862	Snedden Ranch	1864
Lavers Ranch	1858	Robert Woody Ranch	1862	Carver-Bowen Ranch	1869
Nickel Enterprises (descended from Miller and Lux)	1861	Rankin Ranch	1863	Twisselman Ranch	1887

"Beef Endures the 'Taste' of Time"

The Early Years, Part I

1860: Ferdinand A. Tracy arrives in Kern County. Born in Wilkesbarre, Pennsylvania, in 1829, he was attracted to California in 1850 by the lure of gold. In the early 1850s he was a commissioned lieutenant in a company of United States troops sent to quell an Indian uprising in northern California. Later he resumed his gold mining activities in Kern County, but this soon lost its luster and he turned to livestock raising. Becoming a resident of Bakersfield in 1862, he joins with Wellington Canfield under the firm title of Canfield and Tracy, becoming a leader in the cattle industry throughout the San Joaquin Valley.

1875: Ferdinand Tracy unites in marriage with Mrs. Ellen Baker, the widow of the late Colonel Thomas Baker, founder of Bakersfield. Ferdinand would go on to occupy positions of trust in county affairs, but he generally shunned public life. The marriage yielded two children, "Baby" and "Ferdie," who both died as infants.

William Tracy, was born August 8, 1866, near Galt, California. In his teens he had purchased an outfit and engaged in teaming on the large ranches in Colusa County, meanwhile, saving his earnings with frugal forethought to the future. As a teenager William was a teamster in northern California's Colusa County, ultimately driving the large teams hitched to the giant grain harvesters evolving in the 1880s. It was perhaps this exposure that kindled his passion for "beasts of burden," namely mules and the Belgian draft horse. Made a legend as a Persian War Horse, the handsome Belgian was revered for its calm demeanor and especially its short, powerful legs that gave exceptional pulling power and agility. (The mules had eclipsed oxen, the unglamorous brute of the westward movement, and large draft horses due to their pulling power and controllability by means of the mouth bit.)

1891: Will Tracy comes to Kern County. He takes up a homestead located one mile east of Wildwood Road, near Seventh Standard Road, on the range of his uncle Ferdinand. He joins his brother Theodore as superintendents of the Canfield and Tracy herds. Although "Uncle Ferd" was a cattleman, nephew Will's aspirations leaned toward the breeding of draft animals. It was an equine epoch that created ever-increasing demand for such animals in agriculture, commercial drayage and eventual service in WWI.

It is the year 1898. Tracy Ranch is a fledgling thirty-six years old; the horse-drawn carriage is about to become a "horse-less" carriage; Teddy Roosevelt and his Roughriders have defeated Spain in the Spanish-American War in Cuba. The enterprising William Tracy moves his quaint seven-year-old homestead cabin one mile west (on Wildwood Road) toward the setting sun to its final resting place… today's Tracy Ranch headquarters.

After settling into his cabin, his next priority was the construction of a horse barn followed by a nearby blacksmith shop (where horseshoes were forged from scratch). Built in the expedient, functional, California box "balloon" construction (no foundation, no framing) out of redwood, the barn and blacksmith shop would become the hub of a breeding enterprise that included jackasses, mules and the venerable Belgian draft horse.

"Ultimately the Tracy Belgians grew to over two hundred in number. At the head of the drove were two prized stallions: Predominant weighing 1500 pounds and Silver Tip, tipping the scales at a mighty 2000 pounds… both fine specimens of their breed."-Wallace Morgan, Kern County History, 1914.

Frances Selvidge recalls:

The Old Barn, which was built in 1898, was our dad's first building after the homestead house, and he was very proud of it. Dad raised mules and horses and needed a jackass for breeding with mares. He kept the jack confined in a corral to the side of this barn. In 1915 a new barn was built. With WWI the demand for mules increased; a second jack was purchased and confined in a corral on the north end of the New Barn. Occasionally one or both jacks would escape which resulted in great consternation, as they would literally attempt to kill each other by aiming at one another's jugular veins. Loud cries would reverberate through the grounds. "The jacks are out! The jacks are out!" Then someone would approach with a shotgun loaded with buckshot. A slug in the hip would do the job, and both jacks would be escorted to their quarters.

Because of the scarcity of metal during WWI, dad could have sold the galvanized roof of the New Barn (still intact today) for more than the whole rest of the building. Over the years the Old Barn has been the subject of many an artist's brush and can be seen in various interpretations, according to the eye of the beholder.

JB Selvidge, 2000:

In 1978 with its roof half gone and its walls on verge of collapse, restoration began. Care was taken to preserve its character, reminiscent of a period when "patch and make do" was the essence of survival. During restoration, many of its original features were carefully preserved while structurally enhanced with new wood bracing, brackets, cables, and a new cedar-shingled roof. After twenty years, restoration was completed in 1998, its 100th birthday.

1900: William Tracy completes the first phase of a ranch house that would later become home to a wife and family of six children. It is here that his ranch hands have their meals prepared by a Chinese cook. (The current home of JB Selvidge, built in 1984, sits on the site of the original home.)

The horse barn (1898) that would witness 105 years of ranch history unfolding in the shadow of giant eucalyptus trees, stands today restored. Called "The Old Barn," it is an enduring tribute to its builder, William Tracy, in the Tracy Home Place Centennial year.

Tracy Ranch barn under restoration in 1995

Will Tracy's Blacksmith Shop

In due time William would acquire the partial holdings of Canfield and Tracy and the holdings of his brother Theodore. With these additions to his original homestead, the ranch would grow to 3,080 acres.

April 3, 1904: William Tracy marries Fannie C. Rowlee, daughter of nearby rancher (north of Lerdo on Rowlee Road), Charles W. Rowlee. Born in San Joaquin County, she moved to Kern County as a child. She entered Chico State Normal and later graduated from San Diego State Normal. She would become a teacher at the Wildwood School (one mile north of Seventh Standard on Wildwood Road). As the pioneer matriarch of future generations, she would teach her children and grandchildren the valuable lessons of life that she had experienced first hand. Denied access to the ranch kitchen by the Chinese cook upon her marriage to William, she takes to the yard, planting eucalyptus seedlings (the largest standing within the current grounds). She hauls water to them in buckets carried by a small wagon. One of these trees would grow to become the specimen Longbeak Eucalyptus in the United States (173' tall in 1983).

Within a year after William and Fannie's marriage, a son, Cecil, is born. He is followed by Darrel, Frances, "Teddy," Martha and Tilton. (By 2002, all were deceased except Frances and Martha.) Through the years the Tracy family stood together on a ranch torn apart in the '20s and '30s, but perhaps no more daunting than that experienced by thousands of Forty-niners who braved the Great Plains, the Rocky Mountains and the High Sierras in search of their pot of California gold.

"The little tractor that could." The ranch's first little caterpillar stands beside a Cat D8 that was in use in 1983.

Ostriches Arrive at Tracy Ranch

The year is 1906. It is the flamboyant Edwardian era of the early twentieth century, an era epitomized by optimism and innovation: Henry Ford introduces the Model T, the Wright Brothers succeed in flight, Benjamin Holt demonstrates the first successful crawler tractor and the United States is completing the Panama Canal, having taken over the ill-fated French venture. Founded in 1862 as a rangeland cattle operation, Tracy Ranch of Kern County had expanded, c1900, into the breeding of mules and Belgian draft horses and is now forty-four years old.

It is the gilded age of high fashion, lace and corsetry from San Francisco to New York City's 5th Avenue. Wealthy industrialists give their ladies carte blanche to purchase opulent clothing that matches the glory of their own achievements. Fashion designers respond by creating accessory regalia that would adorn the nation's high society, designed around a common element... the rare and exotic ostrich plume.

The ostrich feather legacy dates back thousands of years when it was used as an amulet in Egyptian religion signifying truth, justice and order; with mythical gods portrayed while wearing the ostrich feather as part of their head ornamentation. The feather's first tangible use was as plumage atop the polished steel helmets of European medieval knights. (A plume is an arrangement of several individual feathers.) In 1794 the feathered plume was first introduced to the United States but this died out in 1797.

Over a century later it would make a grand comeback, creating a demand that spawns ostrich farms in South Africa, Australia and the United States. Suddenly the ostrich is the source of exotic pure white feathers which, when several are constructed into a top grade "willow plume," could fetch $100. As a bonus, its speckled hide provides soft fine-grained leather most popular in stylish cowboy boots and handbags.

On the not-so-exotic Tracy Ranch, where fencing pliers and a lariat are of more use than a plume, it seems paradoxical to consider raising birds that are the source of such paradisiacal adornment but Fannie is enchanted with the prospective ostrich business and rises to the challenge. She and William form the Tracy Stock and Ostrich Farm in 1906, thereby becoming the pioneer ostrich ranchers in Kern County. Little did they know what the irascible ostrich had in store!

A mated pair of ostriches, a female named Tempe and a male named Phoenix, are shipped by rail from Phoenix, Arizona. (*Struthio Camelus* is the scientific name for the ostrich species; the word camelus denoting its ability to go for days without a drink, made possible by water stored in cellular tissue throughout its body. The nine-foot-tall ostrich falls in a class of flightless birds along with the rhea and the emu, called ratites. A ratite is characterized by a flat, raft-like chest as opposed to the deep-V breastbone and proportionally larger breast muscles of a normal flight bird... perhaps an evolutionary phenomenon of survival in that a flightless bird requires much less to eat than its flying cousins. However, for what the ostrich lacks in flight it makes up for in "punch." With Herculean force it can splinter corral boards, kill a man in one swift downward kick, or propel itself in twenty-two-foot strides at forty miles per hour, easily out-racing a horse.) The ostriches arrive at the Bowerbank siding three miles east of Buttonwillow. Phoenix has practically destroyed his shipping crate and is fairly thrashed, with feathers everywhere. (He would later die from infection and be replaced with Phoenix II.) The two are off-loaded and driven four miles northward across alkali and sagebrush to their new home on the Tracy Ranch that is perhaps as arid and dusty as their native Africa.

By 1910, the happily mated pair of Phoenix II and Tempe has increased to a flock of 174. The Tracy Studio of Design and Manufacture at 2323 Chester Lane (across from William Penn School) in Bakersfield is turning out an array of multi-colored plumage, satin-backed lap robes, boas, pom-poms, fans, etc., all creations of artisans Fannie and her sister Hazel. Because of this successful enterprise, Fannie is soon noted as "the Bird Lady of Buttonwillow."

1915: With the expansion of the Tracy Stock and Ostrich Farm, William builds a new and larger horse barn 300 feet south of the original. His expanding Belgian draft horse operation, on its way to a peak of around 280 head, has by now gained a far-reaching reputation for quality in commercial and military applications. (Later referred to as the "white barn," it has undergone phases of restoration since the 1960s.)

1917: William and Fannie give a Red Cross fund-raiser picnic attended by 5,000 people. A dance hall is moved from Burbank Road to the ranch, and the party continues around the clock for two days and nights, with two bands switching off. The building would later be converted to a commercial dairy with the product picked up by the Wasco Creamery. (In 1951 the barn was converted to a shop, which is the mechanical shop on today's ranch.)

Ostrich Tales On Tracy and Tejon Ranches

"The Bird Lady of Buttonwillow" and her husband William were not the only ones to stick out their necks. Harry Chandler and General Otis of the Tejon Ranch joined the then-profitable bird-raising business a few years later, providing ostrich plumes for the fashionably well-dressed Fifth Avenue ladies and to those living on remote arid ranches of Kern County.

Able to out run a fast horse and out-kick a Missouri mule, the lanky, flightless ostrich can be a formidable foe. With minimal pectoral muscles its wings are of little use, albeit effective for high speed steering and braking and, as a bluff, intimidating when flapped and thumped. Fannie Tracy, recalling the ostrich's kicking ability in a 1949 interview, was asked how they got the birds to stand still long enough to be plucked. She explained that they were first lassoed, then blindfolded and placed in a special pen where the plucking or clipping was performed. This operation caused no pain to the bird, and the feathers grew back. During the interview she was asked the obvious question, "Did you ever see an ostrich hide his head in the sand?" She and her sons Cecil and Darrell looked at each other and smiled. All concurred, they had never seen this performed by a captive bird.

One moment strutting with dignified countenance, the male ostrich—during mating season or while guarding a nest of eggs—can suddenly turn in tempestuous ferocity. With giant thighs each swinging two long toes tipped with sharp toenails, the ostrich is a fighting machine that can slice open, knock out, or trample its foe to death. "When I was a young girl," Frances later recalled, "daddy came in from the fields one night with the front of his coveralls ripped open top to bottom. In those days the ostriches grazed in the same pastures as the milk cows. That particular evening as daddy went out to gather the cows an aggravated male ostrich assaulted him. With its characteristic style of first kicking forward in an upward arc and then slamming downward, the bird's sharp toenail luckily just missed daddy's flesh!"

Upon his release from the armed forces at the close of the First World War, little did Perry Sprague know what a hand fate was to deal him when he accepted a job by Tejon Ranch as caretaker of their troop of 186 ostriches. The year was 1919. Like those in the Tracy family, Perry never saw an ostrich hide its head in the Tejon's sand, dirt or alkali. When interviewed in 1949, he took great relish in relating how an infuriated male that he had foolishly struck had chased this "tough" marine veteran of WWI out of a field. If he hadn't climbed in between the horses of his plow team the ostrich would undoubtedly have killed him. "Those big Nubians often weigh 350 pounds, and when about to strike they rear up to a height of ten to twelve feet. They kick forward and then down with a terrific force. I've seen them shear off one-and-a-quarter-by-six-inch corral boards with a single stroke. You can imagine what that would do to a man!"

"What about their speed?" Perry Sprague was asked. "Can they keep up with a horse?" Recalling a humorous circumstance, he replied, "Frankly, I don't know just how fast they can run, but I chased a Nubian for forty miles one day with a car that would do slightly over fifty miles per hour. He could run faster than that car at top speed. As for matching a horse! Why, on one occasion I saw a fast horse with a rider overtaken. Then every few strides the ostrich would leap into the air and try to kick the fellow out of his saddle. Chased him a mile and a half clear into Rose Station before the bird would give up!"

Cecil and Darrel related how they had been forced to shoot a pursuing bird on one occasion, or face the consequences.

The post-WWI era marked the waning popularity of the ostrich feather. With the increasing popularity of automobiles, ostrich-plumed hats would die out and, by the early 1920s, so would the entire industry. The birds of Tracy and Tejon ranches were dispersed; many of those on the Tejon Ranch died of greasewood poisoning, some were sold to carnivals, and all except two of the Tracy troop were donated to the Hart Memorial Park. A few birds were kept at the Tracy Ranch as curiosities and the original pair, Phoenix and Tempe, lived out their lives here. In 1946 Tempe died of infected wounds after dogs chased her into a barbed wire fence; Phoenix died of pneumonia in 1952 after an exceptionally wet winter.

So ended the drama and the trauma of ostriches on Tracy Ranch; an era spanning the turn-of-the-century opulence, two world wars, the Great Depression, numerous floods, and a struggle for survival on the ranch kept alive by an elusive dream.

1920-21: William, against the advice of Fannie, invests heavily in the newly formed Imperial Livestock and Mortgage Company and in the Verdun Parking Company, both represented by persuasive slick professionals. Both firms fail, plunging the ranch into a twenty-year struggle for survival.

Cecil Tracy, the adventurous eldest son of William and Fannie, strikes out on his own to become a range cowboy for the giant cattle barons, Miller and Lux. He would later ride broncos and bulls at the local rodeos.

1926: William sells off property to generate cash flow. Section 5 (southwest corner of Seventh Standard and Tracy Avenue) is one of the parcels to go, bought by a consortium of affluent sportsmen from Los Angeles. Chosen because of its natural ponds and wildlife, the original forty shareholders agree, on January 28, 1926, to purchase the property for $16 per acre; naming the 660-acre (oversize section) "The Buttonwillow Land and Cattle Company, Inc." The years following would see duck hunters and entertainment personalities such as the Warner Brothers of Warner Brothers Studios visit the ranch, where Fannie Tracy contracts to "room and board" them, generating a supplementary income.

1928: With the ranch reduced to less than half of its peak acreage, the ostriches and the Belgian horses gone, little is left for the sustenance of a large family. Fannie and William sell the home place on Wildwood to a Mr. Napoleon Merritt of Los Angeles, whose son becomes caretaker. Within a few years, Fannie quitclaims the contract after learning of the neglected facilities, acreage and livestock. William's health at this time is beginning to fail (early 1930s).

1938: Much of the ranch is leased out for oil and gas exploration. It is this income that brings hope to a bleak situation. Cecil proceeds to remodel the ranch house, blending several classic appointments, such as a Victorian staircase, hardwood flooring and a giant red brick fireplace with a seven-foot hearth. A born romanticist, Cecil adds the final touch to his love nest by bringing in an Art Deco, 1930s, Wurlitzer jukebox and an eight-foot square plush mattress pad which he places in front of the giant open hearth. His flair for life and adventure is reflected in the Tracy home that, in later years, would see many legendary parties.

William Tracy has become an invalid, and Cecil spends much of his time helping to care for him as well as tending to ranch duties.

Turning Point

A turning point out of the abyss of the 1930s depression era and the beginning of a long uphill climb commences which, by the turn of the twenty-first century, has transcended four generations. Tracyism: "The true meaning of life is in the struggle."

1941: William Tracy, who had a stroke in 1930, is committed to Camarillo State Hospital where he passes away on July 4, 1941, at age 74.

1942: In 1942 World War II threatens to take all available young men to the battlefield. On Thanksgiving Day of that year a family council is held. It is deemed absolutely necessary that at least one male member of the family must stay with the ranch. To assure this objective all members of the family, including in-laws, agree to come to the ranch from their various locations in the hope that at least one might be left for the management of the ranch operations. It is decided that a family partnership would begin on January 1, 1943.

As agreed, family members come from San Diego, Hollywood and the oil fields of Coalinga to launch a new operation of farming. Fannie, who along with Cecil had shouldered the ranch burdens through the Great Depression, spearheads the reorganization. At 1,200 acres, the ranch stands at less than half of its 1922 size of 3,000 acres.

It is in this year, 1942, quoting Cecil Tracy later: "just twenty years from the first stunning blow, the ranch paid off the last red cent that held the 'hangman's noose' together."

1943: Mrs. Fannie Tracy applies for a $23,000 loan from the Federal Land Bank of Berkeley. She receives a letter dated December 7, 1943 stating "we regret our inability to be of assistance to you at this time," citing that too many families were dependent on a certain producing acreage. The solution: develop more acreage. Jack M. Frey coaxes a $10,000 personally guaranteed loan from good friend Mrs. Baker of Baker's Transfer and Storage, Bakersfield.

The ranch's first cotton, under Jack's direction, is planted the spring of this year, with no more than a Farm All 'M' wheel tractor (still in operation in 1994) and a Caterpillar '15' crawler. A two-bottom moldboard plow with an innovative rear wing acts as dual-purpose ditch plow and border-maker. Stifling mosquitoes, floods, bullheads, and rutted-out alkali roads mark the era.

1946: Fannie Tracy applies for a loan of $25,000 from the Federal Land Bank that, for the second time in a letter dated January 10, 1946, declines the loan, implying that the ranch is further out on a limb than in 1943. The ranch's first new tractor, a model 'LA' Case wheel tractor of 55 DBHP (drawbar horsepower), along with a 10' offset disc, is purchased from E.O. Mitchell of Arvin, California. This is a step toward big-horsepower tillage.

1949: Long-awaited mechanical cotton pickers are introduced on the ranch. Called a "Red Okie," one machine replaces forty hand-pickers. The machine is the result of pressure put on International Harvester Company by San Joaquin Valley farmers who, beginning in the 1930s, had been experiencing labor problems. Cotton farmers were blamed for mass unemployment after picking season, which degraded into the "slum" camps—nicknamed "cardboard cities"—that were described by John Steinbeck in his book *The Grapes of Wrath.*

Tracy Ranch, Inc. is formed out of the original 1943 partnership.

The Fabulous '50s

We saw how the ranch was rescued in the early '40s from twenty years of hard times. The

Field of Dreams is written as a tribute to Jack Frey who came to the ranch in October, 1942, with his wife Martha (Tracy) Frey, and to those deceased who, since 1862, have put their heart and soul into Tracy Ranch.

The diversity traditionally marking the Freys, Selvidges and Tracys has been the ranch's strength… where the bonding philosophy has always been, "When in doubt, do what is best for the Ranch." And, as Fannie Tracy once said, "Don't worry about day-to-day inequities, for in the end it all works out."

This was the spirit that ushered in the memorable 1950s, when unprecedented expansion began… something new, something old, as ground lost in the 1920s was eventually regained along with the new. Abundant water was cheap, mechanical cotton-pickers were revolutionizing the cotton industry and farming in the Golden Empire was on a roll. On a roll also was the mighty Kern River, unleashing a small part of its fury through the ranch, where our next story begins:

1950-52: The Great Floods

On November 19 and 20, 1950, heavy rains in the High Sierras raise the Kern River to an all-time high 37,000 second-feet record flow (officially, since recording of flows at first-point near Hart Park) nearly flooding Bakersfield.

The ranch's usual problems are suddenly trivial compared to one rising problem: waters behind the historic, wooden Bellevue weir just upstream from the current Stockdale Highway bridge are about to overflow into Jerry Slough (historically, Gooselake Channel), threatening farms through Rosedale and, twenty miles further west, the Tracy and Wegis ranches.

There is no safe haven for dozens of dozers and scrapers as they are rounded up on a Sunday. A lone International TD-18 dozer with a sign 'Rexroth and Rexroth' on top the hood sits on a mound of dirt near an oil well sump. Unable to locate its owners, the tractor is borrowed that Sunday night. The following day its angry owners track its disappearance several miles north, where they find it building a levee. They soon calm down after observing the situation, leaving with a handshake and permission to "use the tractor as long as needed."

With the combined horsepower of Cat 'D8', International 'TD-24', and Cletrac mammoth crawlers, it is possible to channel, just in time, the Kern's flood waters through valuable cropland with minimal damage. Later, in May of 1952, a second flood (ranging 5,200 to 8,300 cfs—cubic feet per second—at "first point") flows down the same channel. However, a forty-foot-wide break in a levee north of Lerdo Highway floods acres of cropland. The break is ultimately plugged with the aid of several earthmovers and dozers working into the night.

Isabella Dam was completed in April 1954, greatly reducing the danger of flooding.

The Field of Dreams - Brushfield 1952

While Jack Frey is a pragmatist and a man with a mission, his brother-in-law, Cecil, is a romanticist and a man with a vision. Born in 1905 on the ranch, Cecil's adventuresome and colorful life and character could no doubt be turned into a Hollywood movie. He has a keen sense of nature that ultimately leads to the development and later acquisition of some of the finest farming ground in California.

Cecil was always captivated by the lush mesquite and Buttonwillow trees and the hundreds of acres of cocklebur brush lying five miles north of Buttonwillow on Highway 78 (now 58) in the Gooselake Channel flood plain. He is personally acquainted with the landowners, the "Kern County Land Company," with offices at 19th and H Streets, Bakersfield. After years of coaxing, Cecil finally gets Kern County Land Company to agree to a development-lease in 1952, as it has a standing policy to never sell ground (until years later, when they were bought out by Tenneco).

Tracy Ranch has little capital to work on, so a crop-share agreement is formed. Initial survey prior to clearing is accomplished as Cecil anchors a stepladder in the bed of his 1950 Ford truck where, from his perch, he directs the driver through ten-foot-tall brush. The fine, sandy silt soil, unlike any other on the ranch, necessitates running on half-deflated tires. The simple name "Brushfield" is coined, and it would become synonymous with a cornucopia of high yielding crops ranging from cotton, wheat, rice, alfalfa and, later, carrots, potatoes and tomatoes.

Today, the multi-thousand-acre Brushfield is a "food and fiber machine," the flagship of the ranch, proving the premise that "not all soils are created equal." Family member Wes Selvidge inherited its management in 1967 and, later in 1986, promoted and developed a highly successful potato operation. Growing in size each year, Wes's vision, like Cecil's, is coming to fruition as he ships direct to all of Frito-Lay's west coast plants, to Eagle Brand in Visalia, and to In-N-Out Burger plants in Los Angeles. Jason Selvidge is following his father Wesley's footsteps, assuming more and more day-to-day duties.

Skip Selvidge

Built around 1900, the old Tracy home went through many changes over the years. This is a 1937 photo looking east toward the house and grounds. The "New" Barn was built in 1915.

A Note from Nancy (Frey) Trahan, First Editor:

As I was completing the first transcript of mother's book, I needed to look through some of the genealogical records handed to me for safekeeping from my Aunt Frances—Frances (Tracy) Selvidge. I had filed this particular envelope without examining the contents. It turned out to be a delightful discovery, and I called mother excitedly.

"You'll never guess what I'm holding in my hands!" I exclaimed. "They are on old, yellowed paper about to fall apart, and they have Grandma Tracy's notes compiled by a lady named Mary Gregg."

Like the Dead Sea Scrolls found in a forgotten earthen jar, these yellowed notes add new insight to known history. The notes date from the late 1950s to 1961, and Mary Gregg collected them during the first attempt at setting down the history of Fannie Tracy's colorful life. In them, Fannie relates many anecdotes not included in her book, *The Tracy Saga.* I have interspersed some of them into the text because they offer a more intimate view of Fannie, and others because they provide a glimpse of life in the Bakersfield era starting twenty-five years prior to Fannie's birth.

Many are fragmented; most have at least some of the pages missing. Some are Mrs. Gregg's typewritten notes, executed in a kind of "shorthand." I have tried my best to extract the information faithfully. My favorite passages, however, are the ones written in Fannie's own hand, as they reflect her fine sense of story telling.

Among the notes were transcriptions from *The Morning Echo,* an early Kern County newspaper that was a precursor to the *Bakersfield Californian;* Ellen (Baker) Tracy—Aunt Ellen to the family—was a frequent contributor to this newspaper in her later years. (Although she was not a blood relative, Ferdinand Tracy's wife, Ellen, was the first family member to settle in Kern County, and her memories set the stage for those who followed.)

Mary Gregg and Grandma Tracy's first attempt at a book went very slowly, as is alluded to in some frustrated communications between them. Mrs. Gregg's apparently failing health eventually stalled the attempt altogether, and Fannie turned to local historian Glendon Rogers to help her write her autobiography.

Ellen Baker Tracy

Ellen M. Baker Tracy ("Aunt Ellen")

Excerpted from Kern County Historical Society Publication Number Seven, May, 1944:

Ellen Alverson was born in Ann Arbor, Michigan, on December 21, 1837, the daughter of Dr. Labon and Charlotte (Graves) Alverson. Her early life was spent in Michigan where she married, at the age of fifteen, a Mr. Whalen. As a bride she came westward overland by ox team via the route of the Donner party. She gave birth to one child, a daughter named Mary Emmaline (Mrs. H.A. Jastro). Not long after the birth, she was widowed, and in 1857, she married Colonel Baker. In 1863 she showed her courage and devotion to him when she accompanied him to the present site of Bakersfield, where she aided him greatly in establishing that city.

By her marriage to Colonel Baker, Mrs. Baker had three children, Ellen, called Nellie (Mrs. C.C. Cowgill); Thomas Alverson; and Charlotte Esther (Mrs. J.M. Jameson). The story is told that when Colonel Baker first contemplated his move to "Kern Island," he hesitated to take his young wife and family to that wild country until he had first prepared a home for them. Mrs. Baker heard of his scheme, and quoted to her husband Ruth's answer to Naomi in the bible story:

> *For whither thou goest, I will go; and where thou lodgest, I will lodge; thy people shall be my people, and thy God my God: Where thou diest, I will die, and there will I be buried; the Lord do so to me and more also if ought but death part thee and me.*

And then she added with an emphatic tone: "If you are going into that God-forsaken country full of Indians, bandits, and Heaven only knows what else, I am going with you." And she went.

At the time of the death of Colonel Baker on November 24, 1872, he asked his friend, Ferdinand August Tracy, another leading pioneer of Kern County, to look after Mrs. Baker and to take good care of her and her family. He performed this duty faithfully by assisting Mrs. Baker in her business affairs. About two years after the colonel's death, the Widow Baker became Mrs. Tracy, and the two lived happily together until Mr. Tracy died on January 10, 1908. Mrs. Tracy then went to live with her daughter, Mrs. Charlotte E. Jameson, until the death of the former on September 6, 1924. Mrs. Ellen Tracy published the following passages in *The Morning Echo* around 1900:

> While there is much to be endured in this testing-time of the foot soldier of civilization as obstructions are removed, and the way prepared for the onward march of time, yet the compensations are not a few. The pioneer gets closer to the heart of nature, and through this communion feels the omnipresence of the Great Father. Words are inadequate to describe the peculiar charm of the unsettled country.
>
> California—wave-washed on the west with her noble mountains and beautiful hills; her fertile plains and flower-decked valleys; her picturesque lakes and streams; her dense forests, home of the giant cedar; her fair groves of oak, pine and other members of the tree family; indeed, every variety of scene and climate; the paradise of bird and beast—California in her virgin beauty was well-fitted to exert that indefinable influence upon the spirit of man and cause him to exclaim: "Truly, this is God's country." But due to the "gold excitement," man, as a rule, sought California with the muckrake in his hand and eyes bent upon the sordid. Now and then one came, however, whose soul responded to the wonderful power, and whose eyes looked down through the vista of years and saw visions.

Kern Island

> It may interest you to know that at some time Kern County was a part of Tulare County, and Bakersfield with its surrounding country was known as Kern Island by reason of the river on the north and large sloughs on the other three sides. In 1865 Tulare County was divided, and the southern division was given the name of its principal river—Kern. Havilah, a

mountain town, was appointed the county seat. The mountain sections were the most popular, since the valleys were considered unfit for human habitation, and were given over to great droves of cattle that fattened upon the luxuriant vegetation. Finally a few venturesome families braved the malaria of the rich damp soil, and settled in the valleys, and found that death was not "lurking in every passing breeze."

Colonel Baker

When Colonel Thomas Baker came in sight of the Kern Valley, he was delighted at the prospect. Pointing to the distant line of timber that fringed the river, in the spirit of prophecy, he exclaimed, "This is God's country! Some day it will be filled with happy homes. The largest town south of Stockton will have its site here. Three or four railroad lines will come through those mountain passes and center here. The place is rich in future possibilities."

How improbable did it all seem then. We had come to this place with teams and camped out on the journey, for there were no houses this side of Porterville. It was a tiresome, three days' journey with the loaded wagons. Of course there were no good roads, as now—only uncertain trails to be followed.

Colonel Thomas Baker, founder of Bakersfield

The Early Home

When we reached the desired spot, we found not shelter that could sustain the dignity of the term house. There was no lumber within a hundred miles, and windows were an unknown luxury for these parts. An old shanty built of poles planted in the ground, and thatched with tules, or cattail flags, windowless, doorless and floorless, must be our home for the time being. We moved our belongings into the one room, and then built on a brush shed to serve as a kitchen and dining room. There were two or three other families here at that time and they occupied similar abodes; but for this, we could not have thought it possible to exist amid such surroundings. We had sold all of our furniture save a few of the most necessary articles, and our housekeeping was a very primitive type.

One hundred miles to the nearest store made shopping an impossibility. If our provisions ran short, we must wait for the freight team, although sometimes a spring wagon was sent out to meet the slow-moving freight, and a few pounds of the most necessary provisions brought on ahead. What a delightful change from several days of the potato-bill-of-fare! The freight team consisted of a large wagon drawn by six mules that brought our supplies from Los Angeles, taking two weeks for a journey that is now accomplished in a few hours. Freight was three cents per pound.

We were not able to afford wheat bread when flour cost $10 per ten-pound sack; therefore, we used corn instead. This we ground in a small mill. Many and many a time, when about to prepare a meal, have I first shelled a panful of corn, then gone out and ground and sifted it. The fine part was used in making corn bread, and the coarse part cooking it for hominy. Sometimes our coffee consisted of sliced sweet potatoes browned in the oven, and it was not a bad substitute. Honey took the place of sugar. We had moved a couple of hives of bees from Visalia, so we were well supplied with sweetening. Honey cakes and candy were some of our luxuries.

High Water – 1864

Speaking of high water in the river reminds me of another little incident that happened in the year 1864. We were returning from Visalia and when we reached the Kern River we found a stake on the bank with a note fastened to the top, saying that the river was dangerous, and not to attempt crossing in the night. As it was late in the evening, we heeded the warning, and camped to await daylight. When morning dawned, we found a small boat tied to the willows. In this we crossed, and in the same manner brought the goods over; then swam the horses across the river, and finally long ropes were tied to the wagon, the horses hitched to the ends of the ropes, and the wagon pulled through the water.

About twenty-five years ago, the channel of the river was ten or twelve feet deep, but now the riverbed is filled with sand to a level with the surrounding country.

Mail Service

Being without a post office, of course we received neither letters, nor papers, and often for weeks we heard naught from the outside world. At the time of the assassination of President Lincoln we had received no mail for six weeks and knew nothing of the nation's sorrow. One day a Mexican swam his horse across the river—the banks were running full—and told us that the President had been killed. We supposed it to be the president of Mexico until some friends in Visalia sent us a bundle of papers with the belated news. Often letters that were directed to us went the round about way to Los Angeles, and came to us with the edges all worn and ragged. I made a solemn vow that when the post office came to us so that we could have even a weekly mail, I should take all the papers and magazines that we were able to afford to make up for lost time. At the present time we take about twenty periodicals.

In those early days there were no churches, no schools, no doctors in the place. It was four years after we came before we had the privilege of listening to a sermon, and about this time the few settlers determined to have a much-needed school. A log cabin was built in the center of the settlement about three miles from the present site of Bakersfield, a lady installed as teacher, and about one dozen barefooted children attended. The school-room was fitted out with rude puncheon seats and homemade furnishings. One warm day, when the children were busied with their lessons, a large rattlesnake crawled into the room between the logs. The frightened children ran out through the door, and the brave teacher summoned the larger boys to help dispatch "his snakeship." For some time thereafter the little folks were afraid to sit with their feet on the earthen floor, so without let or hindrance, they perched Turk-fashion on their puncheon benches, until with the passing of time their fears ceased.

Home Remedies

There was more or less sickness in the few homes, principally chills and fever, but with medical aid a long journey distant, the people became somewhat skilled in the use of simple home remedies. Quinine and calomel occupied the biggest part of the family medicine chest.

Stage Line

In 1868 a stage line was started from Visalia through the mountains, by way of Glennville to this place. A post office was built and the mail came in three times a week.

By this time it was discovered that the country was adapted to fruit raising. The finest peaches in the world grew here, and nearly everything in the form of fruit, vegetable and grain grew and flourished. The jungles were cleared, orchards planted, and grain and alfalfa cultivated. It now became a healthy country, and people from other places heard of our wonderful resources and came to make homes. Houses, schools and churches were built.

New County Seat

In 1872, the county seat was changed from Havilah to Bakersfield; one of the three or four continental railroads was built, and from a howling wilderness, patient industry brought peace and plenty.

Colonel Baker, after whom the town was named, did not live to see all his wishes fulfilled, but all his expectations, I believe, will in the course of time come to pass.

Oil Discovery

Years before the sensational discovery of oil in the Kern fields we had a certain knowledge of the oil deposits, but the expense of development was beyond our financial ability; also there was no available market. At that time we did not know all the various uses to which the oil could be put, especially that of fuel, nor that an oil well could become as valuable as a gold mine. I think that the development of the oil fields is yet in its infancy; and I believe that Bakersfield is situated over an immense lake of oil, and that at some future time deep wells will be bored to tap this source of wealth.

Street Railway

The Bakersfield of today, compared to the Bakersfield of twenty-five years ago, shows wonderful improvement. Now we have fine brick blocks instead of the humble wooden structures that served at that time. Gas takes the place of lamp and candle in some homes, while the electric light has superseded gas generally for illuminating purposes. The Kern River has been harnessed to run our street railway system, a thing not dreamed of twenty years ago. We have more telephones in use in Bakersfield than in any town of its size on the coast; today anyone may have his choice of a church home; our school system is the equal of any; the sick and infirm are cared for in an excellent hospital; we have two daily papers, a free library and, according to the times most modern, a Woman's Club.

Martha Martin Rowlee, Pioneer Woman

Early Beginnings

My maternal grandmother was Martha "Mattie" Ann Martin Rowlee. She was born July 18, 1850, in Philadelphia, Pennsylvania of Lydia Moore Martin and John Martin, both of prominent Philadelphia Quaker families. Martha had two older brothers, David and John, and a younger brother, Benjamin. One of her grandfathers was a tailor by trade and is remembered by the family to have been a friend of Benjamin Franklin. He is said to have made suits for both Benjamin Franklin and Daniel Webster. Until the Tracy home place burned with most of the family relics in it, we had the shears that were used to make their suits. Still barely surviving is a weather-beaten table where, according to Fannie Tracy, Ben Franklin used to take tea with Martha's grandfather.

When Martha Martin was still a young girl, her father John Martin thought that Philadelphia was becoming too "citified," and was a bad influence on his boys. He traveled to the Kansas Territory with the thought of resettling his family on some of the land that had just been opened up. He apparently found an opportunity to his liking, for he got word to his wife, Lydia, to send him her $2,500 cash dowry, which she dutifully did. When she had heard nothing from him after many weeks, Lydia, with her two youngest children (Martha and Benjamin) and her brother, re-traced his route in a covered wagon.

Kansas had not yet become a state; it was considered "open territory," and was a pretty wild place. When the family arrived, they learned that their father had been murdered. They were warned that evil forces prevailed, and that it would be best for them to leave the area as quickly as possibly. That night, as they lay terrified in their beds, the team of horses that pulled their wagon became highly disturbed. Grandma Martin hid little Martha and baby Benjamin inside cracker barrels and, although she was a Quaker, she took up her rifle and went to the front of the wagon to defend her family.

During those frontier times there were bandits that would prey on the pioneer families. One ploy was to steal their draught teams at night, then come around the next day offering to sell the hapless family its own team back again! This extortion was usually effective, because the family would prefer to buy a familiar team rather than taking a chance on buying a new team of questionable training.

Apparently someone was trying to steal the Martins' team. However, when they saw a woman in her white nightdress and flowing hair emerge from the wagon with the long gun in her hands, they decided to take no chances and cleared out fast! The little family survived the night, then fled early the next morning without trying to locate and claim the body of Martha's father.

Pioneer Wife

Eventually, Martha's widowed mother Lydia married the kindly John Bullock, and the family migrated by covered wagon to Iowa. Because of her strict Quaker upbringing, Martha was well taught to perform all womanly duties. She learned the preservation and preparation of food, was taught to knit and crochet and, learning the art of quilt making, she would become a fine seamstress. In addition, she had an inherent gift for nursing. These qualifications were to be severely needed during the pioneer life that lay ahead.

Martha also had a remarkable singing voice, and many compared it to that of opera singer Alma Gluck. Her uncle offered to sponsor the training of her beautiful voice at the Chicago Conservatory of Music, but her parents deemed this unseemly for a young Quaker lady. Instead, her talents were directed toward the practical arts of wifery, motherhood, and sound moral values.

Martha had the first glimpse of her future husband when he came to her parent's home to be interviewed for a teaching position. Curious, Martha pierced a hole in a window blind with a pin, and peeked at her husband-to-be through the tiny opening.

Charles W. Rowlee (b. 1848, D. 1933) and Martha were married on December 24, 1869. Their two eldest daughters, Nellie and Dolly, were born while they lived in Cedar Rapids, Iowa, and Martha was already pregnant with Fannie, my mother, when they migrated to California. Grandfather had given up school teaching to become a wheat farmer. They settled in the San Joaquin Valley, and here Fannie, Gilbert and Hazel were born. They lost an infant boy at birth. The young family began a long and arduous search throughout the San Joaquin Valley for suitable land to homestead.

While Charles searched for more productive land, his wife and children endured very primitive circumstances. Time after time the family was uprooted and moved from one desolate sagebrush area to another. Grandmother Rowlee had an inherent love of flowers, and she valiantly tried to carry her plants with her as they moved from one deserted cabin to the next. Finally, the family settled permanently in the Jerry Slough, about ten miles north of Buttonwillow. (Also referred to as Jerry's Slough, this was a channel cut by the floodwaters of that day. It simply took its name from Jerry Bush who lived upon its banks.) Here they finished raising their family.

Years later her daughter, Fannie, would illustrate the spirit of her brave little mother and the difficult pioneer life they led in her book *The Tracy Saga*:

> In later years when my mother saw a big dust [cloud] that seemed stationary on the horizon, it indicated to her there was a bunch of cowboys working cattle, so she prepared immediately to feed them. This meant putting on a large pot of beans and plenty of good bread and coffee. When the sun indicated it was midday, the cowboys rode in pairs to eat while the others stood guard. The "chuck wagon" always followed those rodeos, but most often the boys would be many hours ahead of it, and would have no food unless they were encamped nearby and welcomed by a settler. The hospitality of my mother became so well advertised, Miller's foremen ordered the Miller and Lux chuck wagon to carry an extra quarter of beef and drop it off every time it traveled our way. This continued for many years and was indeed a great help to our family.

By now my mother's eldest sister, my Aunt Nellie, had married George Tilton. She bore him four sons in rapid succession. Worn out by child bearing and difficult circumstances, Nellie died while her children were still very small. Fannie and Dolly grieved for the loss of this sister whom they loved and admired so much. During this period they helped care for the four little Tilton boys until their father remarried. Because of this close association, Fannie would choose "Tilton" as the name for her youngest son.

Quilts and Carpet Rags

In early days the ingenuity of pioneer women was illustrated by their use of fabric scraps. In the beginning, quilt making was a necessity, but in time it became a creative outlet to lighten the burden of their dreary lives. During her lifetime Martha Rowlee created many quilts, one or two of which are still in existence. Less popular was the pastime of weaving "carpet rags." Here again, nothing was wasted. Clothing worn beyond the point of repair was cut into long, narrow strips and tacked together with needle and thread. These multicolored strips were then rolled into balls and became the weft of a carpet loom.

Carpet looms were a yard wide, and fortunate was the lady who owned one. After the yard-wide piece of weaving reached the proper length, it was removed from the loom and another took its place until there were enough to make the desired width. These were then joined together to form the finished carpet. I can remember seeing one such loom in my grandmother's home, but I can't recall ever seeing a finished carpet. By the time I was a small child the hardships of her life had, no doubt, left little time for a luxury such as rag carpeting.

Educating the Girls

Due to the family's migratory life and the poor schools of the day, the Rowlee children's education was sadly neglected. Dolly had managed to graduate from the 8th grade by "boarding out" with other families, but Grandma Rowlee had produced a rebel in Fannie. Although she was a dutiful daughter in every other way, this horseback rider of the sagebrush did not take to the confines of formal education. More of her time at school was spent in creating mischief than in studying.

When she was fifteen, Fannie suddenly realized the error of her ways. Barely able to read or write, she decided she wanted to become a teacher. The nearest school, however, lay six miles from their home. She persuaded her parents to allow her and Dolly to move into a deserted little shack that was closer to the Wildwood School. The two girls lived here, alone in isolated wilderness, for the next four school years. With barely enough food for survival, they studied religiously under their teacher, Mrs. Stinson.

At that time it was possible to achieve the equivalent of a grammar school diploma by passing a county examination. The two girls, undernourished and in poor health, traveled to Delano to take the dreaded examination. To their surprise, they both passed the test. Spurred on by their success, Fannie appealed to their parents to send them to a Normal School where they would prepare themselves to become teachers. They received little encouragement from their father who lacked funds to send them. But Fannie, undaunted, went bravely ahead, plotting a course.

Now Martha's nursing skills came to the assistance of her girls. By necessity and through practical experience, she had delivered many babies. The giant cattle and farming operation of Miller and Lux had imported many Italian families into the Buttonwillow area to carry out their farming activities. Martha became a midwife to these women as they delivered their babies. She went into their homes, attended the births, cared for mother and child, and carried out all household duties until the new mother recovered. With the money she earned this way, she was able to keep her daughters in school.

Visiting My Grandparents

My grandmother always smelled of wintergreen, a fragrance emanating from her body because of the liniment she rubbed on her swollen ankles. From the earliest I can remember, she could only walk by slowly sliding one foot in front of the other. She called her infirmity "rheumatism." Grandma was short of stature, and her heavy body impeded her movement even further. But it did not impede the movement of her agile hands, for her hands were never idle. Her manner was kind and gentle. She had endured the roughest that life could offer, sometimes serving her fine bread and preserves to harvest crews when her swollen limbs could barely move, yet she never complained or lost her sweet and amiable disposition.

The Mountain Lion

I was too young to recall this incident, but I clearly remember my grandmother's recounting of it. Grandfather ran some cattle on a small range in the foothills above McKittrick, which must have been the setting for the story. Implausible as it may seem, grandma and I were left alone in a camp in this wilderness mountain area with only a tent between us and the darkness of night, when the roar of a mountain lion wakened us. Frightened, I began to cry. Grandma was, no doubt, as frightened as I, but she thought of a way to calm me. She told me that if the lion came into the tent, it would eat her first because she was bigger. She said that he would then be so full that he would no longer be hungry and therefore I would be safe. She later claimed that this allayed my fears, and I went peacefully back to sleep. She seemed to enjoy telling this story. I think she found it amusing because of her excessive weight.

The Sting

Another time, when we children were attempting to turn cartwheels on Aunt Dee's lawn, my hand came in contact with a vicious-looking black insect. (We always knew mother's sister Dolly as Aunt Dee.) The insect had a powerful sting! A day or two later my swollen finger itched to the point of distraction. My grandmother, perceiving my discomfort, coated my finger with salve then took a table fork and gently scratched the offending finger. I shall never forget the merciful relief or the kindness of her act.

Grandfather's Gold Watch

By the time I was in the eighth grade, Grandma Rowlee and I were living together in Bakersfield in a small house behind Aunt Dee's. We called this house "the shack." I was soon to graduate from Emerson School, but I had no dress for graduation or money to purchase one. Grandmother, once again, came to my rescue. She handed me the only article of any value that belonged to her, her husband's beautiful engraved gold pocket watch. She asked me to take it to Wickersham's, our local jewelry store, and take what they would give me for it. When I returned with only $2 for this precious keepsake, grandma seemed a little sad. But it was that $2 that purchased the green silk for my graduation dress.

Martha Ann Rowlee (1850-1948) and Charles Wesley Rowlee; (1848-1933). True pioneers, they migrated from Iowa and homesteaded in several places in the San Joaquin Valley before settling in the Jerry Slough area. Martha sits while Charles keeps time to the music from the radio at their Chester Lane home, circa 1930.

Grandpa Rowlee

Grandpa Rowlee had also survived a difficult life. Charles Wesley Rowlee was born in Tioga County near Oswego, Pennsylvania. The family moved west to Illinois when he was nine, and he attended public schools and completed his education at Upper Iowa University in Fayette, graduating as a teacher. At age nineteen, he taught successfully for seven years in Linn County, Iowa—some of his students were older than he was!

In 1876, the Rowlee family traveled westward to seek their fortune at foothill grain farming near Stockton, California, in the little town of Clements. They migrated to Kern County in 1884 and, time and again, Grandpa Rowlee would stake out a homestead and put his plowshares down in what would prove to be infertile soil, only to pull up stakes and move on. Eventually they homesteaded 160 acres on Gooselake Slough, and they finally sank an artesian well to 560 feet, providing the good water that was critical to farming the area. He installed a reservoir for irrigating his crops, and he grew alfalfa and raised cattle under the "half-circle V" brand. Charles built his ranch to 880 acres and, after the age of sixty, the family had also acquired a mountain ranch and the home at 2323 Chester Lane in Bakersfield.

When I made my first trip to my grandparent's home on what is still known as "Rowlee Road" I could not have been more than four or five years of age, but I remember it as clearly as if it were yesterday. From my first remembrances of this distinguished looking man with a quiet manner, my grandfather's mind had begun to fail. No longer capable of earning a living on his own farm, grandfather had taken a job filling chuckholes on the county roads. He drove an old Dodge touring car, and he never drove faster than ten miles an hour. Unlike other cars, the Dodge had yellow headlights. As evening drew near, grandma would light a kerosene lamp, stoke her big old kitchen stove with sagebrush, and place a few potatoes to bake in the oven. Grandpa must have worked long hours, for it was dark when we began to look for those yellow lights. When they finally appeared, it seemed like it took him forever to reach the house. In the meantime, grandma would start frying the bacon. When we sat down to the dinner table, grandpa cut his potato in half and then slowly poured bacon fat over it. "My," he would tell grandma, "this hot bacon fat on a baked potato makes a mighty fine dish, Mama." This ritual never varied in the slightest degree, nor did the menu. Night after night we listened to his appreciative comment. The scene made an indelible imprint on the mind of this small child.

When I was five or six, mother had asked grandma to sew a dress for me. By now there was little doubt of grandfather's senility. Even at my age I noticed that the buttons on his shirt did not always meet the corresponding buttonholes. As grandma sat and sewed on my dress, grandpa would pick up a newspaper, select the most prominent headline, and read it aloud. With mounting emphasis, he repeatedly read the headline until he reached a resounding climax by rolling the paper up and slapping it sharply against his knee. He then put the newspaper down as if he were through with the whole business, only to pick it up again a minute or two later and repeat the entire ritual. This production went on for hours at a time, and although his behavior seemed strange to me, grandma continued with her sewing and seemed to accept it as perfectly normal.

Martha Rowlee, in her last known photograph

Farewell

Grandpa Rowlee passed away in 1933 while Grandma Rowlee lived to be almost one hundred years of age. She died in mother's home on Jerry Slough in July 1948, and I am still saddened when I remember how she spent the last several years of her life totally blind and bedridden. She, who had served so many with loving care, now lay in a world of darkness. The hands that had never been idle, now lay useless by her side. The beating of her strong and courageous heart had far outlived the usefulness of her body.

Frances Remembers

Excerpts from the Writings of Frances Tracy Selvidge, September, 1987:

Martha met the tribulations of her new pioneer life with great determination. She served as midwife to many families in the remote areas where she lived. Much of the money she earned went toward sending her girls to normal school. Her bread and cake baking were never matched by any of her four daughters.

Grandpa Rowlee also had a good voice, and I remember them singing with each other, evenings. It was a great joy to us all when grandpa came home with a new Edison phonograph, with those thick records. We had no toys as such in those days, so when I was visiting I would while away hours creating dances to the voice of Harry Lauder or to orchestrations of John Phillip Souza marches.

When they retired, they bought a home at 2323 Chester Lane in Bakersfield. Grandpa later died, and grandmother's rheumatism became so severe that she spent most of her last twenty years in a rocking chair. She had a grand sense of humor and was loyal and proud of all her family. We never heard a word of complaint from her.

Back to the Future

From The Bakersfield Californian, January, 1994; By Frances Selvidge:

> The recent article in your Accent Section on the resurgence of midwifery took me back to my sense of place. First to the memory as a child, of sitting on the front porch waiting for it all to be over—while my grandma brought another little brother or sister into the world.
>
> Then my mind fast-forwards to all the tests and costs of present hospital births and I think—WOW! I rewind a bit to 1941 and review the original statement of my eldest child's hospital bill in West Los Angeles as follows: ten days at $6.25 each, circumcision $2.50, mother's sutures $5, add to this the doctor's prenatal care and delivery of $125, for a grand total of $195 for one baby boy! What a changing world we live in! I note the present midwife fee for delivery is $995. With my mother's ingenuity in the late 1800s, she would have stretched this amount to cover her whole college career!
>
> From 1905 through 1919 came Cecil, Darrel, Frances, Teddy, Martha, and finally Tilton, the nucleus of a new branch of the Tracy farming family. I was number three on my grandmother's list. Out of six, she had birthed five of us, having been short-changed by my mother getting caught thirty miles away from home in Bakersfield when my sister Martha was born.
>
> Tracy Ranch was our home and grandmother lived six miles to the north by horse and buggy on Rowlee Road. Having born six children of her own and seeing the need of other expectant mothers in that desolate, treeless sagebrush plain to the West, this courageous pioneer met a need and thereby acquired her birthing talent.

By the time her own children approached motherhood, she, without the aid of stethoscope, plastic sheet or gloves, would tell grandpa it was about time to hitch the buggy to go South, especially to Tracy Ranch where she would bide her time to ply her trade.

In earlier times when one of her daughters decided it was time for an education, grandmother was by then recognized for her midwife expertise, and a little cash began to flow in. With this trickle of anticipated income plus the sale of some poultry and small animals, daughter Fannie dragged a reluctant sister onto the train to arrive at Chico at night in a rainstorm; cold, hungry and with one dollar left in their purses.

It is to her credit that she, with husband and growing family, left a secure life on a prairie in Iowa to follow the lure of the West and to conquer the travails of a Western plain. So, from knitting lace in Lincoln's time, she took us back to the future in helping birth a new generation.

Martha Ann Rowlee's life personified her namesake "Martha" of the Bible. Martha, a tireless workingwoman who faced life head on, gently left us in 1948 at the age of 98.

WILLIAM TRACY

A Wife's Tribute

In December of 1962 at the age of 85 my mother, Fannie C. Tracy, published a book with the aid of Glendon J. Rogers. Entitled *The Tracy Saga,* it was published by Cardon House Press. In it she relates the struggles of her pioneer family on the desolate sagebrush plains of early California. Later, she tells of her hard-won education, her engagement and thirty-seven-year marriage to William Tracy, the rearing of their six children, and her venture into the ostrich business.

The Tracy Saga:

> Mr. William Tracy, with us no longer, was born August 8, 1866, within a few miles of my birthplace at Galt. We never met in our childhood, perhaps because he was ten years old when I put in my appearance. Orphaned at the age of ten, he went to Sacramento to live with a married sister whose husband owned a grocery store under the name of Kilgore and Tracy (Ferdinand Tracy's nephew).
>
> My husband-to-be attended school there, working in the family store mornings and evenings doing janitor work. After completing the eighth grade, he hired out to do farm work plowing some of the lands of Colusa County. In those days the gangplow was much in use, pulled along by four or six horses. Those plows were composed of five to six shares, leaving as many furrows turned as they moved along. The driver rode on the plow as he wished, for no special seat was provided. Several of those plow teams often followed one behind the other in the same field. In Colusa County grain was the chief crop grown, being highly dependent on the yearly rainfall.
>
> Before he was twenty-one, Mr. Tracy had saved some money from his farm work activities. Deciding to make a change, he sold his plow teams. This action, plus the money from the sale of his mother's estate, gave him some six hundred dollars. At the suggestion of his uncle, Ferdinand Tracy, he moved southward to start a new life with his brother, Theodore, now a homesteader along Jerry Slough. At that time the brother was having difficulty in financially "proving up," so William's help was appreciated…
>
> I am sure that you realize by now that the boundaries of what is now known as the Tracy Ranch Incorporated [later the Buttonwillow Land and Cattle Company] were first shaped in 1886, and some even a little before, as Tracy and Canfield thought to leave the Sierra mines and put cattle on the vast grassy plains of the central valley. Later when these men grew past their prime, they retained William Tracy, Ferdinand's nephew, to look after their remaining interest. Young William remained on the

Canfield Ranch as manager, until Mr. Canfield lost his holdings to the Kern County Land Company [later purchased by Tenneco] then under the direction of H.A. Jastro. This was brought about when Canfield sold his water rights outright to the Land Company in return for extended credit. The company sold him water for one year. At the end of that time no more water was put in his ditches, and the company then took possession. This land is still known in 1962 as the Canfield Ranch.

At the time the land carried a debt of nine dollars per acre, though it was considered to be some of the finest land in the county. William Tracy had been offered the ranch as a gift if he would assume the debt. He did not accept, for he had some doubts about the possibility of being able to redeem it under the circumstances, and the questionable water rights.

In addition, a partnership existed with an older brother, Theodore, on a Jerry Slough homestead adjacent to the Canfield-Tracy land. There they had started a herd of cattle of their own. Canfield and Tracy still owned a small herd, a remnant of their partnership. The older men had requested these be maintained under young Tracy's management. They were marked with "91" which had been the brand used on their cattle when they ranged through the valley. At the time of my marriage to William Tracy this arrangement still existed. However, my husband decided that things within his own partnership were becoming too complicated, so he bought out his brother's interest for ten thousand dollars. This debt was to be paid off in installments. However, his brother in turn sold the note to a bank at a discount, thus obliging him to pay all the money immediately. Theodore then moved to Bakersfield, quickly falling in with an early-day crowd of good fellows, and soon his money was gone…

In later years, Mr. Tracy often spoke of his love for this location, and what a satisfaction it was to farm here because there were no troublesome rocks or boulders like those that hindered the work in the northern county. In fact, he had a real antagonism for rocks, while I had a fondness for them, sometimes bringing a small supply home just to have around. Strangely enough, they always disappeared when my husband had an occasion to work with cement. My precious rocks were sure to be a part of the mix!

He liked it here, too, because of the use of irrigation rather than the dependence on rainfall. He often remembered, too, the long days and weeks put in waiting for the weather to clear, sometimes a delay of three to four weeks at a time. Those were not the more pleasant memories. However, Mr. Kilgore once said, "I think Willie made a better choice by coming down here."

Even though in the process of establishing a ranch along the Slough, Mr. Tracy went to work on the Canfield Ranch. He did general farming, including dairy and vineyard activities. While working there he rode on horseback, a distance of twenty miles, to his homestead each Friday night returning to Canfield's the following Sunday. It was on one of these trips that he stopped to chat awhile when my sister and I lived in one of the cabins before going away to college. He was riding a horse called Friday, which he did not dismount. He sat steadily in the saddle talking quietly. Soon afterwards I went away to college, and we did not meet again until several years later. At the time of our first meeting, I was still in grammar school. When we saw each other again, I had graduated from college.

The occasion of this meeting was Mr. Tracy's house-warming party to dedicate the new six-room house he had built. Before this, he and his brother lived in a cabin on the part of the ranch known as Section 32. By now he was looking after the Tracy and Canfield cattle as earlier agreed to, as well as starting a herd for himself and his brother? Eventually, the two men, Tracy and Canfield, as they grew older, dissolved their partnership, and deeded their land in the Slough to my husband with the understanding that he would permanently look after their cattle. This he did and, as a result, that land became the nucleus of our present ranch.

At one time Mr. Tracy courted a minister's daughter in Bakersfield. The girl, very interested in marriage, wanted to elope. The minister said he did not want his son-in-law's boots to smell of the barnyard; he didn't want any farmer in his life. I don't think the romance was very serious on Mr. Tracy's part.

At the same time, he was working on the railroad then being built to McKittrick. Working for one dollar per day, he also received one dollar per day for each of his two span of horses. He worked with a scraper and was the last one to pass over the roadbed as a finisher. Camping his horses at the point of operation, he rode his saddle horse back and forth. It was interesting that the land in Jerry Slough was valued at three dollars per acre; that was my husband's income for one day. Eventually, he bought out his brother's land interest, and then spent the rest of his entire active life acquiring and developing his interest along the Slough. It was to this principle that our children and I have dedicated our lives.

Being a man of generous impulses, and a born gentleman with an inherent gift of fine manners, he had an unusual directness of method in dealing with men and the early events of this region. He was void of vainglory, and a good neighbor in any way you wished to take him. He held more than a local reputation for his fairness and honesty. Holding warm affection for his family, he shared equally with me the concern for and complications of bringing up our six children.

Mr. Tracy carried fond memories of his mother's teachings and a warm affection for these memories. During the early period of his life he had been impressed with many elements of good character. He was uncomplaining and recognized by the men in the early events of this region as a man of justice and generous impulses. He fully realized that he could not bring the acreage he had accumulated into full fertile and productive status in his time. He appreciated the thought that some day his boys could continue its complete development.

I have often seen him stand by his shovel, just dreaming, as the water spread over nearby fields, and I am sure he was thinking far into the future. Indeed, his was a life well lived and enjoyed. He was fond of poetry and carried his book of poems into the fields that he might, refresh his memory, until there were no forgotten lines. Scott's *The Lady of the Lake* was his favorite, and he enjoyed quoting long passages from this beautiful work. It seems almost sure that he must have been introduced to and learned his fondness for this poem while in school in Sacramento because, as a boy and in his later life, he carried this little book with him and could quote most if not all of the poem. He was a great reader throughout his life and was exacting in his memory.

He was courageous in meeting the buffets of life and tolerant of those who threatened to do him harm. He seemed to have a way of taking such things as floods, diseases and poor finances in his stride. It was his wish that our children should absorb the responsibilities as equal participants and complete the expansion of the ranch to the undeveloped area of its boundaries. This they have done far beyond his dreams.

It was to the lands of Jerry Slough that my husband devoted his strength, intelligence and drive to build a solid foundation for a future Tracy Ranch Incorporated.

Will Tracy, age 25, in 1891

My Father, William Tracy

A Trip to Town

I could not have been more than four or five years of age when I accompanied my father on a long journey to a place called Bakersfield. I have no recollection of how we arrived there; perhaps it was in the old Willis Knight for that is the first car that I remember.

One thing I remember clearly is our walking along the streets of Bakersfield and having other men greet father warmly. They seemed glad to see him as they shook his hand and called him "Will." It came as a great surprise to me that anyone would know my father in this place so far from home. It made quite an impression on me and elevated my father considerably in my childish estimation.

Grab Bag

A happy memory from my childhood involves the times when our father would drive a team of horses and wagon to Bakersfield, some twenty-three miles away. Such trips were rare, but when he returned he would bring with him a whole stock of bananas, a few coconuts and a little bag of candy. The older children soon demolished the bananas and the coconuts lasted a little longer, but the bag of candy was for Martha, alone! This little bag of candy was called a "grab bag," and it was made of manila-colored paper and decorated with narrow purple, yellow and green stripes. These colors appealed to me almost as much as the candy. Inside the bag were a few pieces of candy but, best of all, each bag contained a small surprise! Father was a very hard-working man who had little time to devote to each of his children, but the memory of this single little act of thoughtfulness has remained with me all my life.

The Stroke - 1931

Although I treasure these childhood memories of my father, it was not until much later that I appreciated his full worth as a man. Mother was ten years younger than he, and since I was number five in a family of six, he was getting along in years by the time I was in high school—he was forty-eight when I was born. Father, at sixty-five, was somewhat broken in spirit after making some unwise investments and was not feeling well. Always considerate of father, mother suggested that he spend a few weeks at our cabin on Huntington Lake to get out of the heat and to rest.

Grandma Rowlee, eighty-one and severely crippled with rheumatism, my six-year-old cousin Ruth Cooper, and I accompanied him. We had not been at the lake for more than a day or two when my little cousin came running to tell me that there was something wrong with "Uncle Willie." When I went to him, I found him lying on the floor in a frightening condition. In a state of paralysis, he seemed to be trying to tell me something. I finally realized that he was asking for water. When I gave it to him, he almost choked to death.

Ruth ran to summon Dr. Coleman, our friend and neighbor who lived next door and taught at nearby Fresno State summer school. When he examined father, he discovered that he had the measles. Father had caught them from my younger brother, Tilton, and no doubt they triggered the stroke.

At age sixteen, I was now the only able bodied "adult" in camp. Needless to say, this was a pretty awesome responsibility. But Dr. Coleman called from the Lakeshore Resort telephone to inform Aunt Dee in Bakersfield, and somehow the message was relayed to Buttonwillow. (Twenty-three years would pass before our ranch would enjoy the convenience of telephones.) I knew that it would take at least six hours for anyone from home to reach us. That was the longest six hours of my life.

It was pitch black by the time we heard the siren of an ambulance as it made its way up the hill to our cabin. That was the sweetest sound I had ever heard. Mother arrived in the ambulance with the attendant, and Cecil followed in the family car. The load was suddenly lifted from my youthful shoulders. Unfortunately, it landed squarely upon those of my mother whose shoulders had already borne so much.

Cecil drove grandma and little Ruth home, and I accompanied mother, father and the attendant in the ambulance. I shall never forget the eerie sound of the siren as we sped through small towns in the dark of night. It was nearing daybreak when we arrived at Aunt Dee's in Bakersfield.

By morning, father had improved considerably and was regaining his power of speech. He had suffered a severe stroke that left him with some paralysis on one side, and he was diagnosed to have a blood clot lodged at the base of his skull. For the remaining ten years of his life, it would cause him to suffer seizures that would render him unconscious for several hours at a time. That, of course, was a great concern to mother and Cecil who feared he might fall against one of the ranch's hot cast iron stoves or suffer a seizure in the fields as he irrigated with his dogs.

Incapacitated for the remainder of his life, father would no longer be the head of our family. From that time on it would be our brave mother, with Cecil's help, who would shoulder the responsibility of weathering a severe depression and saving a run-down ranch. Since mother was teaching at the time, Cecil became father's caregiver. During most of those years father enjoyed taking his dogs into the field to irrigate, but eventually he had to be confined to the house.

Jack's Bad Habit

A young Jack Frey at Tracy Ranch

By the time I was dating Jack Frey in college, he was spending his summers working for his Uncle Frank on an oil lease near Coalinga. On his days off, however, he enjoyed coming to the ranch to help Cecil with whatever work was to be done. This was a few years after father had suffered his stroke, and Cecil and mother had been forced to take over the management of the ranch. It was during one of these visits that Jack's bad habit got him into a bit of trouble. He made the mistake of using a few swear words in my father's presence.

When Jack related the incident to me I could easily visualize the episode. The face so deeply furrowed by the passage of time and illness would always betray father's emotions. Beneath the shock of white hair with one brow lifted above the very blue eyes, in a manner gentle but firm, he addressed my boyfriend: "See here young man; for more than fifty years we have managed to get along on this ranch without the use of profanity. I see no need for the use of it now."

This did not completely cure Jack of his bad habit, but I'm sure that, from then on, he conducted himself with a little more care when he was in my father's presence. Father may have been disabled by his illness, but his values were still securely in place.

Will Tracy the Orphan

Not long after Will Tracy was born on August 8, 1866, his mother Mary Dicks appears to have been abandoned by her husband, Edgar Vernet Tracy, leaving her with a family of six children to raise with no means of support. After divorcing her husband, she carried on as best she could until her death in 1877. Will, age ten and shattered by the loss of his devoted mother, was placed in a foster home. Although our family knew that our father had been orphaned at an early age, he had not spoken much about his childhood.

The Manuscript

In 1935, four years after his stroke and while Frances and I were attending college at UCLA, mother suggested father write down some of the memories of his life. Not until fifty-five years had passed did Frances and I discover that he had done so. By then we were the only surviving members of the original Tracy family.

The autobiography, written in pencil on inferior paper in father's unsteady hand, could so easily have been discarded as the scribbling of a weary old man. But sometime during those difficult years following father's death in 1941, a distraught Fannie passed the manuscript into the hands of her son Darrel for safekeeping. (I suspect that mother, knowing what was in the autobiography, handed it to Darrel for "safekeeping;" not quite willing to air the family's dirty linen, yet not wanting to throw away the beautiful and sad story of daddy's childhood.) Darrel, in turn, passed it on to his daughter, Marilyn, for transcribing. In 1958, while still in high school, she undertook the task of deciphering the script. With meticulous care, she examined each word to put the autobiography into readable form. (She did, however, censure the passage alluding to the uncertainty of Will Tracy's parentage. Fortunately, she preserved the original manuscript.)

For many years after that the family was not closely in touch with Marilyn. She completed college at U.C. Berkeley, married, lived in Southern California, and we saw her only briefly at family reunions. But in 1989 Marilyn decided to return to the ranch and build a home here. It was only then, fifty-four years after it had been written, that we learned of father's manuscript. Not only were Frances and I surprised that the manuscript existed, but we were also amazed that father, at that point in his life, could have written with such clarity.

Were it not for mother's suggestion that he write, and were it not for the careful preservation of his manuscript, the family would never have known of this side of his life. Nor would we have learned the true and poignant story of a very brave little boy.

The Family Secret

By the time we children were adults we were aware of a family secret. We learned that our father had some question as to his own parentage. His mother, Mary Dicks Tracy, was married to Edgar Vernet Tracy. In father's autobiography he lists the six children that preceded him, one brother and five sisters. He also suggests that someone other than Edgar

Vernet might have been his natural father. The rumor was that our real grandfather was Charles Wellington Canfield, though it could just as likely have been Edgar's brother Ferdinand.

Before reading father's autobiography it is helpful for the reader to know a little of the two men, Ferdinand Tracy and Wellington Canfield, who played such an important part in the course of his life.

Ferdinand Tracy

At the time that Glendon Rogers co-authored mother's book, he could not have known about the "family secret," for father's autobiography had not been unearthed. When he researched the Tracy name, he found very little about Edgar Vernet, but a great deal about his more accomplished brother.

Referring to Ferdinand Tracy, mother's co-author states in The Tracy Saga:

> In anticipation of the forthcoming publication of *The Tracy Saga*, a cooperative venture in writing by Mrs. Fannie and the author of this column, it seemed appropriate to provide some of the background to the Tracy name. It is not necessary to go back into the family tree but rather to look at the name more in relationship to the past hundred years. To do this we can well start with Ferdinand Tracy, a successful pioneer, who was responsible for many of the early day beginnings of Tracy Ranch, Inc. It would be an error, too, to transfer all credit to this man, for the eventual marriage of Will Tracy and Fan Rowlee brought into true perspective what is true today and the major basis for writing *The Tracy Saga.*
>
> On January 10, 1908, one of the largest funerals of all early days took place in Bakersfield as F.A. Tracy, following an illness of four years, then in his 78th year and 46th in Kern County was committed to his Maker. His survivors included his wife; a niece, Mrs. James Arp; and two nephews, William and Theodore. A brother lived in the northern part of the San Joaquin Valley. In looking over the publications of the day, including the local newspapers, one finds recreated for the reader a picture of a man fine and strong in his influences, one held high in regard by his fellowmen, recognized for his leadership, and at all times a gentleman.

From "History of Kern County," by Wallace M. Morgan, 1914:

> FERDINAND A. TRACY—the life that this narrative depicts began in Wilkesbarre, Pa., in 1829, and closed at Bakersfield, Cal., January 9, 1908. The intervening years represented a period of intense activity crowned by the regard of friends. From 1850, when he was attracted to the west by the discovery of gold and crossed the plains with a large expedition of emigrants, he was identified with the material up-building of California and gave to its development an intensity of devotion indicative of his fondness for the commonwealth of his adoption. In the early '50s he was commissioned lieutenant of a company of United States troops to quell an

Indian uprising in northern California. The company was successful in establishing peace, and it was the last time these Indians were on the warpath. When his services were no longer needed Mr. Tracy was honorably discharged. During the period of mining activity he followed the occupation in the Mokelumne region and indeed he never lost his interest in the industry, but after he had left the Sierras he became interested in the mines of Kern County, notably those in the Tehachapi range. From mining he turned to stock raising. During 1860, in search of feed for his herds, he came into the Carrissa Valley and from there he proceeded to the Kern delta, whose possibilities attracted him. Becoming a resident of Bakersfield in 1862, he operated his stock interests from this point on with Wellington Canfield, under the firm of Canfield & Tracy; he became a leader in the cattle industry throughout the San Joaquin Valley. Their herds increased in size until their brand was more frequently seen than that of any of the stockmen in the state. It is said that during the fifty-six years of their co-partnership, never a word of disagreement arose between the two partners, but they remained in the end as in the beginning close personal friends, congenial associates and devotedly attached each to the welfare of the other.

The marriage of Mr. Tracy in 1875 united him with Mrs. Ellen Baker, the widow of the founder of Bakersfield. Politically he always advocated Republican principles. Frequently he was called upon to occupy positions of trust in county affairs, but these came to him without his seeking. Indeed he was so unassuming, so reticent in his own claims, so strong in his dislike for notoriety that he shunned public life and often was superseded by others who had not the moral and intellectual equipment for an honorable career that he possessed. Had it not been for his exceeding modesty he would have ranked with men known in the annals of the state. His manly nature appealed to men in a manly way. Generous to a fault, hospitable in act, attractive in personality and genial in companionship, he made friends of all whether rich or poor. In every sense of the word he was a true gentleman and this was particularly noticeable in his desire to protect all helpless and dependent creatures. The poor had in him a helpful friend; the suffering never sought his assistance in vain. His integrity and honesty were of the kind that sought no personal emoluments, but upheld the highest principles of honor through innate purity of soul. Self-poise guided all of his acts and was apparent in every business detail, but it was not the self-control of the selfish man, instead the natural temperament of one ever ready to make sacrifices for others and one who displayed patience and kindness under every circumstance. Possessed of splendid mind, he developed this through a wise and long-continued course of reading. The masterminds of all the ages became familiar to him in his readings and thus he acquired a cosmopolitan culture. A fondness for poetry did not deter him from delving into the intricacies of science and political economy, while in history he was exceptionally well informed. Of religion too he was a thoughtful student and while with innate reticence he never revealed his thoughts concerning the spiritual life, his own deeply religious nature pervaded his entire existence and made beautiful his adherence to the stronghold of Christianity.

Charles Wellington Canfield

Another who strongly influenced the life of Will Tracy was Charles Wellington Canfield.

According to The Tracy Saga:

> Though certainly not in any real sense part of the family this man who in 1877 was state senator from the district that included Kern County, certainly helped influence and perhaps shape some of the decisions made by earlier members of the family who were county residents. He was a well-informed man, who used his talents wisely as indicated in his work on some of the laws of early day California.
>
> Upon coming to Old River he acquired the Barnes-Harris homestead, one of the first in the valley (1859), and the John Olton homestead, all of which soon became known as the Canfield Ranch. A small townsite was laid out and christened Canfield. Soon he brought his wife west from their home in Boston. The long separation while Mr. Canfield built his way up in the west had led to numerous warm love letters for theirs was a very close association. Mrs. Canfield was a teacher in the Boston Conservatory of Music. While attending Mary Lyon's School for Girls at an earlier date she had traded eggs and other farm products for her tuition. Of wide interest was her father, the well-known Judge Cooley, famous for his authorship of a book on law.
>
> Eventually Mrs. Canfield had her grand piano shipped from Boston to become one of the earliest in the county. That the Canfields had carpets on their floors in those early days is a matter of record.
>
> Mr. Canfield was a cattleman, yet he helped to shape the "no fence law" which is still in effect today. In brief this law required a cattleman to restrain his cattle on the open range from trespassing on another's property. The fact that he helped pass this law gives insight into his regard for duty.
>
> Once established in the Old River area he became one of the first to grow a vineyard and establish a dairy. At one time my husband could have purchased the Canfield Ranch for a nominal sum per acre. However his judgment advised against such a move, and this action was never regretted.
>
> When the powers of these two active men, Tracy and Canfield, began to wane they retired to their first love, "books." Both men had said they liked to work in the mines because they could retire at night and read. Mr. Canfield was content to spend his last years living a quiet life here in Jerry Slough with Mr. Tracy and myself. As long as he could lift a hoe he raised sweet peas. Then driving two small mules hooked to a buggy, he carried the flowers to his wife's grave in Bakersfield. It was only after 1902 that I

came to know C.W. Canfield, for that was the way he signed his name.

Eventually his Old River property passed into the hands of Carr and Haggin and finally to the Kern County Land Company where it is still known as the Canfield Ranch.

Will Tracy on coming to Kern County

Father's Mystery

A quote from father's autobiography states: "Meantime, my uncle on my father's side and the man who was purported to be my natural father had favored me exceptionally, and I found myself the owner of a considerable acreage of undeveloped land."

Depending upon how these two lines are interpreted, he could have been referring to either man as his "natural father." If Canfield were his father, then the man who was thought to be his uncle was of no blood relation to him. If Ferdinand was his father, Will was a "sort of cousin" to his half-sisters and brother. Whichever the case, there is no doubt that these two men who "favored him exceptionally" had great respect and admiration for the young teamster.

Will was born in 1866. Ferdinand married Ellen Baker, widow of Colonel Thomas Baker in 1875. Was Will Tracy actually his own son? If so, then Will was his only surviving natural son, since Ellen Baker's adult children were all from her first marriage to Colonel Baker. No wonder Ferdinand "favored him exceptionally." Or, as family rumor had often hinted, was Ferdinand Tracy's business partner, Senator Canfield, (who married a Miss

Cooley of Boston in 1864, but brought her out to California much later) actually our grandfather?

Careful examination of the Time Line shows that at the time William was conceived, Wellington Canfield was recently married on the East Coast, while Vernet had just been released from the army at nearby Fort Point, California. Ferdinand's whereabouts at the time are still unresolved. Cecil and Frances, who knew of the rumor as young adults, searched their memories and old photographs as clues. They eliminated Canfield as a possibility to their own satisfaction because of their father's strong resemblance to his brother (Vernet's son) Theodore, and attributed the allusion to a "natural father" as wishful thinking.

In any case, there is no doubt that a strong bond existed between Will and Ferdinand. In his declining years, when Ferdinand was no longer able to run his cattle, he told Will that he would deed all of his Jerry Slough property to him in exchange for tending his herds. For the rest of Ferdinand's life father lived up to that agreement. The famous Canfield-Tracy "91" was used for Ferdinand's cattle and inverted to a "16" for Canfields, while our father used a block "T." Eventually father would inherit the "91" brand.

Ferdinand decided to leave all of his property to his nephew since he had no surviving children of his own. On the other hand, Canfield had no children either, and during the last few years of his life Canfield made his home with Fannie and Will. They named their third son, born in 1908 four years after Canfield's death, "Charles Wellington," though they always called him "Teddy."

Regardless of Will's parentage, Mary Dicks Tracy must have held either of the other two men in higher regard than the elusive Edgar Vernet Tracy who abandoned her and left her with six children to rear.

Although her son did not become a "clerk in a store," part of his mother's prediction was true. Will Tracy was no "ordinary man," but a fine and honorable one. I think that Mary Dicks might well have been exceedingly proud of her youngest son.

The Native Son

Reminiscences
of
William Tracy

December 1 to December 27, 1935

Transcribed by his Granddaughters,
Margaret Marilyn (Tracy) McDonough
and (later)
Nancy (Frey) Trahan

Retrospective

At home, Buttonwillow, Kern County, California

Will Tracy in March, 1940

I find myself in my sixty-ninth year, well and strong with a slight weakness in my left side caused by a slight stroke I received on recovering from an attack of measles I experienced in the year 1931. I fully anticipate ten years more of peaceful life with my exemplary family, consisting of a splendid wife and our five splendid boys and girls ranging in age from thirty to sixteen years of age; none of which including my wife and self have ever been addicted to the use of alcohol drink or tobacco, which I think all will agree in this time of the world is unusual. [A sixth, Charles Wellington "Teddy," died in 1920 at the age eight of Bright's disease during a scarlet fever epidemic.]

My mother Mary (Dicks) Tracy [b.1808, D.1877] with her husband [Edgar Vernet Tracy Jr., b.1850, D.1913] and their four children [Alice, Theodore, Emma, Ida] came across the prairie in a covered wagon from Iowa to California in the year 1851 or 1852. Two more children were born before me—two girls [Mary and Sarah]. Father was then called to serve in the Civil War as a Union soldier. I was born August 8, 1866 [in or near Galt, California], after he came home from the war, and there were two more fine girls [Anna and Nellie].

I can never remember father living with us. We all loved our mother dearly, but none of us cared for father. Mother got a divorce and acquired a deed to the home where we lived and were raised before she died in the year 1877.

It took me many years to live down the self-consciousness of this truth of being a son of divorced parents, because the children at school and others reminded me of this when I started to school when eight years of age. But my father was fair with me, and I can forgive him all. God is just and his life will be tried before an impartial judge.

My oldest sister was made administratrix, and all the children got an equal share of the property as we became of age.

After my mother's death I was turned over to a young married couple, living not far away, to be raised. I was just coming eleven years old. I lived with this man until 1882. Then, on account of a little difference, I told him that I wanted to leave and go to one of my sisters. We came to an agreement, and he took me there.

I was then taken care of by different ones of the family for a few years and graduated from the elementary school of our state and given a diploma that would admit me to high school, of which I never took advantage. I went to work on my own account from then.

In 1887 when I was twenty-one years of age, I received from my mother's estate my portion that amounted to something over $600. I had acquired at this time six good horses and was hiring them and myself out for wages. This was a good life for me. And in the year 1890 I was the proud possessor of $1,600 in cash.

I then looked up my only brother, eleven years older. He took me in as an equal partner as he had homesteaded 160 acres of land. My $1,600 came in very handy now in a new country. We worked together until after I was married in 1904. Then he became dissatisfied and I bought him out.

Meantime, my uncle on my father's side and a man who was purported to be my natural father had favored me exceptionally, and I found myself the owner of a considerable acreage of undeveloped land.

My wife was an excellent and practical woman with a good education, having a State Normal Diploma of California. She has borne six children, one of which we lost at the age of eight years. Besides most of the household cares, she has taught for many years and is now teaching our home school. And we are all living happily together.

My life's account has not been so romantic as some I have read in the *True Story* magazine, but I am very thankful for that.

All our boys and girls have had the advantage of high school and our two girls are making good progress in the University of California, Los Angeles.

Our elder boy is manager of our still undeveloped holdings of land.

Remembering Mother

Well, to go on with my story. Last night I did not sleep much for remembering things that happened when I was a little boy, sixty-five years ago. It is remarkable that my mind remembers.

There is a little incident that was related by my mother that I must report here, for fear it will be forgotten when I get busy with the business of the ostrich farming.

When our folks were crossing the plains, some of the men took their guns and went out in search of game. One of them shot and killed an Indian woman. After breaking camp, they had not gone very far when they saw a dust rising in the rear. Finally, it developed into a band of braves on their horses. They approached the captain of the train and demanded the man who had killed the woman. He was given over to them, and the train was in the hearing of his cries as they skinned him alive.

When I was five or six years old, I found an old oyster can in my mother's barn. I hauled it out of its cubbyhole and beheld its contents. It was nearly full of gold coins—five, ten, and twenty dollar gold pieces. I went straight and told my mother about it. She said, "Put it right back where you got it. It belongs to Dick Rude."

Dick Rude farmed our place. He had a yoke of oxen—Nig and Mike—two large steers, and a threshing machine for threshing wheat, barley and such after it had been gathered in the straw and stacked. He had a fine span of mares. One was a sorrel, the other a bay. The sorrel weighed 1,600 lbs. I heard him say so one day after she had stepped on his foot. He had a two-horse wagon and a double barrel shot gun. It was a wonder to me. It lay most of the time across his wagon seat, and I used to feast my eyes on it.

I saw him shoot a squirrel and load the gun. He took his powder flask and poured some black powder in each barrel. Then he took out a ramrod that just fit in some ferules underneath the barrels and rammed the paper down tight and tamped them hard. He then took his shot pouch, turned some shot in each barrel, and stuffed the paper down again, but not so hard. The gun was what we call a muzzleloader. Then he took a small box of caps out of his pocket, pulled the hammer back, and put a copper colored cap on each tube. He let the hammer down on them, pressed them hard with his hand, and the gun was loaded.

Our mother had about a dozen cows named something like this: Reddy, Blossom (for shortness called Bloss), Lilly (called Lil), Old Brin (for Brindley), Spot (of whom we were afraid for she would hook), and Old Brock.

Mother made butter and carried it to a little town three miles away on our little saddle mare. She wrapped it up in two-pound rolls with white muslin cloths around it. One day when she came home there was blood running from the saddle down the pony's front legs to the ground. I thought, "Surely I have lost my mother." The girls and I lifted her down and asked her how she had got on the horse in town. She said the men helped her on and were awfully good to her. In a few days she was well again.

I remember nothing more clearly than the sweet sad smile that my mother wore on her face at times. She told the girls that I would be no ordinary man, but would be a clerk in a store or something. And they all loved me very dearly.

We had one big bay work mare and the saddle mare. One day a Negro came to borrow the work mare "to haul some dirt," I heard him tell my mother. We called the devil the Black Man, and I thought he was the devil and lived under the ground. I wondered how he was to get the mare down there. In a few days I discovered he was a near neighbor and a good kind man. He had a small boy with whom I liked to play. We spent many happy hours together.

But what I liked best of all was to ride a horse. The girls minded the cows and used the saddle mare, and they taught me to ride. They rode sideways, but I rode straddle.

One day someone yelled, "The cows are all going down the road." I ran for the saddle mare that was all saddled and bridled, standing in the corral. I ran for the mare, climbed on, and headed for the gate. The gate was a pair of bars. Did you ever hear of a person having a gate like a pair of bars? Well, that is where it comes from. It is a couple of poles and when the gate is shut, each end is resting on a post at either side of the passageway. Well, the bars were up and I did not notice it. Well, she cleared the bars and I found myself groveling in the dust, my eyes so full of the soft dry manure dust that I despaired of ever being able to see again. But the tears soon began to flow copiously and washed my eyes out, and I saw the little mare running down the road. She turned the cows and brought them back, and I let down the bars and turned them in the corral.

I loved to go in the barn and watch that little mare; I would look her over and marvel that such small little legs—especially from the hock to the fester joints—could carry her body with two of my husky sisters so fast, and turn the cattle without breaking.

We had a big black curly dog that thought it was his duty to protect me. Sometimes, when the girls and my mother saw fit to correct me, they would try and run me down; but Boxer, the dog, ran right along by my side and would not allow them to lay a hand on me. A female hound dog came to our home and I taught her to drive the cows. She became so proficient that she seemed almost to understand the English language. All I had to say to her at any time was, "Flora, go get the cows," and she would get them wherever they were and bring them home for milking. She had little puppies. We raised one of them, and he learned to help her with the cows.

One day my mother went over to one of the neighbors on the little mare, and she came home carrying an old iron teakettle before her on the saddle. The girls all cried, "What have you, Ma?" and she said, "Something for Willie." The girls lifted me up so I could look in the kettle, and what do you think I saw? Two little pigeons in full plumage. I think it was the most thrilling sight I have ever seen. There was a bloom on their plumage like ripe fruit. I was overjoyed. I raised them on bread and milk, opening their mouths with my little fingers to poke in the damp bread for about a month—when they learned to pick it out themselves. I wish people could know what joy can be thrown into a little boy's heart by such a little act of kindness.

One day I found a monkey wrench, and the man who farmed our place at that time gave me fifty cents for it. He had just paid my mother a twenty dollar gold piece. She told the girls to take me in the bedroom to see it. It lay on the bureau. I think it was the first I had seen since I found Dick Rude's oyster can of coins. That evening I could not find my fifty cents, but after going to bed I remembered laying it on the bureau when I had inspected the twenty-dollar piece—and there it was.

Some of our neighbors were Democrats and they called me the Little Black Republican and accused me of loving a "Nigger." I told them I did not think I loved a "Nigger," but that our neighbor Mr. Nelson was sure a fine man. This Mr. Nelson was the man I had taken to be the devil the first time I saw him. One day I asked my mother if our neighbor could be a Republican if he was of a mind to be. Of course, she told me he could be. I thought it a very strange thing a man preferred being a Democrat in preference to a Republican. I never could do that.

The Orphan

When my mother died and I went to live with Mr. L—that is what we will call him for convenience—I missed my dogs. We soon heard that a young dog was at my uncle's about ten miles distant. So Mr. L gave me a horse and saddle to go after him. I bought him and we kept him until the dogs in the neighborhood got to killing hogs at night. Then Mr. L shot him dead. That was a hard blow. I loved the dog although he was guilty, for he had fallen into bad company.

The day Mr. L killed my dog, he went over to one of the neighbors where he knew one of the guilty dogs was, and he told the man his dog was the aggressor and that he must dispose of him, or he would shoot him right there. He had his big shot gun along. The man promised to get him out of the country and we never saw him again.

I had also raised about ten sheep by hand, on cow's milk, from lambs that had been given me one time by the neighbors. Mr. L took these also. There was a ram among them and he would bunt a person, especially me. But I could catch him when he ran at me, and many a time have I led him to the fence and climbed the fence to get away. The school children met to go through the field to school. There was one big fat boy; the ram had it in for him. He would knock him [over] and run backwards for about thirty steps and come and knock him over again until he could get to the fence. The children quit going through that field. Mr. L finally butchered him. I did not like that very well, either; but they were good people.

Mrs. L I loved as a very dear mother and she counseled me very wisely, which has kept me out of many a pitfall.

While living with the L family, I had my earliest sweetheart. She was my rival at school for the head of the spelling class. One day she said, "I'm sorry that I spelled the word; I hate to go above you because," the brown eyes lower fell, "because, I love you!" One day she [had] asked to be excused from the schoolroom and been denied. Presently, her class was called, and in the middle of her recitation she wet all over the floor, then turned and ran as fast as she could go out of the schoolhouse door. We saw no more of her for a week. My heart was very sad for her. Of course, she had the sympathy of the whole school.

She became a teacher in time, and taught at that same school. I had the pleasure once of visiting her school while she was teaching, when I was visiting the L's, but our relations were never resumed. However, I was always fond of teachers and finally, when I was thirty-six-years old, married one and have never had cause to regret it. My advice to young men is to marry a teacher. They are not only more able to instruct your little ones but are a source of income. By all means, marry a teacher.

Had Mr. L treated me with as much tact as his wife, I undoubtedly would have been with them when I became of age; but when he fancied I needed punishment, he would say to me at the supper table, "Willie, stay with me a little while after supper." Then when he had finished his supper he would unbuckle his leather belt, which he always wore, and accuse me and whip me with the belt—I will not say unmercifully, but he knew I was a good boy. His whipping never made me love him [but] his wife's council made me respect her. And then I heard him get into quarrels with other men and use such profane language I was feared he would break out on me sometime.

Two little girls were born to them while I was there, and we were very dear to each other. They separated a number of years after I left, and I cannot help but think that had I stayed it might have been different. They both married again. She married an old sweetheart, who I have heard jumped off a pier and drowned himself. I was dreadfully homesick to go back for a long time and the mother and girls were very sorrowful. I went back to see them every year for many years. They are both dead [now]. One girl married a neighbor boy and lives near the old home. I still visit her.

When I went to live with Mr. L, I took some of my pigeons. I made them a nice box with the door facing out so I might see in the box. One morning when I was attending to my morning work of feeding the chickens and such, I heard a tapping in the pigeons' box. I went where I could look in the door and behold, a pretty spotted and striped little skunk came to the door. It stamped his foot at me—pat-pat-pat-pat. I went to the house and got the old shotgun that I had all ready to shoot. It was a big ten gauge with a heavy load in one barrel. I took good aim, point'n into the door where the little skunk was challenging me and I blew most of him clear through the barn wall into the barn. Prodigious! Oh, what a smell. If it had not been for that barn burning up about forty-five years ago, I would not be surprised to see it smelling yet.

At the time I went to live with Mr. L, my sister just younger went to live with Mrs. T's parents that we often visited, but before I left Mr. L, my older sisters had taken her away. She got a State Normal Certificate and taught school many years and also married a good man and raised a family. The next younger, the youngest of the family, was adopted by a good old couple and received a good education, and married and raised a family.

My sisters, with the exception of one, have passed on. They were all excellent women; married reputable men and each one reared a nice family. They were all loyal to each other and to me, and I have no reason to doubt that we will meet again.

Family Times

Have been out sawing a little firewood and feel a little rheumatism in my right shoulder. Don't know if I can write. Am writing a little every day now on my story. When I began, I said to mother, "Oh, pshaw, there is no use, there is nothing in my life to write about to compare with the stories I have been reading in the *True Story*," but every day now I have an inclination to write. I get almost as much of a thrill out of it as I did out of reading a story.

It is almost Xmas time—the girls are coming home from college to spend the holidays with us—Frances and Martha. Cecil is here, the manager of the ranch, and Tilton, the baby—sixteen years old—and has ridden his first Brahma steer successfully. Darrel, the postmaster—they will all be here; also, their grandmother on their mother's side. She lives with us most of the time; she is eighty-six-years old and infirm.

Mother said to me, "We haven't much money to spend for presents this year."

"What are we going to do?" I said.

"They will get by," she said. "I think it would be a good time to give each of them their portion of the property."

I said, "It is yours absolutely."

"Well," she said, "I always want your cooperation in matters like that."

I told her that when I had deeded it to her, my whole idea was to make things safe for her and the children, so of course I am agreed. So, she had me make out the legal descriptions of five-quarter-sections of land, and she sent them to town to have the deeds made out. To the eldest boy, the manager, she gave the land on which is located the home where all the children were born and raised. I will tell you about the property after the holidays, the Lord willing.

My arm is still a little sore, but I can write.

We raised on the ranch from one hundred to a thousand turkeys as a little side issue to help out with the expenses and the accommodations of our friends for the holidays. Each year it has fallen my lot... this year to take care of these birds. They do the harvesting of the wheat and also the corn in certain fields that is of the milo-maize variety. I take them out to their feeding ground each morning carrying my gun, and all the dogs on the place at my heels, the idea being to keep the premises from coyotes that abound in this locality. It is great sport for the dogs and me.

We are each day expecting the girls home from college for the holidays. There will be great sport for them, riding horseback, running wild and all that. They are all horsemen. The girls and boys and all from the time they were two years old have rode in front of their daddy on the horse and learned to rein a horse. I believe it is natural for children to like horses. My first little boy, when he was four days old, was observed [turning] his head as he lay on the bed and watching me through the window as I passed back and forth plowing the orchard. Little did I know at this time that he was destined to see the day that he could ride the worst horses in the world. In the year 1932 at the age of twenty-seven years, in a contest against all comers at the regular yearly rodeo in our part of the country, I saw him win first prize consisting of a beautiful western riding outfit—silver-mounted saddle, bridle and spurs—and championship of "best all round cowboy" of this part of the country in which we live. He is now manager for the whole ranch.

This year the youngest boy, sixteen, rode his first Brahma steer, straight up, while he rode him backwards. They are supposed to be the hardest animal to ride. The little fellow does most of the riding required in the handling of the cattle on the ranch. You would laugh to see him swing a sixty-foot riata from the left side of his horse, for his right hand is on his left side. The only left-handed one we have. All his horses become left-handed after he has ridden them for a while. So thoroughly has he learned to rein a horse across the neck as we break our cow horses that when I taught him to drive a team to a wagon (it was a flat hay frame) he would take both lines in one hand and run from one side of the wagon to the other when he wanted to make a turn.

At home… The flocks are gathering into the fold for Xmas. We love to have our boys and girls about us. They are about all we will have left after the Xmas day is over. They are all single yet.

In his later years Edgar Vernet Tracy was proprietor of this stable in northern California

Ostriches

Very early in our married life, my wife found a very interesting ad on the ostrich industry and from then on she read everything she could find on ostriches. And of course I became enthusiastic, so I struck out for Tempe, Arizona, where I knew Mr. Pickerel had an ostrich farm. I bought a train of birds for $250 and had them crated and shipped home. One of them, the male, became hurt in shipping and soon died. I had him mounted and gave the specimen to the State Normal School at Chico, California, that my wife had attended, and sent right back for another bird. We kept this train in a small enclosure. They were named Phoenix and Tempe in honor of two towns in Arizona; Phoenix, being the male bird, and they were the wonder of this part of the world. We had raised stock all our lives, but had barely seen an ostrich, so we had lots to learn. They were quite docile, however, and after a year or so the male scooped out a hole in the dirt with his bony chest, and Tempe began depositing large white eggs weighing about three pounds apiece, until she had laid a dozen in twice that many days, and people sure made a road to our place to see them. In about forty-two days, there appeared seven little gray beaks whose harsh covering was about the consistency of excelsior. They were about the size of chicken hens, and sat very flat and close to the ground. The old birds were sure an impartial pair; she keeping nest all the day and he usually through the night, and one was as attentive to the young as the other.

We had two boys at this time—five and three years old. One of them—the younger [Darrel]—had a mania for gathering hen eggs. He would gather his little skirts and go out through the weeds where the hens had formed a habit of laying, holding the sides of his skirt and putting eggs in the front until they would crowd each other out at the back. You might track him all through trails in the weed patch by the trail of eggs he had lost, and about all he would say was, "Hen lay eggs." He was so interested in the ostriches, we were afraid the ostrich would get him and he would so often say, "Look out, Birdie get you!" One day we heard him just screaming, "Birdie get you, Birdie get you!" and went out to the ostrich pen and there he was, about four feet high on the fence, and the old male just parading the fence back and forth and brushing very close to him, and he almost frantic—his little skirts sticking out behind. It is marvelous that we never lost some of our children by those enormous birds. I do not stretch the point when I say enormous. They weigh 350 pounds and can reach the top of a ten-foot wall, and they can stand on one foot and strike a tall man in the chest with the other.

We raised these seven and a fine lot they were. Our next inspiration came when we found the birds of the concession of the ostrich farm, a show place in Idora Park of Oakland, were for sale. We bought them with the right to the concession and shipped the birds—forty-seven in all—to the ranch. It was an exciting day, loading them from the boxcar into a wagon and handling them to the ranch. My wife was helping in this and they picked the diamond out of her ring that I had given as an engagement ring, and she never saw it again. [Whenever Grandma Tracy was angry at one of her granddaughters she would say disgustedly, "I'm going to will her the diamond that the ostrich ate."-NFT] We put these birds in an alfalfa field fenced with netting wire, and they throve beautifully and raised several broods of young besides producing many beautiful feathers. My wife had the younger sister [Hazel], at this time living with us, who was wonderfully apt at any kind of handwork. She sent her to Cawston Ostrich Farm in Pasadena to learn the trade of making up ostrich feathers and believe me, when she came home she was master of the art. The finishing touch of her fingers was magic. My wife set up a shop in our town, and they took in other girls to learn and were sure putting out some wonderful work. They made plume fans, sofa pillows and artificial flowers to beat the world, but the ostrich industry was doomed.

Fannie feeding her birds at Idora Park, Oakland, California, in 1912

With automobiles coming in and a world war coming on and our government discouraging any industry that was not necessary to life, the ostrich business could not prosper. Consequently, ostriches were a drag on the market, but we were so enthused we went to Arizona where the Pan American Ostrich Farm was shipping them off at $10 per head. We took a carload of them and shipped them home and turned them out in our big fields with the cattle and horses. And, with a big open country all around us, they got scattered all over this part of the world. We gave the last of our herd of 175 birds away to our own county park in the year 1929 and resumed our farming and stock raising operations.

I would advise anyone who fancies the ostrich business to keep them under very close control with netting fence to protect their plumage. You cannot mix them up with the general stock of a large ranch. We have had many exciting times with these birds. In the first place, although some of them are so docile, a lot of them are natural born fighters and especially are they antagonistic to man. They do not conduce the temperament of Maud

Muller. When they come over to investigate you when you are raking the hay, and even if you do have a good fork when one makes a lunge at you, you cannot stick it into his breast more than three or four inches. When he turns to run, the fork handle slips out of your hand and you are helpless for an assault with another bird right behind him.

I have never seen anything that could frighten a horse so much as an ostrich. I have seen them drop like they were shot and have seen them run like a wild horse. We live on a big open range, and cattle often break in our fields and mingle with our stock. When the buckaroos come after them, we have to assist them in getting them out, for if they go in the fields alone with a strange horse, as soon as one of those old scarred warriors gets an eye on him, he drives him out, regardless of barbed wire fences. For without a man on his back, he can overhaul him taking twenty-two feet at a stride, and taking them fast. And he can scratch that horse from stem to stern, and man and horse will tangle with the barbed fence whenever he comes to it to get away. But we have horses that I have trained to face a fighting ostrich, and when he attacks me I can grab his head and give his neck a wrap around the horn and drag him around until he is good and can guard the men while they work the cattle.

I have worked them over for the camera, and at the Pan American Exposition at San Diego in 1915 I was shown working ostriches on horseback while my wife and little girl are shown feeding baby ostriches soft granite rock, a thing of which ostriches are very fond.

The ostrich has a peculiar way of mating, and he walks across the field and at intervals lets the air escape through his mouth until the bag of air is exhausted. This performance is usually practiced in the morning hours when the world is quiet—a booming sound resembling the roar of a lion that may be heard for several miles. He does this by inflating a section of his neck with air resembling a large toy balloon.

To supply grit for the birds, we pound up soft granite rock and they will come for it as far as they can hear us pounding.

In handling these birds on horseback, I have on several occasions thrown a noose of my riata over the head and accidentally choked the bird to death, and some have killed themselves by getting entangled in barbed wire fence. Some of these we have dressed and [then we've] eaten a portion. We did not relish the meat, but the skins we have tanned and made into nice satchels and purses for ladies.

I find the best way to protect oneself from infuriated old males is not with a pitchfork or shovel, but [using] a club about the size of a shovel handle, [then] striking the bird across the neck. You may belt him like a beef and he will be knocked out long enough for you to get out of the way, but I have never killed one that way. When they challenge you to fight, they go through some of the most peculiar movements imaginable. They will intimidate the most courageous men, and when in contact they can strike you fair in the chest and lay you flat on your back.

Ostrich nest: the female ostrich lays her clutch of eggs in a nest excavated by the male

The ostrich egg is a good article of food. It compares favorably with the hen egg. Of course you cannot serve them hardboiled for breakfast as it takes one hour of boiling, but you can serve them scrambled. One egg will serve ten persons, as it is equal to thirty hen eggs; or, you may break open one end and use as much as you like and set the rest away in the shell to use later. The shells are thick and strong and make nice jewel cases. With a little paint, you may make wonderful designs on them or you can dress them up with silk or satin into beautiful ornaments.

I have gone out into my fields in laying season with my machine and brought in one hundred of these monstrous eggs. We used to sell the insides to the bakeries and have them reserve the shells, of which we made hair receivers and jewel cases. We sold them as souvenirs when my wife ran the ostrich farm in the concession at Idora Park, Oakland, California.

Afterglow

December 27, 1935

Christmas is over. All the family [members] were home to dinner and a Xmas tree; there, each child received—along with numerous small presents and remembrances—a quarter-section of land with a ranch house fit for a home; a pleasant evening.

William Tracy

Martha continues:

As hard times continued, mother went back into doing what she did best; she secured a teaching position in her community. Throughout her life, our mother was a dedicated and creative teacher—perhaps far beyond her time in her teaching techniques. Cecil, then, was left the complete care of our father, whose condition over a ten-year period had become so debilitating that it became necessary to confine him to the house.

During the last year of father's life, his malady became so severe that his mind gave way completely. In his deranged state, he would overturn all of the heavy ranch furniture. Father then came to look upon his eldest son as his jailer, and turned against him. This had to be a bitter pill for Cecil, who was so devoted to his father. Mother and Cecil struggled valiantly to keep him at home as long as they possibly could, but finally, in May of 1941, they could no longer manage. It was with heavy hearts that, in desperation, they placed him in an institution, where he died within a few weeks. Even though father had been a burden for so many years, both mother and Cecil were devastated by his loss. It was with great sadness that we buried our beloved father. He died on July 4, 1941, when I was just two months pregnant with his second grandson.

"To Thine Own Self Be True"

Several years after our father passed away, Cecil erected a monument in his memory. What could have been a more appropriate symbol than our beloved old dinner bell? The bell was suspended from ten-foot-concrete scaffolding, at the base of which were inscribed the words "To Thine Own Self Be True." This was our father's favorite quotation and the creed by which he lived. This was Cecil's way of paying tribute to this honorable man.

The Dinner Bell with Will Tracy's homestead cabin in the background

FANNIE ROWLEE TRACY

My Mother, Fannie

Fannie California Rowlee was born near Clements, California, on February 18, 1877. She was the third of five children born to Martha Martin Rowlee and her husband, Charles Wesley Rowlee, and she was preceded by sisters, Nellie and Dolly, and followed by a brother, Gilbert, and a younger sister, Hazel.

Her mother was pregnant with Fannie when the family migrated from Iowa to California by horse and wagon. Her father gave up his profession as a teacher to try his hand at wheat farming in San Joaquin County, near Clements, California.

Fannie was the more creative of our parents. There is little doubt that her offspring inherited their somewhat theatrical behavior from her! But above all else, Fannie was a dedicated teacher. Once the school mischief-maker settled down and applied herself, she studied hard enough to endanger her health. At normal school she became the favored student of the school's most exacting professor. She was well educated in the use of grammar, and she saw to it that her children did not commit common errors of speech.

Child of the Plains

In search of more fertile soil, Charles Rowlee moved his family from Clements to one desolate location after another, and Fannie was left to her own devices. No toys or books or neighboring children occupied her time, so she created a world of her own. The living creatures of the desert became her friends and playmates. She talked to frogs, turtles and "horny toads," and she gave them names. In her fantasy world, puddles of mud formed by infrequent rains became the fields in which she planted her beans, wheat and barley seeds, and because beans sprouted easily, they were her favorite. Rocks of different shapes and colors became cows, horses, sheep and chickens, according to size.

At one time Fannie decided to train a pet calf to be ridden. Snubbing him to a manger, she sat on his back for long intervals until he became accustomed to it. Eventually she trained him to accept saddle and bridle, and he became her mode of transportation. She must have been an amusing sight, perched on the back of a calf!

Cowboys played an important roll in the life of the pioneer family. Sometimes they were the only source of communication. Fannie became friendly with these colorful characters who often stopped by their home. To entertain them, she made their images in statues of mud and leaned them against the fence posts to dry. With only mud at her disposal, Fannie used mud as a form of self-expression.

Excerpted from *The Tracy Saga*

Some Memories from Early Childhood

As I recall now, some of my earliest memories include the large green letters across the front of my Sperry Mills flour sack nightgown. My attention was called to these letters each morning by those who dressed me. While they all said pretty, I repeated it as "prettily." Soon everyone said "prettily," and I was very happy for I had become the "teacher." I remember, too, having difficulty with many other words. Vinegar was one that distressed me very much because I could only say "vingar."

My parents had come to California in the depression year of 1876, accompanied by my two older sisters, Nellie and Dolly. Father, at the age of twenty-one, had entered upon a teaching career in Iowa. Within a few years, he gave up a promising professional career in schoolwork to come to California, eventually settling in San Joaquin County. Here he changed his occupation, and became a farmer.

The family finally located in the rolling hills near the small town of Clements about twenty miles northeast of Stockton. Clements was a bustling city in those pioneer days. Father soon became a dry wheat farmer of those rolling hills, and there I was born on February 18, 1877. It was there that father gathered his first serious experiences in farming. Of course there were no irrigation problems in those early days, for the dry hills depended entirely upon the rainfall, according to the whims of Mother Nature.

Some of my fondest memories date back to the activities connected with farmyard life. The chickens, being the easiest to find, quickly attracted me. Soon too, I began to do a great deal of exploring. More often than not, I ran into some trouble, which usually sent me back to the house in a hurry. I recall that late one afternoon while on one of my prowls, I saw a terrible looking animal coming slowly toward me. As it advanced, its head kept turning from side to side in a most peculiar manner, as if to feel its way along an uncertain path. This "thing" seemed to have the body of a pig, and the head of a fish. As usual, I ran for home in a hurry. Mother laughed merrily when she saw me come running. It seems this monster had been created by a small pig rooting his way along into the discarded head of a fish, a large sturgeon. The sight was one to scare any child.

A creek ran not far from our house. As it was dry most of the year, it was called Dry Creek. Along its banks here and there grew buckeye trees. They bore a seed, or nut, much like a chestnut in size and color. I was taught at an early age that those nuts were not good to eat. However, I found them useful in my imaginative play. On my make-believe farm along the creek bed, the nuts became chickens, pigs, horses and cows at my will. The dwellings and fences as well as the livestock on this farm were all rocks, but not alike in size, shape or color. My favorite pastime was to set the hens on tiny white rocks. These hatched as I wished. Then the mother hen and her little rock chickens were put out in their rock pens. My farm was my very, very own. I lived a wonderful life there with chickens and pigs as real as any that ever existed. I lived in a world of joy and fantasy.

In time, I lost interest in my imaginary farm. Now it became time to train my pet pig to carry me about on his back. I had named him Dick Pig. Though I put him in a corral, he knew where there was a hole in the fence. He regularly ran for that opening and quickly scraped me off. I thought this gave me a fair ride and found it to be great fun, so I rode Dick Pig in this style until he went the way of all good pigs. I begged for his tail, which was short, but the man who helped father said he was entitled to half of Dick Pig. To tease me, he took the half that included the tail. I was very disappointed and unhappy; then, the next day, a messenger brought me a package. Inside it was Dick Pig's stubby tail.

Another time I became curious as to how a hen laid an egg. Deciding to find out, I secured a model from the hen house, which at the time was probably a setting hen and would not lay an egg for many weeks. My mother often called such Dominick hens "old settin' hens." Sitting down beside the house, I pinned my victim close in my lap, and started clipping feathers from the end where I expected to find the eggs. Though I found no eggs, I continued to clip. By this time the hen was getting quite naked and exposed. Upon hearing someone laugh, I looked up to see an audience at the window. I quickly gave my victim her freedom, but she was hardly respectable enough in appearance to reenter the flock.

"Setting" a Hen

I played a lot in the barn, as there was no other shade. One day, I thought our chicken population should be increased, and that I should be the one to do it. Securing a roomy box, I constructed a nice nest of straw, putting in the usual number of fifteen eggs. I then secured a hen of the variety to sit on the nest. She seemed a little dissatisfied with the duty I had assigned to her; so I nailed a few slats across the top. When I left, she was still standing up and cackling her dissatisfaction.

When I returned in the morning, she was still standing over the lot of crushed eggs. She was eating breakfast too, as I had not thought to provide any other food. I was somewhat surprised and disgusted, but not dismayed. Studying the situation, I thought how easily it could be managed. I would confine her to a smaller box whereby she would lose even more of her freedom. I secured a wooden soapbox, which once held one hundred bars, yet was only about six inches deep. Giving her fresh eggs and a new nest, I carefully observed that it was now impossible for her to stand. I left, assured that all would go well. Though she could not stand, she must have wiggled a lot during that night, for my morning inspection revealed a nest of scrambled eggs and shells. In disgust I gave her the freedom she had won and gave up the chicken business until a much later date. The eggs could have been used to more advantage, for times were very poor.

Cowboys

In those early days the cowboy was a very important man who was well liked by the settlers. He was ever ready to help out in any difficulty as he rode about the country. In a way he was the mail carrier. Riding up to the door, still in the saddle, he sat with hands and arms folded across the horn leaning forward in a restful position. So situated, he narrated all the events and news he had been able to collect. He rode ahead of the floodwaters warning those in its path what to expect. The usual phrase was "Poso Creek's down." He was always hungry, so he was always invited to come in.

Mud Cowboys and Barbed Wire

When I was small and we first lived on the homestead, it was a lonesome life, indeed, for a child. The scarcity of things to make a child's life interesting was predominantly great. I had never seen a barbed wire fence, and now that one ran right by our house I was very interested and sought to make some use of it in my play. This fence took the place of trees, for not one existed for miles around. We very seldom saw anyone because there were no neighbors. Miller's chuck wagon passed by occasionally, always leaving us a quarter of beef. Cowboys enroute stopped to chat and water their horses. Between their visits and the time they might appear again, I made mud images of them all.

Moving About Came Often

I still had no thirst for knowledge. I didn't even want to attend, and the new school certainly did not bring happy thoughts. I attended because I had to, for father sent me everyday. Unfortunately, we lived only about a mile south of the school. I did like the ball games and pump-pump-pull-away at recess but, inside the classroom, I spent much of my time drawing pictures on my slate. I opened the books only to close them again. Any distraction I could think of that would disturb the school room and not be too obvious was employed. In spite of my behavior, my teachers liked me and were very patient. One thing they did not seem to realize was that I had no substantial foundation; therefore I had no interest.

In spite of this, I was gradually acquiring a little knowledge of reading. One day a fairy tale book fell into my hands filled with many pictures. These tales seized upon me as nothing had before. Taking this book, I crawled into the heart of the forbidden rose bush and was completely lost in fairyland. Lovely princesses were carried away by handsome princes on beautiful white horses. Ogres and wicked men were punished to my satisfaction. Perhaps the thirst for these stories had more to do with my learning to read than did the schools of that day. The next year a schoolhouse was built, and old-fashioned sage plants were set out to improve the looks of the building.

Sagebrush, School and Me

Eventually, we moved our house from "77" and set it on an unfenced portion in the sagebrush about a half mile north of the red barn. Though the land was not of good quality, Father was quite stubborn about planting wheat. Finally, he did harvest enough to pay a year's grocery bill in Delano. Apparently, too, one did not worry too much about whose land he settled on. Residence was rather temporary anyway.

My father was adamant about my continuing in the Hamlin School, and as usual I fought the plan. To make sure of my daily attendance, he bridled my horse each morning before breakfast, and tied him to the rack. This animal, given to me by an uncle, had been broken as a cow horse, and was young and spirited. When I first became his owner, it took two men to see that I was safely mounted. I rode bareback and sideways carrying a lunch pail but no books. Girls in those days did not ride astride, and we certainly took no books home for night study.

My horse had been named Cocha by the cowboys, which meant "pig" in Spanish. It helped to describe his plump build. We did make a gay pair! For awhile I tied him to the school fence during the day, mounting him by climbing on the fence and jumping on. The minute I touched his back, he was off on a run. However, he did not run away unless something in my lunch bucket rattled. I was always careful to wrap my spoon or anything else that might cause noise.

There was a wide shallow ditch full of water in front of our house, and we always landed with a thud in the middle of that stream. Water splashed high and showered over us. Riding sideways bareback took a lot of skill and did not leave me free enough to properly handle a horse with this spirit. No harm came from it, though. I was thrown a few times, but when this happened he always stopped. As soon as he felt my body across his back, he was off to a quick gallop with me on my stomach, struggling to an upright position.

Pleasing My Parents

As a child I was very sympathetic to my parents. I wanted to please them. They in turn always placed great responsibility on me. It seemed that Fannie was always putting hay in the manger, loosening collars and croupiers, or cleaning the barn.

The kitchen's wooden floor had to be scrubbed with lye, and newspapers pasted on the walls where needed. The yard had to be cleaned or trash raked out from under the house. The wood box needed filling. Fannie was always busy.

I do not remember that my father ever had any socks such as men wear now. Strips of cloth called foot rags were made from a flour sack. One sack could be made into about three eight-inch strips. One of these strips was then carefully wrapped around each foot. It was quite an art to do this smoothly. Beginning at the heel it was crossed over and around, until the foot was encased in a perfectly neat sock or foot rag as it seemed to be.

I was fond of trying to please my parents by performing small services. One of these was washing the foot rags. To be sure that my father always had a goodly supply on hand was a pleasure for me. I worried too about the brown spots. The stain came from the shoe leather about the heels and would not wash out completely.

Mother loved to have me comb and brush her hair. This seemed to sooth her during difficult times. I tried to be of good help to my parents.

Animals Were My Friends

Bunt, our calf, became my mode of transportation as I trained him to be a saddle steer. Breaking a steer to ride and guide is not an easy accomplishment, for I didn't have any instruction as to how it might be done. So, I used a method of my own design. At this time Bunt and I both were a little older. Though still quite small, I was vigorous and spirited. Bunt was of the Durham breed and was quite a hunk of meat. He was gentle, as he had been kept on the end of a rope attached to a picket pin. It was my chore to move the pin each day so that Bunt would have a new area for feeding. In this way he became used to me.

One day I took him in our shed barn and snubbed him close to the manger. This still gave him some freedom. He could not move his head, but could swing his body from side to side with me on his back. Soon he grew accustomed to my being on his back, and I could "stick him," a phrase used by the cowboy bronc riders. However, it was not the same when he had his freedom, as I was to find out. The first day I put him on the road, I fastened a long rope about his neck, and asked my brother to slow him down should it be necessary.

Bunt took off at a pretty good pace. I called out, "Hold him down! Hold him down!" The answer came back, "I can't! I can't!" and the calf was free. He increased his stiff-legged pace, threw me forward, and I found myself just behind his horns astride his neck. Every jump brought his head low to ground. Soon I was sitting in the dirt.

To avoid this and to be seated more securely, I saddled him the next time and used a bridle. In time, I could ride Bunt anywhere I wished to go, but never at a very fast pace, unless something frightened him. I even rode around to the neighbors to show off my skill. It was a sad day, indeed, when Bunt had to be butchered.

Next, I tried two smaller calves. This time, they became a team. The calves had been running with their mothers in the pasture, and so proved much more difficult to manage. I kept them in a corral for several days, mauling them around while trying to fit them with a suitable harness. At last I deemed them ready to be put on the road. The gate was a double one, and I told my brother to open it. Unfortunately, I had forgotten that these gates had short posts or stops where they came together. Mounting myself in a box wagon, I gave the sign to open the gates. The calves, seeing their way to freedom, started for it at a fast pace. Everything held together until my box hit the little post in the middle of the gate. This was very disastrous for me. I had a terrific jolt in the smash-up, and the calves, in a tangle of harness, sped away. I never tried this activity again.

My attention now turned to riding a donkey that ran in our pasture. He was an uncontrollable outlaw. In fact, a bad actor. I had seen him stand patiently until well packed, then put his head down and buck so hard the yard was strewn with frying pans, pots, kettles and provisions. The donkey and I got along very well, though we had our troubles. One of his favorite tricks was to run up on top of a pile of discarded machinery and junk. There, high on this pile, he would stand until I dismounted. However, we grew to be good friends, and I rode him any place I wished to go.

Later in life, I found one of my own small children pinned up against the fence between the horns of a wild cow. She was standing quietly while he proceeded to carefully pluck the cockleburs from between her horns. Then there was the billy goat that never caught me when I was a child. A horse will often stop when a child falls from his back. My horse, Cocha, spirited and free in a large open section, would come to me each day at four o'clock to be ridden home from school. As I have gone along in life, I have come to believe animals respond differently to children. A child can often go in the field, catch a horse, and have no difficulty while, for a man, he would have to be corralled. There seems to be some unaccountable understanding between animals and children.

I nearly got into trouble one day when I hitched our saddle horse, Old Gray, to a piece of draper canvas from an old header. I fitted Gray to a collar that he had never felt in his life. He knew only the bridle and saddle. He had never felt tugs about his legs; therefore he was not too satisfied with my venture. When I got that piece of canvas with the slats fastened to the bottom, loaded with sagebrush for firewood, it was a sight to frighten a bunch of wild cows. They came running, their horns thrown high, completely surrounding me. I stayed still, and let them fully investigate me, before I moved on.

Helpful Catalogues

The most I can vividly remember as a child is feasting my eyes upon the pretty things as I turned the pages [of catalogues]. Often I made lists of things that I hoped in some way I might be able to send for, but always my pleasure came just from anticipation as I laboriously jotted down the items. I remember too, those items that had to be returned, especially my mother's shoes, for she had a very tiny foot, and the catalogue-ordered shoes were always too large. Anyway these catalogue packages always brought a lot of excitement just to get them and see them opened.

One time an early bachelor settler who depended a lot on his Sears Roebuck periodical and who always looked for what might be new and interesting, ran across the item toilet paper. He fully intended to investigate this new item, but by the time he got around to writing the company, his catalogue had gone the way all Sears catalogues went in those early days. So he just scribbled for them to send him an order of toilet paper, paying no attention to the cataloging numbers. He received in return a letter asking him to state number and kind or whatever hieroglyphics catalogues were using at that time to designate their product. The old timer answered them rather abruptly because the order was not filled promptly by saying, "If I had your catalogue I would not need your toilet paper."

A Glimpse of Another World

As the Rowlee girls became old enough, they boarded with other families to assist with the housework. This glimpse of the outer world began to stir an awakening in Fannie.

A Summer Job

One summer as our school closed, I thought of earning some money. Luckily I found a place in Delano as assistant housekeeper with the large family of C.A. Weaver. My job was helping with the meals, doing all the dishes, making the upstairs beds and sweeping and dusting. Of course, there were no vacuum cleaners in those days. I also helped on wash days by running the "wash" machine by hand. It was necessary for me to be the first one up in the morning to start the fire in the stove for breakfast. In the family there were five children, mother and father, a blind grandparent and myself. They were a very nice family, but five growing children can make housekeeping very complicated.

I soon found it very difficult to keep up with my duties. The mother seemed to feel the burden of her part, and took a little rest when nursing the baby, but that was about all. She went upstairs to inspect my work occasionally. I had quite a bit of difficulty there.

One girl about my age used the upstairs with her playmates to sew and dress their dolls. They cut and sewed doll rags, as they were called, always leaving much clutter for me to pick up. The girl was wont to look upon me as a servant, which I really was. The father, having a hardware business in town, was not home much, but when he was, he seemed to realize I had too much to do and saw to it that his daughter gave me a hand.

By now I was about fifteen years of age, of very slight build, being about 5 feet 4 inches tall and weighing less than one hundred pounds. I was getting very weary, but did not have the disposition to make it known, nor did I wish to shirk my duties. However, I did set about to lighten my burdens.

In the early morning I picked up the toys scattered about here and there, clowns, jumping jacks, tops, guns, plus play gadgets of many varieties. Each day as these reappeared, I put some in the wood cook stove. The fire consumed all except the wires. These I took out later with the ashes, and put them down to the outside toilet. The doll rags and a doll or two went the same way with less trouble if they happened to be rag dolls. Gradually, things eased up, and it was surprising how few complaints were uttered about missing toys.

There was one problem left which took more courage than the doll rags. This was the matter of the canary bird. It was agreed that the young lady could have a much-wanted bird if she would take proper care of it and give of herself. Things had gone well with this chore, until it grew to be an old story. Then its keeper grew wise, as young ladies, including myself, often did, and found a way to put the weight on my already burdened shoulders. The bird was to have a bath, his cage cleaned and cups filled once a week in addition to a certain amount of daily care. All was to occur each Sunday morning before Sunday school.

Probably the reason I was not taken to Sunday school too, was because I had no pink or blue sashes. The town girls all wore wide sashes and big bows with their white dresses. Perhaps I was something of a Cinderella, so I stayed home and took care of the ashes.

I was left at home to care for the small baby and prepare lunch, and soon the birdcage problem was added too, perhaps unintentionally at first. It so happened the cage was put on the floor preparatory to a bath, but not in time for the bird to complete his morning toilet before his mistress left for church. Knowing that there was also a pet cat sleeping somewhere about the house, I took full charge of Mr. Canary. He was chirping, even singing by the time the family arrived for lunch. This went on until I was satisfied the love for the bird had cooled to the point of neglect, and its care had shifted fully into my category.

One Sunday morning my duty lay between the bird and a crying baby. I put the baby in its buggy and intentionally took it for a ride. It finally fell asleep about the time the family came back home. Soon there was a great commotion. The cat was not asleep under the bed as usual but was eating the canary there. I was not blamed too much, but the young lady received the punishment she well deserved for shirking her responsibility.

On the house where I worked was a little balcony porch jutting out from the second story. It must have been ornamental, for it was not large enough to be useful. A door opened from a hall onto this balcony. I often went out there, broom in hand and somewhat leaning on it, and gazed far across the landscape where an endless world of gray sagebrush met the eye and then faded into the horizon. I knew too well what lay out there. I had trailed through that brush, some of it so big it looked like a herd of cattle in the distance. It gave me a lonesome feeling, perhaps homesickness. Then I turned my thoughts to the future, something I had never taken time to do. Always before it had been the busy present that absorbed me completely. Out there was truly a big wonderful world.

> The only newspapers I had contact with were those I had borrowed to put on the walls of our kitchen to make it bright and clean. I had seen pictures in them, but they had awakened little interest. I cannot remember having seen a magazine in my early life. However, men from other climes had occasionally dropped by our home and some stayed for the night. Their conversations interested me greatly and I listened intently. Some of these men were very talkative. Perhaps some may have been professional rovers. They seemed to be wise to the ways of the world, at least they appeared so to me.
>
> It was here, leaning on my broom, that I determined to see and get a taste of the unknown life that lay just outside of the circle of my little sphere. This feeling was a new one for me, and it could be the same force that has caused young people to leave home, as they do many times, with no plans for the future. However, nothing like this was in my thoughts. I loved my home, but some way I must go farther. It so happened that by this time a former employee, an older girl who had taken a rest, returned to the Weavers, so I was sent home again.

After returning home from one of these assignments, Fannie heard that a cannery was opening in Visalia and that workers were needed. Although Visalia was a great distance from their home, Fannie made arrangements with a family who happened to be traveling in that direction. After a long, hard journey by horse and wagon, she arrived and spent the summer working in the cannery:

The Struggle for Education Begins

> By this time too, I was beginning to think of attending Wildwood School. Mrs. Stinson in her talks with me had opened a new horizon, the possibility of an education, and what I might do with it. However, my plans did not work out immediately, so I continued my schooling at Hamlin, knowing that I still had four years to go. I knew that if I would settle down at Hamlin and do my best, perhaps in time I could persuade my parents that a move to Wildwood would be a wise one.
>
> It was now a long journey to the top, but to me education was a new discovery and my frustration was brief. To obtain this elusive thing now suddenly became pleasing. I endured my torture because of full realization of what had happened to me, and with extraordinary ability became self-disciplined.
>
> At first, thoughts rushed through my brain like floods rushing through the land unchecked. This I endured. Intrusions came, but stayed briefly, and I renewed the battle of thoughts. To cloud my troubled mind and body appeared some hopeless facts. Paramount was the matter of funds. In my life as I lived, money had never been any worry to me. Galloping over the plains, and herding my father's cattle, along with the puny social life that existed, had heretofore been sufficient diversion to keep my mind within bounds. All the simple activities that had been necessary to keep me pre-

occupied now fled, and a hunger for knowledge tortured me into achievement.

While no one believed that such a transformation could or would take place, it did. We had coal oil lamps, and I sat by these studying many times long past midnight. My mother, fearful for my health, often came and sent me to bed. The walls and floor where I studied were bare. Nothing hid from view the rafters and rough shakes of the roof. Truly a dismal place at midnight to some, but this now went unnoticed, hidden by my new enthusiasm for study.

My mother tried to do something about all this. She started making carpet rags by tearing old clothing into narrow strips, then sewing them together and winding them into balls to be woven into carpets. There were always a few women in those days who owned looms and did this weaving of rag carpets. The neighbors and family helped, as well as myself when I found time. I still had to do some work with the stock as we had some horses to pasture.

A yardage called cheesecloth was tacked on the walls, which when uncovered were rough and ugly. The edges were lapped and tacked together, until all the boards were completely covered. Strips of netting were sewed together and then tacked to the rafters, thus hiding the shakes from view. The floor was covered with about three inches of fresh wheat straw, and the rag carpet stretched tightly by putting nails in the end of a long stick and pushing or stretching the carpet close to the wall while someone tacked it down. My father pitched in, too, with this work. When all was finished, it was a pretty room indeed for those days.

Of course, as always, some folks in town had better. Many had houses papered with printed flowers pasted over the cheesecloth. Hollyhocks and marigolds bloomed in their yards by the gate, but none were so attractive to me at this time as my very own room with its pretty white walls, and the beautiful colorful carpet with stripes of all hues. What a soft touch to the foot as you stepped on the straw padded carpet. When nights grew chilly, a small brushwood stove was added. All this added greatly to my comfort as I continued to study diligently in my drive for education.

After continuing in my home district for two terms, a new teacher, a Miss Adamson, came to teach our school. She was a graduate from San Francisco Teacher's College, and, I believe, a very good teacher. However, I was to take no chances and determined this was the time for me to renew my efforts to go to Wildwood. Impossible as this seemed, I accomplished it by the time school started the following year.

A place to live was the first obstacle to surmount. I had remembered a deserted claim with a squatter's cabin built in the sagebrush some two miles away from Wildwood. I thought if only I could persuade my sister Dolly to come with me, my parents' objections to a change would be more easily overcome. Dolly had been sent to the schools in Delano and had completed the required eight grades. Her schooling had been continued during the time we had lived in various isolated places. She had stayed with a family, helping them with the housework in payment for her board while attending school. Dolly had a good memory and proved a capable student. Her higher education would come later when we both attended the same college.

Since there was no compulsion about her attending school at Wildwood, she kept the house after we moved closer to the school. She was afraid to stay alone with the cabin so far out in the brush; so we rode hard rubber-wheeled bicycles through the sagebrush to school. These bicycles had been given to us by a close friend, the Superintendent of the Cox and Clarke Cattle Ranch. He had taken many meals at our place and had played numerous games of cards. Father, at my request, came down with a scraper and cleared a path through the brush the width of the blade, giving us a smooth yet somewhat hard trail. However, it soon became soft and dug up with gopher and owl burrows. Bicycle riding proved very tiring.

One day I thought how foolish it was to take this physically exhausting trip when it might be possible to have the cabin moved nearer the school. Accordingly, I wrote a letter to Father explaining the situation, and what I wanted. He came down a second time with his horses and moved the cabin, even though it did not belong to us. At the time, our cabin stood out in the brush near the trail between Miller's Headquarters and Jerry Slough. When word got around that two girls were living there by themselves, no one thought such a thing possible. It was not unusual for the foreman or cowboys to investigate for themselves. When we answered a knock at the door, we always got the request for a drink of water, this being the logical reason for stopping.

The cabin was placed in a fenced field that belonged to an earlier pioneer of Jerry Slough, Ferdinand Tracy. Up to this time my sister had been afraid to stay alone, but now she felt at ease and free to enjoy life such as a cabin could furnish. I have always been grateful that she did not become discouraged by my daring prank and return home leaving me to my fate of living all alone in its new location.

Although our cabin stood directly in the line of travel of the tramps or, as they were usually called, itinerant workers, employed by the Kern County Land Company at Famosa or by Miller and Lux at Buttonwillow, there was never any molestation, and these men never asked for any directions to find their way. Those headquarters were well known to the roamers of that day. In a single day seven passed our cabin without stopping.

At last the time came to return home from Wildwood. My elementary schooling had been completed. After a few days rest I went to Delano to take the final examination preparatory to passing the ninth grade work that I had been studying. This test was given primarily to pass pupils into high school, but at this time I was advised it would also admit one to a teacher training school for a three or four year course of study.

Such schools at that time were called normal schools. It was my desire from the first to become a teacher in the grade schools of California. In those days one could become a teacher by preparing or "stuffing" as it was called, or by brushing up and taking a county examination. This method never appealed to me, for it did not seem to be substantial, as the county certificate would only be good for two years before another test must be taken. One reason perhaps for this decision was that I had nothing much to brush up on. Because of the lack of higher training, I deemed it advisable to take the high road and go all out, all the way. In the end this gave me a life certificate to teach in the California schools. About this time I believe, county high schools were coming into being, but I had never heard of them.

When I finished the elementary grades, I was in a very poor state of health from over-study, anxiety and lack of proper nourishment. In our cabin we had kept only potatoes and flour, a little jerky at times and some skimmed milk brought to us by Mrs. Stinson, our teacher. We had tried to keep a cow, but she caught a fever and died. We were grateful for the skimmed milk, but when the Tracy brothers discovered we were using such milk, they furnished us fresh whole milk daily. It was usually delivered by the older brother on horseback. Their cabin was only about a mile from where we were. At the time, Theodore, the older brother, did the chores and kept the cabin while the younger brother, William, worked with his teams on the railroad that the Southern Pacific was building to haul asphalt blocks from the oil fields at McKittrick.

I do not know how long or how far Mr. Tracy worked on this road, but as far as he did, his scraper was the last to go over it. At that time he was living on Section 32 as a homesteader. His cabin was the standard homesteader's cabin size, 10' by 12'. In his cabin were two cots, a stove, a table, and a homemade cupboard. Mr. Tracy had made the table and cupboard that we still have today. In fact, they have stood all kinds of abuse, but show no signs of breaking up. He said a five-gallon can of coal oil lasted them one year. They only used it in their lamp, as they burned sagebrush in the stove.

Finally, the time came to take the examinations. My brother and I drove into Delano each morning for a week to take those final tests. He realized my weakened condition, and the last day he tried to cheer me up. He said he had a little money in his pocket and would buy me a can of sardines. Because of my fondness for these little fish and the fact that canned goods were very scarce in those days, this was a great treat. I think at this time the sympathy was more needed than the nourishment.

Miss Kern County Goes to College

A long anxious time followed as my sister and I waited for those important test results. Finally, the letter came bearing the good news that we had both passed, and were now eligible to enter high school. I had the honest feeling that I had not done well in the test and perhaps did not pass. Did the local officials need a flock of students for the purpose of building up the countywide high school? Had they made the test easy? All these questions ran through my mind. Whether this was true or not was none of my affair, and since I later made excellent grades and graduated with honors from college, it can be said that I made up in diligence what I might have lacked in knowledge.

A greater task now lay just ahead of me, that of obtaining information concerning college. I had already sent a letter to all the normal schools in the state except San Diego. This one was new and I was not then aware of its location. San Jose answered saying that they could not take any more students. Chico said it could admit additional numbers and that registration was still open. I sent to Chico for more literature. When it came, I asked my father to look it over and give me his opinion. Father took the school manual and sitting down in the kitchen in the light of our reflector lamp carefully read it from cover to cover. I kept a watchful eye on him. His interest gave me encouragement.

When he finished, he carefully closed the book and handed it back to me with no comment. I then asked him what he thought. He said, "It's a good school." No more was said.

The publication, *A Precious Sense of Place, The Early Years Of Chico State* by the Friends of the Meriam Library of Chico State, chronicles the history of that institution. The following excerpt gives a good idea of what Chico Normal School was like in Fannie's day. Her father may have read something like the following:

> Students entering the Normal School came from widely differing educational backgrounds, for common schools varied in quality and there were few high schools in California before 1891. They were required, therefore, to take examinations for placement in either the Preparatory Department or the Professional Department, the latter allowing them to complete their training as teachers in two years. Of the first ninety students admitted in 1889 only fifteen were placed directly into practice teaching; seventeen were assigned to the Preparatory Department, needing as much as two years' work to qualify for professional training.
>
> At the start of each day, both students and faculty were obliged to attend morning exercises, whether "devotional or literary," but prayer and readings from Scripture were frequent enough that students came to refer to the start of "chapel."
>
> "We are glad you have come to a normal school, if you desire to fit yourself as a teacher. If you do not so desire, the school would be better off without you." So wrote a student greeting incoming students in the Normal Record of October 1897. Indeed, at entrance students were required to sign a declaration: "I hereby declare that my purpose in entering the school is to fit myself for teaching, [and] that I intend to teach in the public schools of the State of California."
>
> Three features were judged essential to the formation of teachers: a "thorough grounding" in the subjects students were preparing to teach; a "course of study upon the science and art of education," and practice

> teaching in a training school. This was "the normal school idea" – a single-purpose institution for the development of teachers.

The Catalogue of 1897-1898 left little doubt about how the students' lives should be lived:

> *Discipline*
>
> *The regular study hours, from seven to ten P.M. should be unremittingly observed all days except Friday and Sunday.*
>
> *All unnecessary promenading upon the public streets should be avoided.*
>
> *Absence or tardiness, except in the case of sickness, should be entirely unknown.*
>
> *Keeping the company of the opposite sex is, as a rule, inconsistent with strong work at school.*
>
> *The habit of speaking of teachers, pupils, or others in complaining or uncomplimentary terms is harmful to the well-being of the speaker, and should be suppressed in its incipiency.*
>
> *It is the duty of everyone to be cheerful; to avoid worrying; to be just; to be healthy. Hence, each student should be regular in all matters of exercise, diet, sleep, and study.*

The original curriculum of Chico Normal adhered closely to the ideal for such schools. Training in the art of teaching was joined with the study of language, mathematics, and science in each year. To this were added such useful but "miscellaneous" subjects as "penmanship, morals and manners, vocal music, drawing, and calisthenics."

Initially, even the program of the Professional Department was at a pre-college level, but by 1909, work done there could gain college credit for transfer to the University of California. Similarly, studies in the Preparatory Department eventually became equivalent to high school, and when, in 1917, high school graduation was required for admission to normal schools across the state, the Preparatory Department at Chico was eliminated.

***Tracy Saga* continues…**

> I believe now, the probability of my chances to go to this college were so remote that my father gave the idea little, if any, serious thought. There were several problems, weighty ones, to be solved. The matter of money was a most serious one. Clothes entered in, and shelter and sustenance followed too. My father's coolness did not discourage me, though I was saddened by it. However, I took my problems in order.

Again, I set out to interest my sister, Dolly in joining me. She at first did not enter in wholeheartedly, but finally agreed that she would if I could get the money together and some new clothes. I thought first of my small collection of calves. They had all been "leppies" brought to me by the friendly cowboys. A leppy was a young calf that had no mother, usually because she had died. Rather than let them perish, they were given to the settlers whenever possible. I had raised my calves by hand on milk from which we had skimmed the cream to make butter. One of these had grown to be a steer big enough for the butcher, though hardly fat enough. However, I sold him for forty dollars and this became the nucleus of my college bank account. I had other calves of various sizes, but not as yet very valuable. My father later made us some kind of a deal with a neighbor for these that gave me a few more dollars. In addition, I sold some chickens and eggs. A cousin who had come out to California for her health was, for the time being, making her home with us. She was handy with sewing and helped us put together our scanty wardrobe. I had already begun to dream about the day I would graduate and imagined myself draped in silk. Sears and Roebucks sent my mother a large bunch of silk samples from which I selected a piece of corn-colored silk material for the "grand affair," my graduation. I carried it with me when I went to tend the cattle on the range. There I could admire it to my heart's content with no questions asked. I also pictured my family all out in front beaming over my achievement. From this time it would be four long years of hard study. When the great event really came, my dress was of simple organdy, not silk. Not one of my family was out in the audience to witness the "grand affair." How wild can a girl's dream be?

The day came when we finally took the train at Delano for Chico. After paying for our transportation, we had very little money left. A short stopover occurred in Sacramento to change trains.

After lunch, we continued on to Chico. By now I had one dollar and fifty cents left in my purse. That should be sufficient, I thought, as I had answered one of the letters sent out by the school and made arrangements to have housekeeping quarters with our lodging. Upon reaching Chico we found it was raining torrents, and to complicate things we knew nothing of the town. Luckily where we were going proved to be but a block from the depot. The cab driver who took us there charged fifty cents. A knock at the door brought the lady I had corresponded with. It was Mrs. Walker, a pioneer in Chico. Her husband operated a brick manufacturing business. She told us she could not keep us as originally planned. Her son had made a visit home and told his mother that he objected to her keeping any more college girls. They were too much worry and bother. Mrs. Walker said she was sorry, and her door closed us out in a stormy dark night.

Fannie Rowlee at age 19, just before going away to Normal School. Sam McQueen was courting her at this time. A cousin made her dress from pink cheesecloth purchased at 3¢ per yard.

It was seven blocks back to the middle of town where we found a room for one dollar, which was all we had left. Wearily, we went to bed. In the night I woke up with a queer feeling that my legs were going to sleep. I had little use of them. I woke my sister and told her I was dying. She rushed downstairs and located the bathroom. Returning quickly, she took me down and got me into a tub of hot water and I lived to complete this story. I don't remember why I was ill. Perhaps it was the excitement and just being worn out.

The next day was a strenuous one. We had no money and really no place to stay. Taking the sheet of addresses sent us by the school with the names of those who would keep schoolgirls we started out. The more we walked to find these places, the more discouraged we became. We found run-down places, dirty children playing, and disagreeable dispositions. None of the places proved possible or acceptable. Hunger overtook us to the point of our being nearly famished, when I stopped short in my tracks and said to my sister, "That first woman just had to keep us!" and back we went to see her. I told her what a time we had had, and she seemed sympathetic, particularly as her son had returned to his law practice in Ely, Nevada, and no longer stood in the way.

She first showed us a little brick servant's room in the yard. (The Walkers were well-to-do and lived in a large brick house in front.) Then she offered us the use of what had been a pantry. A Chinese cook had used the room for sleeping quarters. She said we could cook on her stove as she and her daughter lived there alone most of the time. The little room was clean and large enough for a table and two chairs. The price was ten dollars a month payable when "I had it." Our bedroom was upstairs. Luckily, Mrs. Walker let me purchase milk on her bill. Now, at last, we could get some much needed sleep and rest.

Thus it was that, fresh from the wilds of the bandit-ridden area of Kern County, I entered the field of higher education. The plan that I had nurtured for so long was coming along. By this time I had become somewhat talented as a persuader, yet was entering this field without the full approval of my parents. They really did not approve of my desires. The ground preparation for my educational venture was so inadequate. As yet, no preliminary schedule existed for our financial support. A forceful reminder of this had squarely faced me that first morning I found myself and my sister on the street of a strange city, hunger and cold gnawing at our bones. This perhaps furnished the push for further efforts to survive. While I was strongly reminded of the folly of this venture, and the fact that our survival would be competitive, this did not deter me in my plans. I thought that like Columbus I must "sail on and on." To this day I am deeply grateful to those understanding people who aided me in my distress.

In Chico Normal we were at once in competition with city-bred young men and women. Having been a cowgirl roving the plains for so many years I was ill adapted for this abrupt change. My courage was unimpaired however, and I was determined to adjust myself and become a worthy participant in this new life. An early highlight was a talk to the girls by Mrs. Bidwell, wife of the donor of the Chico campus. She was a highly intelligent woman. Her talk gave me inspiration to continue.

The lady teachers of my early day country schools wore little white aprons to protect their dresses from the dust and grime. So it was that I appeared the first day of class attired in a dainty white apron I had been so careful to bring. I was not long in correcting this mistake for I found aprons were not a part of a college girl's wardrobe.

Professor Seymour, who conducted [the biology] classes, was a very unusual and wonderful man who was not appreciated wholly by all of his students. This thought was brought about because of his concern for the importance of small details. If any of these details that seemed unimportant were neglected by the students they were excused from class. On these occasions he merely said, "You may go play tennis," and call your name.

To such students he gave what he called a "goose egg," which of course meant zero… I would like to say here that the teaching in this man's classes did me more good through life than all the others put together. The importance of detail is paramount in achieving success and eliminates many "goose eggs" as you go through life.

I discerned at once that this man's favor was important. It was said by some students that you could never graduate from his classes. This was a sobering thought, and a challenge to me. Failure would be a serious interruption to the goal I had hoped so devotedly to attain. He called for much assistance from pupils to help prepare specimens for class projects. I was always a ready volunteer. He called me "F. Rowlee." When there was a difficult assignment that would require much thumbing of books after class I would say, "Professor, I can't find so-and-so." He would often say, pressing his lips tightly together, "If F. Rowlee will look on page so-and-so in Cowes' Key, she might find it." I grew to be quite a favorite encyclopedia for others because of my excellent class notes in biology.

I scored even higher when he found out I was from the vast desert area of Kern County. My fame must have spread, for on the first class visit of President Ritter of the school, he addressed me as "Miss Kern County."

The first semester at college was a very difficult one for me. I had much to learn because of my neglected years of schooling. It became necessary for me to study each day right after school and far into the night. By so doing I made excellent grades. When school closed for the summer we did not have the funds to take the long trip home, so we went to San Francisco instead to stay with my sister Nellie and her husband. George had secured a position with the streetcar company of that city. By now they had four little children and things were still difficult. However, we did see quite a bit of the city life.

Back home my mother was nursing the sick. Miller and Lux had started farming their vast land holdings at Buttonwillow. To aid in this development a great number of Italian families had been brought in to settle. Many children were born and as there were no doctors either near the colony nor Jerry Slough, mother had many calls to work as a mid-wife and nurse. She liked the people and they paid her well for those times. From this steady work, she was able to eventually keep us in school until we completed our education. Sometimes there were long spells between checks but we kept our credit good without too much difficulty. We did receive one postal money order while in San Francisco. However we did not know anyone that might cash it. Finally I remembered the commission house that mother had shipped her turkeys to in the past. I presented the check there and told them my story. I can't remember if they made any effort to define the truth but they did cash the check.

One time my mother had some old turkey gobblers, so old and tough she was afraid to ship them, so she corresponded to find out what to do. The commission house said for her to send them along. They would fill an order they had for turkeys for San Quentin. "Tough people, tough birds," as it were. It may have been this story that refreshed their memory and helped to get our check cashed.

School continued very much the same the second year. I worked hard as usual. I had some trouble with my sister who did not like the pressure I put on her to study. She wanted to return home. I argued that there was no money to take us home. She tore a few leaves out of her book in anger, and the fuss was soon over. We did not fare so well the second semester of this year, for I contracted two serious illnesses. First the measles, and then a bad case of the flu. I was very ill and without medical attention, until I became delirious and things looked serious. The illness left me in a very weakened condition and I soon developed a bad cough. I had missed a lot of school and in some classes was required to make up work and take extra tests. In my weakened condition, I found it difficult to climb the stairs to the third floor to take the biology tests given by my favorite teacher, but most dreaded by the majority of students because of his hard grading. I was passed by all teachers as having made up my work, but I doubted my grades. I must have been passed for my efforts, made in such a weakened condition.

School closed with our first two years now over. My serious cough had started a chain of thoughts in my mind. A change of climate might improve my health. Chico had been rainy and cold. San Diego was known for its milder year-round climate. A new Normal School had been built there. It was now possible for us to be transferred and continue our schooling.

We left for home in the sagebrush to promote the idea, stopping again in San Francisco to see our sister. We found her ill. When we opened the door to surprise her, we found she was trying to manage her household from her bed. A toddler, one we had never seen, was pulling ashes and coal out of the cook stove onto the floor. His face and clothes had been well covered with ashes and coal dust. We persuaded her to let us take the baby home with us. To this she readily agreed. While riding the boat to Sacramento, we were unable to get any rest and very little sleep. The baby, we found out later, had a gathering in his ear that kept him restless. We had to find a way to keep him quiet so other passengers would not be disturbed. A miller moth kept flying around and hitting against the ceiling. The antics of the moth amused the baby. I would catch it and turn it loose, catch it and turn it loose.

When we reached home, we found our parents had moved back onto the original homestead intending to farm. They had started to improve it by drilling an artesian well, and bringing in a two-story house with the help of Bill Tracy, to take the place of the old one. They were now back in Jerry Slough where they were to live out the rest of their lives. It was indeed a vicious circle, the life they lived as pioneers of Kern County. By the end of the summer mother decided to keep the baby for a while longer. Luckily, our parents had agreed to a change of schools.

We found our youngest sister, Hazel, now twelve and out of school, doing much, as I had at her age. Knowing too well what this could lead to, I became determined to take her to San Diego and put her in school there. Of course, this added greatly to our problems as we were again going to a strange place and had, as yet, no place to stay. Now there would be three of us, one a child.

Normal School Memories

When I entered San Diego, I was coughing so badly it was necessary to carry a man's handkerchief. It was probable that after sieges of measles and flu, all within six months, I was near tuberculosis. At times I was so delirious I would crawl up in one corner of the bed and sell ribbon. All the skin came off the inside of my mouth.

The city of San Diego was new to us, including the school. On looking about, we saw a sign on a house that said "rooms to rent." It proved to be a house in which the upstairs rooms were both unfurnished and unfinished. The price fitted our pocketbook and we moved in. It was close to the school. Soon we were fixing it up.

The normal school had a teacher training school in one of the buildings. It was like any other elementary school except the teachers were the normal students who were called "practice teachers" because they taught there during their last year of training. This school was well looked after daily by two supervisors. Here was where my little sister would attend classes.

By now she had some new dresses and was off to school. I was very proud to see her go on, for I remembered her back home in overalls and out of school. All went well the first day. However, the next morning everything had changed, and she declared she would not go to school there any more. No amount of coaxing did any good. I said, "If you don't want to go, you must be sick, and you will have to take castor oil." I gave it to her by force, then pulled a small board from a cracker box and spanked her. She went willingly after that and became a good student.

We liked the new school very much and the teachers were liked as well. The morning I registered, I sat in the little waiting room of the registrar of the school. On her desk was a stack of letters. The very top one revealed a transferring letter regarding the Rowlee girls. At this time she did not know who we were. It said, "These girls are the kind of young women we would be pleased to have instruct our own children." While this was a cheering note to a girl among strangers, I had a guilty feeling and read no more.

We had no difficulty establishing credit here, and we rode the streetcars downtown for groceries. Several times I noticed a neat little cottage on the corner not far from the school. Always, it was empty. It looked so inviting, I finally made some inquiries. Everyone, it seemed, knew it was a haunted house. As the story ran, a bride and her groom had moved into it, but an unfortunate event followed. The groom, in the end, committed suicide there. In succession, brides and grooms had moved in and out, until the place had become haunted and was known as the haunted house.

This gave me an idea. Perhaps we might, by chance, get the place for a very cheap rent. A call on the real estate office downtown was in order. The agent said he would rent it to us for twenty-five dollars a month. I suggested twenty-five dollars was too much on account of the house being haunted and having stood there empty so long. When we told him we knew the facts and were not afraid, he came to our price. As there were two bedrooms, I looked up two other girls to share the rent. There we lived happily and unhaunted.

When our school closed for the summer, I again called the agency about the house. They wanted fifteen dollars per month to hold it for us. I again persuaded him to hold it free, promising we would not move out. I also told him we needed all our money to get home. He again agreed. Thus ended our third year of college, and we took the train for Jerry Slough. Though I had thrived as a result of the change in climate I did not have much time to recuperate. The change from a school where I was well acquainted with the teachers and surroundings to one of new people and practices was not easy.

The summer passed quickly, and we went back to what would be our graduating year. Things went along well, until the second semester when my sister, Dolly, found the strain of study and supervised teaching too great a load to carry. As a result she gave signs of her courage being exhausted. I did all I could to persuade her that she only had a few more weeks before we would graduate. Thereupon, I got a bottle of wine and attempted to have her swallow a raw egg in a glass of wine each morning. I had heard my mother tell of the wonders that could result from such a dosage. It was necessary to pull my patient up to a sitting position and induce her to try letting the egg slip down her throat, followed by the wine. After a few mornings of spewed egg and wine all over the bed clothing, I abandoned the idea and left her home in bed.

Poking Fun

While a student in the San Diego Normal, I carried on a little sideline all my own. I set out to gather news about school happenings and people during the day. At night I drew that news in the form of cartoons. I would get up very early the next morning, and post the drawings on the hall bulletin board. It was not long before the school as a whole became interested in those drawings, and came to look upon them as a news agency. Though it was just a prank affording much fun for all of us, my identity remained hidden... It was great fun to stand by the board, as I sometimes did, and listen to the comments. Some special, others priceless!

Graduation

I might say here, my graduation was nothing like I had pictured it in my youth. My parents were not present. I was not the spirited young graduate that I had formerly pictured myself. Standing on the rostrum receiving a diploma was a very tired young woman of twenty-four years, exhausted, with her dream of long ago fulfilled. Her only longing was to be home and resting. My sister did not graduate until much later.

Upon returning home, I was in bed for several days, and my meals were carried to me. At last I was a teacher, qualified to teach in the primary schools of the State of California. I had a letter from one of the instructors in the college I had just left, saying if I wished to teach in the city schools, he would be able to help me secure such a position. The county superintendent of schools at that time was indebted to him for past favors. This did not appeal to me as my thoughts and heart turned back to the frustrations and confusions of my own school days. The four years I had been away had not greatly changed the conditions that I had experienced. Many girls and boys would need help as I needed it. My place would not be in the city, but in the country, where existed the life I knew so well.

My diploma permitted me to teach in the State of California, and this was my desire. You have already read of the difficulties surrounding my efforts to become a teacher. When I first returned home, I was physically exhausted, yet vibrantly happy that I would soon be active in the capacity for which I had been trained. I visualized a definite way of life as a delegate to the profession, and began to deliberate as to where I should begin my career.

During the summer I received a letter from my former history teacher, Mr. Bliss, saying if I wished to teach in the Bakersfield City Schools, he perhaps could help me. He said, he had compiled a history outline to be used in the schools of Kern County, under the direction of County Superintendent Doub. He said further, he would be happy to send me a recommendation that would assist in securing such a position. While this offer gave me much elation, I demurred at the suggestion and did not accept his assistance.

Because of the life I had lived as a girl on the plains where I had received my meager education, I was cognizant of the educational and corrective assistance I would be able to administer to those outpost schools. I felt I could give my best efforts in the schools where I understood the conditions under which I would be working.

I was pleased, therefore, to have my home school, Wildwood, offered to me. With much contentment I approached some of the small problems that arose. Here would be a chance to stay with my parents and help them with their difficulties. They had already made great sacrifices to keep my sister and me in school with no thought of their own advancing age. In fact, I was so enthused, I did not visualize in the least the possibility that things to come might demoralize my deliberations. I was thoroughly satisfied I had hurdled the last barrier posed against a life with trouble. Unfortunately this was a delusion. Trouble has followed me through all of life's demarcations. As a result I have not only learned to live with trouble, but to live happily.

Fannie Rowlee graduates from San Diego Normal School, June 12, 1902

Since my elementary graduation, the population of our area had changed somewhat. Even the school stood some distance from Jerry Slough. The families for whom the school was originally built had moved away; so the schoolhouse had been relocated nearer the new developing area. Thus most of the population was new to me. The Tracy brothers, Theodore and William, nephews of Ferdinand Tracy who had explored the San Joaquin Valley and settled in Jerry Slough to become the founder of the present day Tracy Ranch, had moved their homestead cabin and built a spacious house in the new settlement.

My parents had continued to live on the homestead some six miles distant from the schoolhouse where I would be teaching. Where to board a teacher, or who would board her, was a question to be answered by the board of trustees. This time there was no problem, I could live at home even though I would have to cover a distance of six miles twice daily. My father had no horse suitable to hitch to a buggy, and besides there was only a cart and Jake the mule to convey me. Jake was not well adapted to a cart or buggy, or even travel. However, I decided to give the mule a try.

The first morning I started very early. I wanted to have the schoolroom well prepared for the children. Jake apparently did not like the idea, for he continually crowded the fences so closely, it seemed like he was trying to rub himself free of the cart and me. Jake had only been taught to pull in a plow-team and now had to be given a complete new education. By the time I reached school, I had done a full day's work and so had Jake. I hoped the fully stuffed sack of hay in the cart would sufficiently revive him for the trip home after the day ended, for I knew it was going to be a long day.

Sisters Dolly and Fannie Rowlee posed for this photograph around 1902. They fought against formidable odds to obtain an education. Both became teachers.

Fannie the Teacher, circa 1903

Fannie's Romances

Regarding Sam McQueen, excerpted from the Mary Gregg notes by Nancy Frey Trahan:

For a short time Fannie was again employed by the Weavers (where she had first begun to dream of an education), this time to work in their hardware store in Delano. A lady who worked in McFarland invited Fannie and her sister Dolly to dinner. Fannie and Dolly went in Fannie's old horse and buggy. The woman had also invited Sam McQueen.

Sam McQueen was ranch foreman of the Cox and Clarke Land Company, north of Dewey (Wasco). He was a good-looking man of pure Scottish descent, tall and well built, with big blue eyes, a black mustache, and fair skin. He could have married any girl in the area.

Sam arrived for the dinner in a fine buggy and "spanking" team. The two Rowlee sisters got acquainted with Sam at the dinner.

When it came time to leave, Sam McQueen was waiting with his fine buggy and asked Fannie to ride home with him. Fannie replied, "No, thanks, we have our own horse and buggy." Fannie was of the opinion that she should return with her sister, the way she came. So, when Dolly came out of the house, Sam asked her to ride with him, and she accepted. Fannie was angry, thinking that Dolly should have stayed with her. After all, somebody had to drive the horse and buggy home!

After the party, someone asked Sam how he liked the Rowlee girls. He replied, "The older one [Dolly] is pretty nice, but the little one is too damned independent!"

Nevertheless, Sam courted Fannie; they attended dances together and Sam saw her at every opportunity. Sam helped give Fannie's eighteenth birthday party. At some point they became "engaged," and he had a ring made for her with three little gems, a small diamond, ruby and sapphire that Martha Rowlee had removed from the decorative back of a pocket watch. [It was handmade from gold wire with the stones set diagonally. Later, as a young child Frances traded the ring to a playmate for a candy bar. The girl lost it and months later it was found at the edge of a gopher hole, but it was missing the ruby. Fannie had to buy the ring back from the girl's parents. The ring went to Martha upon her graduation from eighth grade and to me at my graduation. Martha had a jeweler replace the missing ruby with one that was synthetic. I passed the ring on to my daughter Tracy for her eighth grade graduation, and she still wears it.-NFT] Fannie, however, had by this time set her heart on a college education, and Sam could not convince her to alter her plans. Fannie entered Chico Normal School in 1896 at age nineteen; no marriage date was set. Sam would have to wait until she finished her education, or there would be no wedding. Sam later visited Fannie at school, and this time tried to persuade her to marry and accompany him to Alaska during the gold rush. Again she refused.

Sam left Cox and Clarke to work as superintendent for a large rancher, Henry Glide, in Sacramento, and Fannie kissed him "goodbye"—the one and only kiss of the relationship. Fannie saw Sam for the last time when she and Dolly passed through Sacramento on the way home from their second year at Chico Normal School. (They would attend San Diego Normal School for their final two years of college.) Sam took Fannie to dinner and a ride out to the Glide ranch. Sam proposed one last time, with Dolly on hand. Again, he urged her to accompany him to Alaska, and one last time Fannie refused, citing her goal of a college education.

Sam finally departed without her. She received a letter from Alaska in which Sam referred to her as "his little sweetheart." Fannie replied, but her letter to him was returned unopened. Fannie intuitively understood that Sam had feared her answer, and had refused delivery of the letter. She eventually wrote to his brother in Scotland who replied that Sam had married a nurse in Alaska.

Fannie Tracy and Sam McQueen, circa 1899

The Tracy Saga:

As I looked back through my album of verse, I came upon the writing of two men who played special roles in my life. One was a near romance, the other became my husband, father of my children, and companion for many years. Each one came along about the same time.

In looking back over the years, I have never questioned my great decision. I wonder if Mr. Tracy did. It is true though that I was a long time in finishing my education, and already twenty-six when I married. Therefore I can say there had not been too much time in my life for June and romance.

In 1895-96 I was keeping company with Sam McQueen, foreman of the large Cox and Clarke Ranch with headquarters near Pond. On one of our occasions together, he chose to write these few lines in my album:

Delano, September 1896

Dear Fannie,

These few lines to you are tended
By a friend, sincere and true,
Hoping but to be remembered
When I am far from you.

Sam McQueen

Sam had come to the United States when he was a boy, and had been with the company for many years. He was a good companion, for he drove the ranch's finest teams. He owned a fine trotter too, which could do a mile in three minutes. For some reason the company later transferred him back to Sacramento where they had large holdings.

On occasion Sam drove his fine trotter hooked to a one-man sulky to visit me in Delano. My friends objected to seeing me sit so close to him as we went out for a drive in his small two-wheeler. I must admit there was really only room for one, but this was a near romance. He asked if he might give me Sir Walter, the trotter, but I would not accept such a gift. "I would have looked fine out in that god-forsaken sagebrush with a Janey Trotter." Had I not at this time determined to go to college, a romance might have developed.

However, Sam did meet my sister and me when we passed through Sacramento on our way to college in Chico. He took us to the stable where he kept Sir Walter. It was a two-story building, and Sir Walter was upstairs. When Sam called him by name he came down the ramp. I thought this was quite wonderful. Sam asked me to marry and accompany him to the Klondike during the gold rush. I refused his proposal. This was not my first refusal to Sam, for late one night when some of our cattle were stuck in the mud Sam called out of the dark to offer assistance, which I did not accept.

No other near romance ever developed until I met Mr. Tracy, and I have always been happy about this as I have continued to live in the country I knew the most about, and to carry on and fulfill the dreams that he had.

Will Tracy

In one of our letters from home, mother said Willie Tracy was to be married. This to me was just a piece of news, as I had only met the gentleman twice, once in the flood and once at his housewarming.

I never knew what happened to that marriage until after my own wedding day. My husband still had her picture. She was a nice looking girl by the name of Tucker. I believe she had been raised in Kern County in the Glennville country; then had gone away to school and later was employed by a brokerage firm in San Francisco as a stenographer.

I do not know anything about their romance, but I do know why it was never consummated. She had insisted upon his coming to San Francisco and becoming a stockbroker. Mr. Tracy could not see himself in a business in which he had no training, and evidently his love was not deep enough to take the chance. He was already entrenched in the farming and this in itself presented quite a problem. At last he said "No."

However, my mother wrote to us the very day of the wedding to be. I still remember that night. We were studying, when I looked up from my history book and said to my sister, "You know, today is Willie Tracy's wedding day," and then went back to my history. I was really making my own history.

It is a wonder I did not become an old maid, because of the tenacity I had for sticking to the idea I must finish school before anything else. One sweetheart gave up and went to Alaska, and I had no eyes for the one I did marry, until I graduated and became a teacher. I could not see putting all the years and effort in college and never using it; so when I had the opportunity later, I always did some teaching as long as I could be at home with my husband and family.

Tracy Brothers' Housewarming

The Tracy Ranch house of the present time was built on Section 31 in about 1900. At that time the bachelor brothers were living in a cabin about a mile from the new house to be. When my sister and I changed normal colleges that same year, and while going from Chico to San Diego, we spent the summer at home. During our absence our family had moved back to the original homestead on Jerry Slough, so that now we were neighbors of the Tracy brothers.

They gave a housewarming party to dedicate their new house, and we attended. That fall we left for San Diego to complete our education.

It was in the waters of the Kern River where I first met my future husband, and where our romance was cradled, although it remained dormant until many years later. [Fanny was twenty-six and "seeing" Sam McQueen; Will would have been 36.-NFT] It was fate that the floodwaters should be the medium of our meeting, and I cherish still the memory of the day we made our exodus together out of the swirling waters of the sometimes-wild river.

He was a young man of quiet demeanor, and he politely offered his assistance. The mule he was riding on that day was vaulting here and there in a cautious way as if to keep its feet on firmer footing despite the depth of the water. I was riding a good cow horse and demurred any help. I was then a young minx noted for my good horsemanship, hardened by the rigorous life on the plains. Mr. Tracy had found a cowgirl who was not given to cultivating the attentions of young men regardless how gallant they may have happened to be.

All my early life I had lived in the vanguard of the mighty floodwaters of Kern. Many times before I had found myself caught in a maze of mire and turbulent waters, yet I felt only exhilaration. Now I was not only meeting my future husband, I was within the boundaries of the muddy precinct of what was to become the Tracy Ranch Incorporated.

William Tracy wrote in my album this verse:

Wildwood, April 1896

Dear Fannie,

May your house be full of sunshine,
Your enemies be few.
May your clothes be always warm
and your soles be holy too.
May our loot be just as happy as though
I'd pitched my tent with you.

William Tracy

Eight years later we were married! He was ten years older than I, he being 36 and I, 26. Both of us having reached a sober age, our courtship was not very romantic. Neither had much to debate about regarding marriage. I believed I had not spent enough of my time as a teacher and did not wish to leave the field abruptly. Mr. Tracy was an up-to-date farmer of those times with the up-hill part of getting started behind him. So we settled our variance by becoming engaged with the marriage date still in the distance.

Fannie the Teacher

After graduating from San Diego State Normal School in 1902, Fannie returned to her family's homestead in the Jerry Slough area north of Buttonwillow. There she secured a position in the Wildwood School. She now became the teacher at the school she had attended as a child. Her dream was at last coming true.

On her first morning of teaching she arrived at school and was surprised to find a man's watch on her desk. Soon a tall man entered the room. He was a neighboring farmer and, without introducing himself, he sauntered over to a little cabinet of books. But Fannie tells it best:

> Finally, he approached my desk and, laying a book upon it, introduced himself as one of the near neighbors. Noting a man's watch lying face up on the desk, he said, "Oh! Bill Tracy's watch. He tries to infatuate all the teachers." As a matter of fact the school had no clock. Mr. Tracy, who was clerk of the board, having made an early morning visit, noted the absence of a clock and left his for my use. I could see no cause for such a remark, and feeling no infatuation whatsoever from its lying there, I made no reply.
>
> That first year passed quickly and without incident other than what I have mentioned. The school was small, being only one room, and contained a group of nice, very energetic children. The Tracy brothers, having

noted my difficulty with Jake the mule, offered to loan me a good horse, named Maggie. I bought Maggie for thirty-five dollars. Then I really enjoyed my ride to school. The trustees, we can assume, were well satisfied, and asked me to return the following fall. I visualized another good year with the children so I happily accepted.

About the middle of the summer Mr. Tracy's sister, who was a teacher in the city of Bakersfield, decided to spend part of the summer at the Tracy cattle camp in the mountains near Glennville. She didn't feel brave enough to stay there alone and asked me to accompany her. I also felt the need of a change and agreed. The bear trap we found adjacent to the house called our attention to visitors that could be expected.

About two weeks before school opened I received a letter from Mr. Tracy giving me some startling and surprising facts. Shortly after I had departed, my "early morning visitor" of the first day of school had called the community together, and, acting as chairman, advised them that he felt Miss Rowlee should not continue teaching at the Wildwood School. Since he did not have children in school, and was not a trustee, and since there had been no complaints from anyone, I was greatly puzzled. I had no chance to find out or investigate the matter, for Mr. Tracy had said on my behalf at that meeting, "I know Miss Rowlee would not want to teach this school if you do not want her," and he resigned for me by proxy. Of course, all of this was exactly contrary to my wishes for no valid reason had been given. Lamentable as it was, out of deference to Mr. Tracy and in dedication to our marriage close at hand, I refrained from taking any further notice of how it came about.

From the Mary Gregg notes:

Soon after Fannie came home after her graduation, she and Will formed a friendship and he began calling on her. One day he surprised her by saying, "When we're married, I want all your school friends to come see us." Fannie was shocked, as she had no notion of marrying so soon after beginning her teaching career. She did not know if Will was hinting, joking, or serious in his intent, so she simply ignored the comment. Apparently there were other "hints," which Fannie either ignored or turned down. However, Will visited Fannie at the Wicker School that winter. Their romance deepened into an understanding. As he was preparing to return home, Will proposed again. Though not totally convinced that she wanted to marry at all, Fannie "recognized that Mr. Tracy was an outstanding man, and when I came to tell him 'No' for good, I couldn't do it. I guess I'd fallen in love with him." Fannie at last agreed to marry, and they decided on a wedding date for the following spring:

The next year was a pleasant one, as I taught at the Wicker School near Glennville. A young man named Sorenson, just over from Switzerland, was attending the Hogan School. (He eventually opened a livery stable in Bakersfield.) Though already twenty-one, he was handicapped by being unable to speak much English. At the time we had gone to many of the same dances, but not together; so I was most surprised when he came up to the Wicker School to see me. We went out for a drive, and much to my astonishment he tried to give me a large diamond ring. When I finally told him I was already engaged to Mr. Tracy, he remarked, "If you marry Bill Tracy, you will spend the rest of your life living out in the sagebrush." I made no reply. Thus ended another near romance.

That spring I was married, and as a bride moved to the Tracy Ranch. Two years of teaching had taught me much about people, and of a community subject to personal desires.

The Wound That Would Not Heal

It is hard to imagine a more crushing blow to the young woman who had struggled so desperately to obtain an education and who, as a young girl, had the courage to make her home in a deserted cabin, with barely enough to sustain herself, in order to acquire a grammar school diploma. In retrospect, it is easy for me to see something that Fannie never fully realized. The problem had nothing to do with her ability as a teacher. It was simply a matter of envy over the young man who was courting her.

Will Tracy at age 38 was a man of integrity and highly respected in the community. The orphan who recited poetry as he followed his team and plow had garnered respect from two influential men of the day. One was his Uncle Ferdinand Tracy, the other was Senator C.W. Canfield, who was his uncle's partner. As father reports in his autobiography, these two men "favored me exceptionally, and I now found myself the owner of a considerable acreage of undeveloped land." According to his photograph, he was a handsome young man as well. He had a thorough understanding of horses, and throughout his life displayed a great fondness for them. This was a trait he would pass on to his sons.

By then he was pretty well established as a farmer and stockman. It would take such a man to capture the independent and high-spirited Fannie Rowlee. Perhaps this was enough to make another man envious.

The injustice she experienced after her first year of teaching left a wound that would not heal, and resulted in a tug-of-war over the election of school trustees. The unfortunate incident cast a shadow over our lives and would lead to a pattern of petty politics that lasted forty years. But Fannie was a fighter. Just as she was to fight to win a trial that could have caused the loss of our ranch, she fought for the right to teach in her own community. And from time to time, depending upon which faction controlled the school board, she did teach at Wildwood.

Marriage and Family

On April 3, 1904, following a yearlong engagement, Fannie Rowlee and Will Tracy finally took their wedding vows, and before long, Will's ranch house was overflowing with children, and there was no time to grieve over politics. Cecil Foster was born on January 14, 1905; William Darrel on December 3, 1906; Frances Fay on June 27, 1909; Charles Wellington "Teddy" in 1912; Martha May on May 7, 1915; and Albert Tilton on July 16, 1919.

Fannie Rowlee weds prosperous young farmer Wm. Tracy in April 1904

Fannie and Will were a compatible couple, and the course of their love ran true and deep. It would weather many storms. Mother was never idle. She loved to garden and to can the fruit and vegetables she grew. Having learned the skills of canning as a girl, she took pride in their preservation. In addition to rearing six children and teaching for almost twenty years, she often raised a brood of several hundred baby chicks or turkeys. Like her mother she was an excellent cook. When electricity finally came our way she cooked for the crew that installed the power lines. Later she spent weekends cooking for wealthy duck hunters. But the most memorable of her enterprises was her venture into the ostrich business.

I was very unsettled, almost unhappy as a bride because I had no preparation in making the nest and everything appeared to run smoothly in spite of me. Seemingly I was just a boarder. I needed to have an interest to be happy. My husband was very busy with his farm development. There was no social life, and no place to go that did not take a day. I did have a nice horse and a new fringed top buggy and often went down to see my mother.

My first days of married life were a bit lonesome. The China cook did all the food preparations and kept the house. Mr. Tracy was very busy looking after his farming, which he did on horseback. He had by now accumulated five thousand acres of land. With the equipment that existed in those days, we were unable to cultivate large tracts, and outside of one hundred or so acres of crops, mostly alfalfa, the land was used for grazing. Level land at that time was any land you could get irrigated. There were low places that received too much water and grew nothing, and high spots that did not get any water, and grew nothing too.

One day Mr. Tracy left his horse and sled standing while he came for supper. I took advantage of this opportunity by putting a barrel of water on the sled and watering some of the young trees I had planted. Today the eucalyptus trees I nursed as a young bride are past fifty years old and are the same huge trees one may see at the old home place. Earlier, while Mr. Tracy was still a bachelor, he had planted poplars and cottonwoods. They deteriorated very fast because destructive borers made holes in them and drained out the sap.

There were two or three neighbors near, but I had already told them I would not accept any invitation to call, and I told them why. It was because of the stand they had taken in the school matter. This was perhaps hypersensitive on my part, but neither do I have time even today for those who are hypocritical. If I cannot be a good neighbor, I do not want to be a neighbor at all.

Some who take the other way, perhaps lead a happier life as they go along, but I have a host of good people as my friends who are honest and sincere. Our friendships may seem humble because there is no blare of social trumpets, but real friendship comes with simple humility. Friendships gained in a hurdy-gurdy fashion are usually not substantial or sincere. I do not mean that we cannot be helpful to all those that might need our help, for that also must be given in humility.

We had little recreation in those days. Mr. Tracy relaxed on Sunday and often played games of horseshoes with his men. He never played cards with them at night. This horseshoe business I could never understand. In fact, I was a bit jealous of it. I recall, too, the way I felt. Why would a man want to play this foolish game and leave his wife sitting in the house by herself? Being in the house with a Chinese cook is being alone. However, I never complained about this to him.

Mr. Tracy raised horses for sale as well as good teams for his own use. In those days if a team could go to Bakersfield in three hours, a distance of about thirty miles, they were considered to have made good time. He had a span of leaders for his work teams of which he was very proud. One of them named Jack was an especially fine animal, a Cleveland Bay. Cleveland Bays were noted for their versatility, good looks, and fair speed. They were useful for both work and buggy animals. When I came to the ranch as a bride, Mr. Tracy wanted me to have my own special horse, and suggested that I take Jack. Knowing how very fond he was of Jack as a leader in the team, I refused to accept him for my personal use. He remarked, "Well, you would not object to my using him occasionally, would you?" I answered, "Yes, I would object."

I was then given a very satisfactory Cleveland Bay, called Beauty. Beauty, when hitched to my new fringe-topped buggy with its yellow wheels, pleased me very much.

Starting Out

My husband by this time had a good start in life and previous to our marriage had built a large dwelling. However, there was no bathroom or any closets in the house. The cooking and housekeeping were done by Chinese who also processed the meat down to the oxtail soup. Not much wood was consumed between meals as the drafts on the stove were kept closed. Unfortunately, the kitchen was very hot as the north window was small and high, shutting off any breeze that might enter.

When it later fell to my lot to manage the house, I set out to make some changes by having a large sliding window built in the place of the smaller one. On the back porch was located a hand pump and wooden sink. This porch was an open one, so the flies congregated on the kitchen screen. It became very difficult to keep them out. Finally I had the porch screened.

We used the dining room and a long table where our family and men dined together. This was particularly true when the number of hired hands was small. During hay or harvest time I never sat down at the table until the meal was over. There was always much to be done in seeing that all had plenty to eat. At times it became impossible to obtain dependable Chinese cooks, so we used colored, Indian, Mexican or white cooks as well as Japanese. The colored did very well, but all others gave some trouble, and I was very often filling in between cooks.

When our family increased with three children at the table, we decided to furnish a cookhouse for the workers. However, I still kept my family in the house. This also offered problems, especially when a change of cooks became necessary. At harvest time the women cooks seemed unable to bake enough bread. Generally, they provided the men with hot bread, so I decided to make a fifty pound sack of flour up into bread each day, until the cooks had a head start. My baking helped to solve the hot bread problem. These were very difficult days for me too, as there were small children who required my time as well.

We decided to make another change by building another house, and turn the ranch house into a cookhouse. However, about that time Mr. Tracy wanted a modern barn. I thought his wishes should come first, and the proposed house was never built.

Ranch Matters

At the time Mr. Tracy and I were married, he was farming alfalfa and barley as well as raising cattle and horses. As yet there were no tractors. He employed three men steady, and a crew at hay time, which was during all the summer months. His brother managed the irrigation. All of the men ate in a cookhouse. "China cooks" were hired, he said, because they always had something hot on the stove to eat. The fire never went out entirely. When a new man, usually a tramp or itinerant, was hired, he was brought to the house and fed, regardless of the time of day.

Mr. Tracy never allowed any man to keep a bottle of whiskey or to swear loudly enough in the barn to be heard at the house. Smoking in the house was also taboo. However, at that time it was not the custom for men to smoke cigarettes in the house. Cigars were not smoked by the workers, and women, except for a certain type, did not smoke. The women who did smoke cigarettes were those living in the "red light district" and were not often seen in public.

It was understood most of the men would go on a drunken spree periodically. That meant going to Bakersfield where there were many saloons. No man was taken back if not completely sober. Most men stayed away until all their money was gone, and they were sober and broke. Mr. Tracy never investigated any trouble the men might have among themselves. He gave the dissatisfied man his check and sent him on his way down the road.

However, when three or four crops a year were in stacks for winter use, the fields at a distance looked like a field of huge mushrooms. If it were a good year for wild feed, some hay was sold. If there is anything that causes a man loss of sleep, it is to have no feed to carry his stock through the winter.

In the early years after a mad rush of the Kern, the receding waters often left a disease that greatly diminished our herds. Shocks of hay floated off in the swirl of high waters like ships at sea. Swarms of mosquitoes like black clouds darkened the sky, and stacks of hay became torches from spontaneous combustion. The plague we suffered after the overflow waters left our ranch reduced to muck and mud. Those were forbidding days indeed, yet my husband always showed great forbearance even in the most distressing times. Many difficulties he did not even mention to me.

During the high water he went by boat to visit neighbors who had temporarily moved to high ground. On these trips he rescued small animals and many chickens that had taken refuge on top of woodpiles and in trees. He sailed along over the barbed wire fences to these places. At times he was gone so long on his excursions, I became worried for fear he had been carried away.

After each flood, good forage grew for the cattle, but they soon sickened and died, until about half the herd remained. Though they were kept out in the brush until they seemed normally healthy, later when they were sold to the butcher a few gave evidence of their illness, and we lost those. We attempted to replenish our herd by buying another hundred head. Fifty of these we later lost. As far as I know there was no known remedy at that time for such a disease. Even Miller and Lux, with its great cattle empire, made the mistake of using flood diseased grass to feed their stock. Attracted by the high grass that was luxuriant, their men brought a band of cattle to pasture. They too contracted the disease and were moved away. Thus the dreaded anthrax took its ugly toll. Cattle would move up to the horse trough to get a drink of fresh water, then walk away ten or fifteen feet, lie down and die. One morning I counted ten not far from one of our troughs.

Ostriches

With Will's help, Fannie imported the first pair of ostriches in California. I have often thought that my father must have been a very patient man to agree with some of mother's industrious activities! Combining these capricious, erratic and sometimes vicious birds with his livestock would cause him a great deal of trouble, but he was always very supportive of mother's projects, and indeed seemed to enjoy taking part in them.

World War I would put an end to the luxury of ostrich feathers, but during the years of the ostrich population there evolved a veritable library of ostrich lore. Tramps, or hobos, were prevalent in those days, and they often stopped by our ranch for a handout and a place to sleep. One day mother looked out the window and was surprised to see two of these characters scaling the pipe that led to the top of the tank house. Then, with amusement she noticed one of our pet ostriches sauntering below. Another time, as mother was transporting a load of ostriches, one of them plucked the diamond from her engagement ring. This did not seem to bother mother very much, for she always enjoyed the telling of it!

The Tracy Saga:

The stable boy had instructions to groom Beauty and hitch her to the buggy for me at my request. In case of any delay she was left tied to the hitching rack. On one of these occasions, Mr. Tracy took her for a short trip to look after something in a nearby field, and left her standing too long unattended. She became restless and started for home at quite a fast clip, which some of his neighbors would call a runaway. Such an act was considered quite a joke because of Mr. Tracy's good reputation as a horseman.

During the afternoon one of these neighbors drove to the ranch, accompanied with a friend to see about some duck hunting. He was bold to joke my husband about a buggy mare running away, and added, very much in a bragging manner, "Will, you should have a gentle mare like this one I am driving. Now, I can drive her right up to a train, and she will stick her head into the door of a box car, if I say so, afraid of nothin."

He then proceeded to drive off in a direction that took him by the ostrich pen. The big birds were in the shed eating, but just as the horse and buggy with its two occupants came near the shed, Tempe, the female ostrich, came out. Well! The mare took a different turn as if not knowing what confronted her. She decided to run for her life, and turned to do so in such a manner as to turn the buggy upside-down on top of the two men. Her quick turn broke her loose from the shafts, and she fled to the far side of the field, snorting such a succession of snorts that she outdistanced the boom of an ostrich. As you can imagine, two most astonished and humiliated men crawled out, and the owner of the animal kept saying, "You old fool! You old fool!"

The horse, with great difficulty, was brought back and tied to a hitching rack where its owner had full intentions, in his angry moments, of giving her a thrashing. A colored man of muscular build who had been a witness to the happenings noticed the kindled ire of the animal's owner and the verbal bombs exploding in the direction of the still trembling horse, and he intervened. He was not only concerned about the resulting damage to the horse, but he also wanted to enforce justice. "Mister," he said, "if you lay hands on that 'air hoss, I'ze al beat ye up plenty good! An I'ze means it." Then he offered to help repair the damage, and tranquility reigned again.

The San Francisco Exposition as recalled by Frances (Tracy) Selvidge:

Martha was born on May 7, 1915 but this did not slow mother down. A few months later, she deposited three-year-old Teddy and baby Martha, still unweaned, with her sister Dolly in Bakersfield. Accompanied by my father and my two older brothers, Cecil and Darrel, Fannie set out to take a group of ostriches to the San Francisco Exposition. Aunt Dee must have had a hard time with her, because she often complained about it to Martha and related how she tried to placate her with a "sugar teat."

In preparation for our trip to the city of San Francisco to attend the 1915 Exposition, mother and I climbed the steps of the Bakersfield Tegeler Hotel to buy her a hat at their exclusive millinery shop. I was placated by getting one, too, albeit a little too adult for me but simple in a pretty rose velvet. Off the family went, my parents, my two older brothers and "me too" in our open Studebaker touring car.

We started our tour of the city with lunch at an open-air beer garden. While waiting for our waitress, I noted all the mugs being served at other tables. They looked so pretty with a yellow liquid and lots of foam on top. I learned from surrounding conversations that it was called "beer." Our father never allowed any form of wine or alcoholic beverage in our home, so naturally none was mentioned as our order was being taken.

"And what will the little girl have?" asked our waitress. In my most proper six-year-old persona I said, "I'll take beer!" I need not describe the looks of surprise and embarrassment on my parent's faces!

Our next stop was a little theatre where my parents seemed to already know what would be on the Pathé News. To my surprise, there I was, much younger, sitting amongst a bunch of eggs in an ostrich nest at the ranch!

From the tall exhibit building, our parents sent us off by ourselves. It seems my brothers had wanted to go near the seashore. Having arrived at that viewpoint, nature called. While I was at the restroom the lure of the waves and rocks and sea-life below were too much, and I went exploring. When I returned the boys were nowhere to be seen! Among the large crowd, I began to panic. I saw two people from Bakersfield, friends of my parents, standing by but I was too proud to admit I might be lost. Hoping that my parents might still be in the building where we had departed, I found my way there down the roped-off streets. But they were not to be found.

Fannie Tracy in 1915, eight months pregnant with Martha

Back down the street to the beach I went again, head down and feeling ever so lost. A nice lady stopped me and asked, "Are you lost, little girl?" And I, of course, burst into tears. She turned me over to a policeman who called someone on a phone, and soon a paddy wagon arrived. I was transported to the "Lost and Found" department to join three or four other lost souls.

Time passed and hunger set in. Fortunately, one girl was prepared as she came out with a bag of fruit; enough to allow the other three of us an apple, orange, or banana. I still remember what a hard choice it was to decide on one, trying to figure out which would provide the most nourishment for a long spell.

At long last dad and mom came through that door. Oh happy, happy day! I was overcome as I ran and grabbed mother around the knees, burying my face in her full skirts.

Through it all I still wore and treasured my pretty hat, but as we left for home in our open car we passed a Gypsy camp in the outskirts of the city, and right there the wind whipped off my hat! By the time we could stop, a flock of Gypsy kids was running to get it. They saw my brothers coming at them at the same time, but they got a big kick at it anyway, so the boys recovered it in a cloud of dust.

In my ninetieth year I still vividly remember: being in the news, being lost, the paddy wagon and the beer!

The Tracy Saga:

The Bible speaks thus of the ostrich, "God hath deprived her of wisdom, neither hath he imparted to her understanding." At one time I was known as the only woman ostrich farmer in the world. Amazingly enough, I had never seen an ostrich, until I ventured into the industry in 1906 with the arrival of two birds, a female named Tempe and a male called Phoenix. Incidentally, ostriches had first been brought to the United States in 1882, so I was quite behind in learning about them.

Until this event ostriches were book material to read and pictures to see. My knowledge was brief and related entirely to a pamphlet put out by our federal government. As it has always been throughout my life, I had to be busy. I had to keep my interest going, be doing something. There was little at this time for me to do on the ranch, and I was no longer teaching. One day I was reading an article about ostriches in Arizona, so I said to myself, "That's what I want to do! There is a new business to add to our ranch." Finally, after many letters and contacts I managed to initiate the purchase of a pair of young birds. To date it had been difficult to find a pair for sale anywhere in the United States. A hatch of chicks at that time was valued at one hundred dollars each, and the mated pair were just not for sale. In fact one famous pair in Arizona had the record of having produced chicks valued at $2,200 during one season. It was obvious, therefore, that I would have to settle for two very young birds.

Mr. Tracy had fully sympathized with my wish to raise ostriches. He was anxious that I do whatever I wanted. He agreed further to make the necessary trip to Arizona to secure the birds. Unfortunately, floods prevented his reaching Arizona on schedule, in fact forced his train to return. This brought him back through Bakersfield. He found himself highly tempted to get off and come home to the ranch, but feeling somewhat silly and not wanting to disappoint me, he continued again via San Jose and the [California] coast to Arizona where the purchase was completed, and arrangements for shipping got under way.

Shipping an ostrich is about the same as shipping a horse, only an ostrich kicks forward. I found when Tempe and Phoenix, named for two towns in Arizona, arrived, I had certainly entered into an industry with all that I had bargained for. Of course, there had been some publicity about the coming of our newest entry into Kern County activities. Most people were as unprepared as we were. We were expecting to find two birds when we went to the depot. Only one had arrived in her strongly built, but already battered crate. She was just about to kick herself free. Her mate had been sidetracked for repairs to his crate and himself. When he did arrive later, he was so injured, he soon died.

A label tacked on the crate said, "I am no camel! With my feed I need a drink of water." This presumably summed up to an extent all I needed to know about an ostrich. We could see that the bird had had some rough handling, her back was bare from the loss of feathers. Her big thighs, on which no feathers grew, were bruised as well as her back.

It is well to remember that throughout this period, I knew nothing of the ostrich or his habits. Though I was certainly willing to learn, I could not foresee what changes would take place. From the time the first complete pair arrived to establish their habitat on the Tracy Ranch, ours was no longer the same quiet place of the past. The name was even changed to the Tracy Stock and Ostrich Farm. Although the new letterhead of the ranch stationery displayed a pair of peaceful loving birds standing side by side, this was not the case. Their feathers, being worth their weight in gold, may have dimmed our eyes to their malicious character, but this was only temporary. "All that glitters is not gold."

Naturally, I thought a little freedom would help these birds after their long journey. However, a second thought said that, "to be safe, I must shut them up tight." This we did in a little old cabin. We felt that if they were confined they would not scare our horses into running off the ranch. After all it could have been quite a task to cover the west side and retrieve a bunch of scattered stock. Very quickly we found out that the working men were also afraid of the ostriches. Yes, our problems were both immediate and ahead of us for some years to come.

If there is anything an ostrich likes, it is his freedom. Freedom to get out and go! This we learned quickly as we later watched them pace up and down the fence of their pen, traveling for what would be miles in a desert. However, if there was a gate for them to go through, they would go only as far as the gate, not through it. An ostrich's silly brain could not seem to comprehend the least change in affairs.

I found there were many tribulations and responsibilities essential in building up our ostrich farm. Having absolutely no basic knowledge upon which to build an industry, nor any information that was readily attainable, any successes would depend largely upon self-achievement. There was no question as to the expanding field, both in size and in expenses, into which I had entered nor the tedium of the long wait to follow.

The idea of unfolding and discovering the mystic life of these giant birds impelled an interest that could not be easily destroyed. Upon becoming fully absorbed I began to add my own discoveries to their culture. At the same time I wondered since this was a real personal project, was I endowed with enough ability to carry on? I soon found out that ability was not the only issue. The entire activity had to be supplemented with an enormous lot of personal courage and fortitude, otherwise, I was doomed to destruction. "No withering flower had any business among those pugilistic warriors."

My earlier reading had advised me that an ostrich was called a chick, until he was quite a bird. A full grown ostrich can reach up about ten feet. I learned he was capable of growing one foot a month, until he reached full size, yet did not fully mature for about four years. The females matured six months earlier. On the other hand this bird could live to be one hundred years old. A young hen might lay a clutch of ten to twelve eggs and as she grew older, this could increase to even greater number. Old birds had been known to lay as many as fifty eggs in a season when not permitted to brood. She seldom laid fertile eggs until at least three-and-a-half years old.

The male of our first pair never recovered from his injuries and finally died. The next one shipped to us came through in good shape. We immediately named him Phoenix II.

We kept our first two birds in an old chicken yard about 200 feet long and 150 feet wide. We also put them inside a building at night.

By now a shelter and special enclosure had been built for our two chicks. Because ostriches are not fed on the ground we used an old homesteaders cabin which was open on one side with a manger built into it. They could go in and out of the area as they wished. On stormy nights or in real bad weather we closed them in as they never went into the shed at night to roost by their own initiative. Always they rested sitting down, apparently not desiring to seek shelter.

I was never too sure of my own security with these birds, nor could I get a man on the ranch to help me. My husband did assist some, but he was busy in the fields most of the time looking after the ranch. This project, in its entirety, seemed to fall my way. Finally, the men ventured their services in a small way. They would go into the enclosure, but when a bird turned and went the wrong way they would not head him off. They instead would turn the wrong way too, so I had both men and birds to handle.

I was aware from the first that I would have to wait at least three-and-one-half years for Tempe and Phoenix to grow from adolescence to maturity. Certainly, it seemed three-and-one-half years was a long period to pause for a hen to lay an egg, but wait I had to do, learning all the time. "What time she lifteth herself on high, she scorneth the horse and his rider." Their feathers protected them well from the cold, but rain was far more damaging, for they did not have any down.

The feed trough had been built inside the shed. It was of high construction for the ostrich not only had a long neck, but he had long legs too. One side of the shed was partitioned for storage of feed and a safety zone for the men attending them. The space below the trough was boarded up but above was open to the ceiling. This allowed the ostrich neck room that he needed when eating. In this operation, he curved his neck to his convenience, but he stretched it upward to let the ball of food that had formed in his neck go down propelled by muscular action.

In show places they fed the birds whole oranges so that spectators might observe this swallowing operation. When confined, they were usually fed chopped alfalfa, mixed with bran and grain, but in the pasture, they did their own chopping. They required in addition large quantities of bone and pea gravel, somewhat smaller than a walnut, to aid their digestion. At the World's Fair in 1915, I fed mine granite rock.

The ostrich is dubbed a foolish creature. They take fright at small animals like dogs. If not hindered by fences, they can be miles away in no time. When running, they can make their getaway twenty-two feet at a stride, easily outdistancing a horse. When they are injured or damaged, it is their own fault, for many times they ran directly into a fence. A camera with a black hood would drive them out of their wits and into a stampede. If confined, they might tear down a fence and in so doing occasionally break a leg or their neck. They cannot resist anything bright and shiny, and will pick at it for hours. Fortunately, like most animals, the ostrich became accustomed to any unusual spectacle that gave it fright, if exposed to it long enough, and they become accustomed, in time, to anything in their surroundings.

The years passed slowly by, and the time grew near when housekeeping might be expected in the ostrich enclosure. The excitement of the chapter one had faded. Now it was to be awakened again when the first signs of a nest should develop. At last I would be able to take one more step in my venture. The most important, ostrich husbandry, to increase the flock (or in this case the troop).

The first signs of a family to appear in the Ostrich Temple were made by Phoenix. He was proud of himself that he had become the head of the house, and everything must stay away from his domain. When people came close by, he tried to get at them to drive them away. He stood up high and pushed hard against the fence making a hissing noise. The characteristics of the ostrich seem to impel him to claim a certain area in a field or pen and to fight everything that trespasses.

Henry Miller's cowboys found this out one time in a very dramatic way. They had, for years, standing permission to enter our fields and take out their stray cattle. One time they were scattered in every direction, chased by a single male ostrich who had been placed there to recover from an injury. In the process he laid claim to the field. The horses developed speed that their riders never knew they possessed as they felt the swift punch of a sharp toenail at every step.

One day, Phoenix was down on the ground making an ostrich challenge like two males do when they were about to have a battle, which incidentally becomes a very fierce affair. His head was weaving from side to side, his wings were extended flapping in rapid succession, until the dust was flying as if he might be stirring up a whirlwind. Then suddenly he jumped up dancing a bit in short steps. As there were no other male birds on the ranch at the time, he must have been addressing himself to Tempe who was paying little attention to him.

A short time later he was down on the ground, again this time scooping a hole with his breastbone and wings. As the dirt loosened, he pushed it out by twisting his bulky body around as if to fit the cavity beneath him. When it was completed to his satisfaction, he appeared ready to begin a new life. Up to this time Tempe had taken on no responsibilities and showed no interest in the cavernous hole. She drifted near the cherished spot occasionally, but displayed no avid interest. She looked but showed no signs of approval, agreement, or participation.

Nature's plan however, was at work, so Phoenix decided to make it clear that he was growing weary of her boycott of his ground preparations. It seemed about time to give her a little discipline as a reminder, and persuade her to agree to participate and cooperate in his plan. Resolutely, he went to work on his poorly adapted mate.

It seemed now the confines of their enclosure were doomed to destruction. Tempe was equally determined she was not ready to dedicate their abode by depositing in it the first ostrich egg to be laid on the Tracy Ranch. Their apartment was no longer the quiet little place it used to be. Feathers flew! He chased her and kicked her as she ran up and down the pen, in and out of corners, clawing frantically at the fence, hoping to escape his fury. With his great toenail he prodded her at every step. Feathers filled the air like autumn leaves on a windy day. He rent the air with his hisses like an angry serpent. At last, the participants worn out, the battle ended. The attacker advanced no more objections. Both stood panting, their wings spread to invite a cooling breeze.

The episode was, perhaps, just a preliminary in an ostrich's life, indicating there must be agreement among birds as well as man or beast in a well adjusted home. Up to this time no one had directly entered the ostrich pen with the birds, and after this demonstration, no one had an inclination to do so. Tempe began to show a little interest in the nest, even setting down by it and picking at the dust. One day, lo and behold, she was on the nest. Soon Tempe laid the first ivory-shelled beauty that was to become the nucleus of the original troop of ostriches in Kern County. This egg became the center of much attraction. Tempe dutifully deposited an egg every other day, until the clutch reached twelve. One dozen big eggs weighing a total of fifty pounds to be exact. It took one hour to boil one, and twelve men could be fed on a single scrambled egg or omelet if used for food.

As the days went by and the weather grew warmer, the pair would stand over the eggs to shade them. At night they sat on them. On the hottest day they would fan them with their wings.

When they decided to incubate them, the two confined themselves devotedly to the nest. A six weeks period was needed for the eggs to hatch. Phoenix covered them at night, Tempe by day. The male's black feathers afforded him protection by the dark of the night. The female's feathers blended with the gray of the day.

Artist's rendition by Martha Frey

"The male's black feathers afforded him protection by the dark of night. The female's feathers blended with the gray of the day."—*The Tracy Saga.* Both shared in protecting their nest and caring for the young.

About eight o'clock each morning they changed shifts, so Tempe could settle on the nest. She sat steadily disregarding the hot sun. When the days grew hotter, she stood up over the eggs to shade them. Eventually I built a shade over the nest. However, this practice I did not follow after the troop grew large in number. While it was a good idea, it was extra work on an already busy place, and they seemed to do well without it. The pair never left the nest in any kind of fright or to take refreshments. Their housekeeping went on unimpaired, each keeping his schedule without delay. Other than changing of shifts, nothing unusual went on, until hatching time neared. It was easy to observe Tempe's behavior because of her daytime actions. I never watched Phoenix at night because of the time, and I had a little family of my own to put to bed. Therefore I speak only of Tempe now.

She practically sealed herself on the nest by packing a levee of dirt all around the contour of her body. She even picked up some dirt with her bill and sprinkled it carefully over her feathers as if to stop all the circulation of cold air.

I suspected quite a bit had been going on for a day or two beneath Tempe's body, but I never witnessed the change of shift to see how they managed when hatching was in progress. At no time did I ever see a broken egg or a crushed little bird in an ostrich nest. How they managed to settle their heavy bodies and long legs with their big feet down on the eggs and not break them was hard to define.

One day when I went out to the pen I saw one cute little head sticking out from under his mother's feathers. As the day warmed up, another and another, and so on were looking at me with their big bright eyes. Actually, the shell is very hard, and by the time the chick has picked his way out, he is quite exhausted, so he takes a much needed rest by stretching his long neck out flat on the ground. About the third day they came out from under their mother and sat near her. I put some mash of wet bread close by and soon they were eating. I kept an eye on Phoenix, too, but his disposition had changed somewhat. Now he walked proudly about trying to encourage his offspring to follow him.

In a few days the nest was completely vacated. I could see at once the little ones were going to have very good parents. The two adults walked carefully with their long necks folded; so they could keep their heads close to the ground among the chicks. The body of the chicks seemed too heavy at first, for their little legs that were so short. No one could dream that some day those same legs would be like a pair of stilts. The chick navigated himself in a peculiar way. He seemed to tumble along, then take a run, then another tumble, and so on, until such time as his legs became strong enough to hold up his body. Then he did very well. When a baby ostrich stops to rest, he sits down and stretches his neck straight out on the ground as if it is tired too.

Ostriches are exceedingly good parents. They seem to know every second where every chick is, and some way manage to keep their big clumsy feet out of the way. I never saw them step on one of their chicks. Our baby ostriches were very cute and interesting, yet so difficult to describe. They seemed so unafraid as they watched me with big staring eyes. For the first time I stood and watched Tempe and Phoenix and their little brood leave the pen and march slowly out to their new home in a broad alfalfa pasture. A dream of mine for a long time had come true. The parents moved slowly along nipping at the alfalfa as they went. The little fellows did the same as they tumbled along.

The little birds grew rapidly, at least a foot a month, and before I realized it, they could reach as high as my head, and with their long necks were nipping at my ears. I was encouraged and felt by now I could manage a greater number, so I purchased from Cawston of Pasadena, forty-five mature ostriches for my troop. The original field we used for the ostriches contained about fifty or sixty acres and was fenced the same as for cattle. At one time we had 174 birds in the Tracy Ranch troop.

Now with this many birds, I deemed it would be more profitable if I manufactured my own feathers into plumes, or whatever the fashion of the day demanded. Now that my troop production was large enough, I believed my birds could support a factory for reworking the feathers. I knew I could sell a plume for ten to one hundred dollars and that a willow plume, the best available, was sure to bring the highest price. Though ostriches were normally plucked every nine months, we went over ours every six months for the raw feathers sold for so much a pound. There were about twenty-five large white feathers grown in each ostrich wing. As yet I knew little about processing the feathers, and that enormity itself paralleled the propagation and production of the feathers.

This was a venture, however, that did not need to take me away from home. I did not wish to erupt my family life by putting myself in a distant factory to learn a new trade. I figured the best way to learn was to dig in, learning all I could about preparing the feathers. My sister, Hazel, was making her home with me at this time. She was extremely gifted, excelling in ornamental creations of handiwork. She had become interested in my new venture, so she applied for work in the Cawston Ostrich Farm. Cawston had imported dyers from Germany, so there was an advantage in being trained there. They accepted her application, and she entered their factory, quickly becoming an expert and doing specialized work. This gave her a chance to learn all phases of the profession of feather processing and manufacture.

I went to work at home seeking to learn the art of washing, bleaching, and dying the feathers. I was much surprised to find this was done very much the same as cloth. Hazel, later of course, was of great help in advising me along this line. The feathers while still wet were rubbed in dry starch and then whipped on a board or table, until they were dry and fluffy. The fronds of the feathers were very durable. Later, they were hung in the sun to further dry and bleach. After drying, they were ready to go into the factory and under the knife for building up a plume or into the dye pot. “Raw feathers” was the trade name of feathers as they come from the bird.

I soon became proficient in the dying and matching of colors, except black. The latter I gave up as others were buying black feathers already dyed, so I could do the same. A number of chemicals were used in the processing of feathers, and there are also a number of professional ways in the factory to use damaged or under-standard grade as first class feathers. When a feather was still on the bird and developing, it emerged from a sheath that came off as the feather grew. Sometimes it would not shed readily, and when it stayed in one place too long, it pinched the feather, changed the fluff of the fronds, resembling a scratch.

A plume was built of layers of feathers three deep in the body and five in the tip. The under layers could be in several parts of pieces and might be barred. A plume was seldom if ever made of full-length perfect feathers. All perfect feathers were kept separate for tops. The ostrich feather was said to be the only feather that had even balance on each side of the shaft. For this reason, the ancient Egyptians had used it as a symbol of balance.

When Hazel returned from Cawstons, she was able to turn out work equal to the factories in San Francisco or New York. She quickly originated new designs and new uses for feathers besides those already in style. A specialty was our limousine pillows, which were at first all made at our feather factory on the ranch. For help I had brought girls in from San Francisco. Unfortunately, they did not like either the climate or the lack of social activities, so I gave up on importations, and began training local girls. About this time I bought out a concession at Idora Park in Oakland, California. I didn’t want to be a failure at home, so thought I would try the north first!

I arranged to ship a carload of birds to Oakland and open a concession in connection with a display and sales room. Shipping ostriches was quite a delicate operation as we soon found out. They were first blindfolded with a canvas sock-like hood, then loaded into a wagon or van. The hoods made them more docile and manageable. However, they were unhooded enroute. Having had no previous experience in the moving of ostriches, I planned to ride along in my Model T touring car with an extra man in case of trouble. Our first stop was the railroad loading point at the Bowerbank siding.

We quickly learned the birds had a tendency to shift all to one end of the wagon bed, presenting a hazardous situation. To equalize this balance, it became necessary for someone to ride in with the birds and keep them evenly distributed. My hired man did not relish the idea of placing himself in such a precarious position, so I took his place among the birds. This move proved very costly to me, not in injury, but in loss of property.

I separated the birds carefully, for they were closely crowded and towered far above my height. To keep them separated I extended both arms in a scarecrow fashion. If one of the ostriches could have talked, he probably would have said: "Ah! Here is what I like, and I'll take it right now." He reached over four or five birds and with one mighty nip plucked my wedding diamond loose from its setting, and with a gulp it was gone. Since he was from Africa, he probably knew its value better than I did. I have lived my life unadorned ever since. The world's number one pickpocket cannot equal the ostrich's dexterity in stripping his victim of all ornamental splendors. A gold watch can disappear as quickly as a kernel of golden corn. It wasn't even safe to wear a pair of glasses because of their glitter.

Once in a boxcar with the horizon and the blue sky above shut off, I could anticipate a reasonably safe journey with a carload of healthy birds. Several little precautions attended to in advance further enhanced the keeper's safety and that of the African canaries. Cleats nailed across a sanded floor gave the birds safe footing. A space opposite the wide door, large enough for feed and a man, divided and separated the birds equally. The car paneling had to be of substantial construction. Sleeping facilities were not considered wise. A bed as downy and soft as their gorgeous feathers would avail you no safety. Because of their watchful eye and inquisitive nature, they could have you picked to pieces or even stripped of your clothing by daylight. It seems they meant no bodily harm, but just wanted to exercise their bills. Perhaps, you have never tried to sleep with an ostrich antagonizing you by pecking all night at you or the bedding. Well, one man did, and he never slept the long night through.

When the train reached Oakland, I was there to claim and transfer my birds to their quarters in Idora Park. This activity took several trips by a furniture van, which was certainly ideal for it was covered. The troop was unloaded without injury. They were put in a large enclosure back of the display and sales room. A door from this room permitted pay customers to enter the area occupied by the ostriches. The lot was divided into several pens by wire fencing. All was enclosed by a ten-foot-high board fence. In connection with the showroom were housekeeping quarters and a factory room. I had placed Hazel in charge of this operation.

Two of my children had gone with me; Frances and Cecil who had brought his pet monkey. They soon had the run of the park, which they enjoyed very much. Imagine two kids turned loose among a bunch of concessions. I felt safe to forget them and not worry. Frances enjoyed the outdoor stage-show theater people led by Francis Hartman, and their nighttime productions. She and one of his children played together during the day. They were both of tender age, and one day after I had supplied them with a big dish of ice cream I listened to their conversation as they ate outside near my window.

The one little girl said to Frances, "Let's take little bitsy bites." Frances answered, "You can if you want to; I'm going to take great big bites."

Cecil had the time of his life, as a nickel then bought what it was supposed to buy. I never knew much of the time where he was, but be never caused any trouble. He was given free rides on this and that, and the monkey he had taken along to the park paved the way to many friendships. To capture an ostrich in the wilds was nothing compared to a monkey loose in a park. Jocko once took refuge in the trestle scaffolding of a Race Through the Clouds. He had first visited lesser attractions, but when he found the Race Through the Clouds, that was his Utopia, and there he camped refusing to vacate for three days. Finally, one morning he must have become hungry, for he was found climbing around the ostrich pen. From there on it only took a banana to bring him down.

In the center of the ostrich pen was a very small building just large enough to store a sack or two of special feed and for a man to enter. Visitors never seemed to disturb the birds as a rule, but one day a camera was set up for some special pictures for an Oakland newspaper. When the picture man threw the usual black cloth over his head pandemonium broke loose. A wild stampede took place in which cross fences and enclosures were torn asunder even before I could have the man and his camera out. I had remembered there were four or five visitors in the enclosure when the fuss started and wondered could anyone have been injured? Looking around I could see no one. They could not have vacated or escaped except by climbing the high board fence. This seemed a sound conclusion. A minute later I found every one of them, men, women, and strangers alike all in the little feed shed, crowded in like sardines. Self-preservation came first!

One summer season run at the park ended my city life. I enjoyed the people of San Francisco and Oakland, and I enjoyed the constant flow of people in and out as they marched to the tune of the husky barker I had out front. Some of the vainglory and veneer he put into his flow of rhetoric gave me the feeling at times, that I should wrap a snake about myself and go out front. The people of these big cities were receptive, but fundamentally I was not a show woman, and when the season was over, I shipped my birds home to pasture in Buttonwillow.

The only trouble I had was not with the birds but a variant train crew. My boxcar had been placed on the siding and operations were readying it for the birds. The train crew for some reason wanted to change the schedule and make up the train ahead of time. This would not allow adequate time for safe shipping preparation. It would be hazardous for the safety of man and bird and a foolish venture to undertake. They were deterred only when brought to realize the danger involved and that a man could be killed.

You're Invited to an Ostrich Plucking

Mr. Tracy told me one day on his return from Bakersfield that many people had asked him to let them know when he plucked the ostriches. He said, "I told them, I couldn't do that. Furthermore, I couldn't even remember which one I had told, so I finally put a notice in the paper."

I knew right away what to expect, or at least I thought I did. And when you read this story, you will agree with me…

By eight o'clock the morning of the day announced in the paper, our yard was a living sea of picnickers. No special arrangements for such a crowd had been made. Even though they were just milling around, they seemed to be enjoying themselves. I was bewildered and felt thoroughly helpless. I tried to find Mr. Tracy, but could not. In fact, I never did find him the whole day long. In our brief planning we had made a large caldron of coffee, and barbecued one whole beef. Toward lunchtime, white tablecloths were settled and spread all over the place. People sat back to back, until every speck of a large territory was just like one big checkerboard. As far as I could see, no one could get near the table where the meat was served, but I guess they did, for it was all gone in a hurry.

Knowing I would have some hungry men and a family to feed that night, I had put away in the pantry that morning a large piece of meat. A little past noon, Tom Burke, then manager of the Kern County Board of Trade, in company with a guest, found me. True, I was not too hard to find because I was mostly in one place spinning around like a top, not knowing what else to do. I was very much embarrassed because we had no control over this affair. I kept bowing and gasping, "Hello there!" but hardly knew to whom I was speaking.

Mr. Burke, of course, had hoped to find some kind of scrap left to treat his guest, so I took him into the pantry and proceeded to slice off the big roast. The meat was picked up as fast as I sliced it off, and I commenced to wonder how much Mr. Burke could eat. I looked up for the first time since I started to carve, and found our house was full of people. I lay down the knife and disappeared. It was finished by whom I will never know. That night we had bull's head and nothing else to eat. The men of the ranch had dug a pit and barbecued the head; so we stood around the pit like a bunch of Indians might have, and I ate for the first time that day. I have never tasted anything since that was so good. That bull's head saved the day, at least for the family.

It took six men to clean up the place after all the visitors left. All the moveable furniture had been taken out of the house. The milk separator was found about two hundred yards from the dairy barn, and there were five acres of picnic debris to pick up. Having perpetrated such a disorganized affair, it helped us to see the need for serious planning if ever another picnic was not to be a fiasco, too.

The Red Cross Picnic of 1918

In the spring of 1918 Fannie and Will Tracy decided to sponsor an enormous picnic as a benefit for the Red Cross. After the embarrassment of their disastrous first picnic, Will had suggested they try again. He asked Fannie to take charge of the affair, in order that it not interfere with his farming operation. She did! In her book Fannie refers to the event as a barbecue, but in the memory of her children, it was always called "The Red Cross Picnic." Whatever it was called, it had a tremendous impact upon all of us.

I was three years of age, and I must have been an obstinate child, for I remember someone trying to place a Red Cross nursing cap on my head for a group photograph. I would not allow them to do so, yet I clearly remember that in my heart I wanted that cap on my head in the worst way! Apparently I had made a stand, and I was going to stick to it! Who knows what goes on in the mind of a child?

World War I was about to come to an end, and patriotism was running at a feverish pitch. "The War to End All Wars" was no Vietnam. Who can ever forget the stirring melodies and rhythm of George M. Cohen's famous songs, "It's A Grand Old Flag," "The Yanks Are Coming," "Over There," and "Yankee Doodle Dandy!" The songs are as compelling today as they were eighty-some years ago.

With sentiment running so high, Fannie had no problem securing help with this gigantic undertaking. But she describes it best in *The Tracy Saga:*

> Mr. Tracy and I did not discuss our picnic difficulties very much, though I think he realized what a mistake he had made by his announcement through the paper. By now several Red Cross benefits had been given throughout the county. We attended most of them, and I observed each program very closely. At the time I was not thinking about ever giving one, but was interested in noting why some things went wrong. One day my husband said to me, "I would like to give one more picnic and have it a success before we quit." I agreed with him. We would both be "happier ever after" if we could overshadow our bad affair with a successful one. This 1918 picnic would also serve as a benefit.
>
> As Mr. Tracy was very busy with the farming operations, he told me, if I would manage the affair, he would give me all the men on the ranch to ready the place. Like a general I laid my plans and went to work. I remembered clearly our previous picnics and how I was swamped. I remember too, appealing to some of my best friends for help, and how they looked at me as if they had never seen me before. Scarcely excusing themselves, they went on by. They had the look and probably felt that way too, "What do you mean? I am here to relax and enjoy an outing!" This was true of everyone I entreated. So just like a good housekeeper, people have to have their plans in their minds when they start the day. The people who attended our picnic this time had prepared themselves for an outing.
>
> Realizing this, I took my car and called on the clubs of every town and city in the county at that time. I found these people to be very cooperative. They helped me form committees to take charge of different divisions. It was to be a combination carnival and sideshow, with amusement features, auction sales, roulette wheels, speeches and bands playing. I knew before the big day just when certain towns and certain people would be in a booth. It was a marvelous sight to see the way things clicked that day. On each booth I had painted the exact hour each was to take over. Exactly when it came time, women checked under the railing and a fresh relay took over. Only one hour was allotted to each group, then they were free. This plan was used in each booth throughout the weekend.
>
> During the occasion thousands of people attended. Of course, I do not know exactly who was there, but the registers around under the trees were full of interesting names. Not all could get to the books because they did not want to stand in line. It was a great day to go down in memories of

those who were there. The Tracy Picnic! I remember! I was there! Were you? Seventy-five thousand dollars of phony money was printed for use that day, and a main bank was set up in the cookhouse, the old-fashioned kind on wheels that moved around with harvest crews. It was in the charge of the Bank of America, then called the Bank of Italy. They brought out their regular cashiers and kept the books. It was made easy to change regular money into the phony kind. There were many branch banks under the trees, controlled by the "Cook House Bank."

The key concession and attraction was a professional barker from San Francisco. He had pre-offered his services and took over. He was a marvel! I know nothing of his history before that day, and I think anyone who saw him perform has not forgotten. He never ate a bite all the time. It took six new girls every hour to handle the tickets and crowd that was banked around his booth. The bank kept a separate account on every booth. He started with two hundred dollars worth of "Baby Dolls" and when the booth sold out, the till showed six hundred dollars.

As near as I can remember, we barbecued seven beefs, all donated. Each bakery in Bakersfield, I had learned, took out their bread at two o'clock in the morning. Then they could put a whole beef in their ovens. Each butcher shop prepared the meat for roasting. The barbecued meat was put in large paper-lined dry-good boxes, which were picked up by a truck, and brought to the ranch by six o'clock in the morning. Men had been working throughout the night at the ranch, and had the pits all fired and smoking. The bakery-barbecued meat was stored in our blacksmith shop close to the pit. As the meat was used, it was easy to replenish. No one noted that the meat had not really been in the pit, for the smoke supplied the usual flavor. Another group attended the five hundred pounds of dry beans cooked in big kettles. Two hundred pounds of coffee were brewed, stacks of bread made available, and two hundred cases of pop were on hand.

There was only one hitch, and it will be well for those who are giving benefit affairs to remember this. It was political times, and the politicians asked for an hour of time to make short speeches. Their request was granted. Every booth closed except the "Baby Doll" man, who just lowered his voice and kept right on. At the close of the day, the banks had kept track of the rate the money had been rolling in and knew exactly what those political speakers had cost us. It was plenty!

Two orchestras played, one by day, and one for the dancers all night. Benches and chairs were provided for tired people, and they enjoyed the music and the excitement while they rested. Ostrich riding and fighting went on at one place, and bronc riding at another. Some professional riders happened to be passing through town, so they came out, gratis, and put on an exhibition, which added much to the day.

According to Mr. Tracy, we made one other mistake. He said it took a long time, too long, to move six thousand people from one area to another. However, it certainly was an exciting day, and a gratifying one when compared to the earlier "flop." One thing I had noticed on my inspection tours of picnics was when people turned away with their plates, they seemed at a loss as to where to go. We had a remedy for this by having inviting places ready. At the close of the day and night, our picnic enterprise had $2,500 to turn in to the Red Cross. This, I am sure was a great help, and we were happy over the results of this two day affair.

The next day a call came for a certain gentleman's coat. When found, it was loaded with ostrich eggs. They were stuffed into the sleeves which had been fastened at the bottom by tying with stout string; however the owner never came for the coat!

It was a wonderful day too for all the children, especially my own, three of whom were old enough to be unattended, and three of whom were too small to remember the day. At the close of the day there was pop, gum and candy left. Cecil, the oldest of the children, secured a large tool chest and went into business by storing quantities of these leftovers. To be sure of safe keeping, he included in with the delicacies a very large, five- or six-foot gopher snake. His sister was about as well versed in tricks or subterfuge as her brother, but the big old snake did afford quite a problem. Although she knew he was not poisonous, she did not relish his attitude of defense, and the way he had of showing his tongue.

End of an Era

After twenty years, a war intervening, the world's feather market went to pieces. Women forgot self-adornment and responded to the call of their countries. The once fabulous plume and boa went out of style. There was not much interest left in an outmoded product.

It can only be for the sake of history that I write of this industry, and since it consumed part of my own life, I have included it here. It was a challenge, and as I view it through the vista of time, I hope that it had something to do with building up fortitude, strength, and wisdom to meet the trammels of life that I encountered on the way. Raising a family of six and a troop of ostriches at the same time, and not being vanquished entirely, required one to lead a versatile life and one in which there is no room for vacuity.

Some of my birds I sold to carnivals and circuses, some went to parks, and some became vagabonds by running away. A few were shot by trespassers who entered Jerry Slough to fish or hunt. The dead bird was usually found close to trees where his victim had hoped to escape. The tenacity of the bird being greater than that of man resulted in the bird's death by the blast from the trespasser's gun. I am sure each one of these men could tell an exciting story. Ostriches mixed with general farming were a disturbing factor and did not combine well with animal husbandry.

The First Great Loss

The year of the Red Cross Picnic we lost my brother "Teddy." He died of scarlet fever at the age of eight, and mother was desolate. Adversity she could handle, but death was a different matter. Aunt Dee once told me that "Fan" made it very hard for Will at that time, but father did his best to console her during this difficult period.

Charles Wellington "Teddy" Tracy and pet steer circa 1918

The Swamp School

In 1920, after the birth of her sixth child, Fannie secured a teaching position at what was known as the "Swamp School" southwest of Buttonwillow and about six miles from her home. Barred from teaching in her own community, Fannie was determined to make use of the education for which she had striven so hard. She put Cecil, Darrel and Teddy in school at Wildwood and, taking eleven-year-old Frances with her, she set out to teach the Italian children whose families worked for Miller and Lux. Fannie and Frances usually drove to school by horse and buggy. During rainy seasons the wheels sometimes became so laden with the swamp's adobe mud that they were almost as large as what were later known as balloon tires. The horse could barely pull the load. During flood season they occasionally forded swollen streams, but if the route became totally impassible our father would crank up the Model T and they would take the longer route through Buttonwillow.

At the Swamp School Fannie taught the Italian children to speak English and do arithmetic. She and Frances formed long-lasting friendships with these hard-working pioneer families, and they often sent our mother home with a loaf of the delicious homemade bread baked in their brick Dutch ovens.

Above all, mother was a mentor. If she thought that a person had the least amount of talent she would encourage it. When I was a child I once drew a profile of a horse with both eyes on the same side of the head! Mother liked it, and from then on I became the family artist, with or without talent! The following passage from *The Tracy Saga* reveals a bit of her philosophy as an educator:

> I found all children good children when they were kept interested with enough new ventures to keep up some excitement. I never thought of any of them being bad. Of course, some correction was needed, but not often. To me, all children are interesting, whether they have been branded as good or bad, and I have worked with both. I have never found that a really unruly child was damaging to me if I looked for more strength within myself. Children really have a broader understanding than they are given credit for. Our duty as parents and teachers is to do all that is possible to separate from them what appears to be bad.
>
> When children once see and feel what is good, to separate them from it is almost impossible. All methods of correction, I believe, should be done through activity and not by ostracism.

Despair

In June 1940, it was necessary to confine our father in the State Mental Hospital at Camarillo, California. At this point mother's spirit reached its lowest ebb. The adversary she could not fight was waiting just outside her door. Mother revealed the depth of her sorrow in a letter that she wrote to Frances on May 23, 1941:

> *Dear Frances:*
>
> *If it were not for this great sadness that overshadows all, I would be to your party. We returned home last night where we left Daddy in the hospital, and now I sit here in this tomb-like house. No living sound, only the birds, as Cecil went to town with Jack to have some welding done. The house is that of living death, only perhaps worse than death if it were not for the hope. I am going out in my garden now as it is the only companion I have left. Cecil and I have heart-sinking spells of grief as we tried so hard to stave off the terrible illness that kept creeping up, until as last we had no more power to battle on as we had to save Daddy. Cecil, Darrel and I are going into the hospital tonight to see him. We expect to take him over to the Coast Hospital for treatments. He will know you and be glad to see you. I think this is the first time in my life I was alone here entirely. You know the rest.*
>
> *Love, Mom*

Will Tracy passed away five weeks later at Camarillo. I was just two months pregnant with his second grandson. The orphan who recited his favorite poems as he followed the shares of his plow existed no longer. Gone was the young horseman whom Fannie had encountered in the floodwaters of Jerry Slough. Gone was the kind neighbor who had provided her and her sister with whole milk in place of skimmed. Gone, too, was the young rancher who had left his gold watch on the desk of the new schoolteacher. Now there remained only memories of the gentle husband and father of six children—the one who had comforted her in sorrow and helped her fulfill her dreams. Even the elderly irrigator with his shovel and pack of dogs had vanished from her life. Fannie had reached the depth of her despair. It was difficult to imagine then that she would lead a full and productive life for the following twenty years.

A New Beginning

Although Cecil had been at no loss for female companionship, he had remained single for fifteen years after the divorce from his first wife, Florence. He had dedicated those years to helping mother hang onto the ranch. Now, in fairness to Cecil, mother decided to vacate the home she loved to make way for Cecil should he choose to remarry.

Around 1945 Fannie selected a site on the banks of Jerry Slough about a quarter of a mile east of Tracy Avenue and a mile south of 7th Standard Road. The site would eventually become the office headquarters of Buttonwillow Land and Cattle Company. There, for the next several years, she would garden in the fertile soil and entertain her eleven grandchildren. She resurrected two old buildings and joined them together to create a modest home. By this time our grandmother, Martha Rowlee, was totally blind and bedridden. Fannie would care for her ailing mother until her death in 1948 at age ninety-eight.

Soon Fannie planted trees and an abundance of flowers. Fannie's grandchildren loved to visit their grandmother who always had projects for them to do and would take the time to play. Forever the teacher, she drove the children to Uncle Darrel's post office in Buttonwillow where they would set up a stand to sell the vegetables from her ever-expanding garden. Once she ordered 500 seedling pines from the Department of Forestry, and the grandchildren planted them in one-gallon cans, then transplanted them in Jerry Slough with visions of one day opening a Christmas tree farm. Although the trees did not thrive, it was an exciting plan, and earned the children the astounding salary of twenty-five cents a day.

The Pet Parade

It was not always work at grandma's, for she had a fun side, too. Sometimes she engaged the children in playful fighting, just as she had done with her own children. Once she organized a "Pet Parade" which I recorded on my new movie camera. Three pretty little girl cousins on horseback headed the parade, but the star performance of the show was a "bull fight" staged by cousins Larry and Wesley. No elaborate costuming was used for the event—a white dishtowel served as a flag. Larry the matador flaunted his "flag" with grace and vigor while Wesley, on all fours, portrayed the enraged bull, lunging and plunging at the detestable flag.

Camellias

Fannie was far from inactive in the years following the loss of her husband. She became friends with Mr. Howard Asper, the camellia expert then in charge of the Desconso Gardens and, later, manager of the Huntington Gardens, both in Pasadena. She studied the culture of these beautiful flowers (after her first encounter with mother, Aileen, Cecil's second wife, referred to camellias as "those nasty little flowers"), and before we knew it Frances and I were growing them too. We became active in the Camellia Society, and were soon exhibiting specimens and making floral arrangements at the annual camellia shows. Fannie learned that she could propagate a camellia from a single leaf, and she later taught the skill to my ten-year-old daughter, Nancy. Never one to do something halfway, Fannie acquired hothouses, and before long they were filled with boxes of sand and hundreds of baby camellia plants.

Fannie with her camellias

Frances (left) and Martha at the camellia show

By 1957 Fannie's hothouses were overflowing with hundreds, perhaps even thousands of camellias she had propagated. It was there among her camellias that, unfortunately, she fell and shattered a shoulder. Undaunted, she worked hard to restore the use of her arm, but it became my sister Frances's lot to take over the care of the camellias. (She would turn the handle on the old meat grinder to strengthen her arm, and she told me, "When that gets too easy, I put a little piece of bread into the grinder.")

Under her grove of eucalyptus trees, Frances installed hothouses and additional shade to provide for their care. She fulfilled mother's dream by establishing a business called "Camellia Grove Gardens." Mother attended the opening in a wheelchair.

Fannie's Book

No longer able to carry on her regular activities, Fannie resumed a project she had begun earlier. She would finish writing her book. Time being too valuable to waste in bed, Fannie had always been an earlier riser. Now she awakened early and devoted her time to telling the story of her life.

The day that the proof of her book came off the press, a second fall cost Fannie a broken hip, and with this accident, her health began to truly fail. For the next four years Frances dedicated herself to the care of our mother. This required the attention of two people around the clock, so it became necessary to phase out the camellia business. Frances set up a hospital bed in her dining room at the center of her house, and here her children and grandchildren had easy access to come and visit. From a large bay window, mother could witness the rotation of the crops and take comfort in the fact that Will Tracy's ranch was still alive and in good hands. Fannie finished proofing her book from her bed, where she would be confined for the remainder of her life. *The Tracy Saga* was published in December 1962. It was rewarding that Fannie lived long enough to see her book in print.

Fannie Tracy, co-author, born February 18, 1877 near Galt, California, became one of the early settlers of the Buttonwillow area, even before the age of 10. She has resided there since, having married and raised a family of five, while helping develop a large ranch devoted to stock raising and general farming. Some 25 years of her life were spent teaching in several of the small elementary schoolrooms of Kern County. This book is a result of her efforts to provide her family and the general reader with more than 80 years of personal philosophy gained through "events along the way". She lives now with her daughter Mrs. E. J. Selvidge on a portion of the historic Tracy Ranch.

AUTHORS

Glendon Rodgers, co-author, born in 1911, is a longtime teacher at Bakersfield High School and presently serves as Dean of Student Activities. An enthusiast in the field of Kern County history since 1947, with several publications already to his credit, he is rapidly moving toward other writing efforts. His recently organized "Cardon House" as a family publication venture is proud of its first contribution to local history. The opportunity of working with Mrs. Tracy and her family proved both inspirational and rewarding toward developing a clearer understanding of a time called pioneerhood in the west.

In December, 1962, Fannie and Glendon Rodgers co-authored the book, *The Tracy Saga*

Reflections of Mother

For her book, Fannie invited each of her children to contribute a chapter. Following are some passages from Cecil, Darrel, Frances, Martha, and Tilton.

Cecil, 1961:

> The character of Mrs. Fannie Tracy is a very definite one. Only a few people in her long life have measured up to what she thinks a person ought to be. She is the finest judge of character I have ever known. While my father was taken in on many occasions by well dressed, fine looking and smooth talking persons, never once was mother ever fooled by the same people. I remember mother warning him repeatedly in a gentle way (she was always gentle to dad, and he to her) about the advantages certain people might take of him. Dad did not always heed, but mom was always right.
>
> It was, in the long run, father's disregard for mother's warnings that plunged our family into a twenty-year struggle that shortened his life and placed a mighty burden on the little round shoulders of a great and gallant lady. There is a strange sense of responsibility and resistance of defeat when one is in charge of the accumulated morsels of several lifetimes of honest hard working forebears whose work is yet not quite done. Life itself seems to take second place in importance, and this I will say welled up in all its mighty splendor within mother's heart, and it was contagious.
>
> Mother is eighty-four now, and I am nearing sixty. I am convinced, since I've had time to look around, that I have known intimately one of the great persons of our time.

Darrel:

> Mother delivered me into this world early on a December morn in 1906, on the home ranch a few miles out of Buttonwillow. Since that time we have come a long way together. She has always been close to our family and intensely interested in the welfare and well being of each of her children. Always quite some distance from a town of much size, considerable thought had to be given to provide such things as our schooling, music appreciation, and a few social functions. These she managed as best possible under the circumstances that existed and they were not convenient. She has always been a dedicated teacher of children…
>
> On a number of trips across the sagebrush and plains she drew my attention to rabbits and coyotes and such. There began my interest in wildlife. As we traveled from day to day our attention was drawn to a litter of baby coyotes that played about their den alongside our roadway. Mother said, "We will have them dug out by the men and you may have one of the little ones for a pet." I was most hopeful and pleased over such a prospect. Sure enough, the day finally came and the men began to dig out the canine pups. It was an exciting time for me. The hole began to go deeper and

deeper as they dug, and now and then a question would arise as to which direction the coyote hole would go. At that point I recall so well that a buggy whip that we carried was employed to determine the course of the digging. The whip was very long and the men would push it far down the hole to determine which way the hole turned next.

After considerable effort and due determination the quarry was unearthed, and there at the end of the passageway was found a litter of little fuzzy baby coyotes. I was allotted my share, and we took them home for pets. I believe that such efforts as this on the part of mother and dad have sparked within me an everlasting appreciation and respect for so many of the things that lie in nature's vast realm of wildlife. From that day to this I have enjoyed the out-of-doors.

Frances:

It is difficult to delve into the past and choose from memory any one incident that stands out more than others. When one was raised in a large family on an early western ranch with active parents, and I might say especially an active mother, this effort becomes even more difficult.

She always trusted us, and responsibility was put on us at an early age. At the age of twelve I was doing all of mother's banking for her, and since I could write like her, she trusted me to write checks on her account when I needed or got caught in town without money. When I was at the early age of twelve she would put me at the wheel of our car, and off we would go to Los Angeles over the old Ridge Route and into Los Angeles traffic. Never once do I recall her getting nervous or excited over my driving. I shudder to think of it now, but that was my mother.

The author:

"Mother": a word that has so many different meanings to so many different people. To me as a child, the second daughter of Fannie C. Tracy, it meant a pair of arms to hold me and to rock me when I was ill or discontented. On the other hand, the same two arms could be quick and firm to discipline me when the occasion demanded. Well do I remember once, during a naughty spell, of being quickly snatched up and having my head thrust under the spout of the old hand pump on the back porch, while mother pumped vigorously with the other hand. I thought that she surely was going to drown me as I fought and struggled to get free. But such occasions were rare, and for the most part I think of her as radiating a certain kind of warmth. This warmth may have been, at least in part, physical. Having been born cold blooded physically, I seem to have spent a good part of my life trying to keep warm. My childhood was no exception.

During the winters on the ranch, good old cast iron stoves heated our house. These were a joy and a comfort when their bellies were loaded with eucalyptus wood, and I remember with pleasure the sounds they made as they hissed, and roared, and crackled, and popped. However, they could not heat more than one room. As bedtime approached, the fire was allowed to die down, and soon we were in another world. We would quickly jump between the icy sheets and seek the brick that we had heated on top of the stove, and then wrapped carefully with newspapers. As the night wore on, no amount of covers could keep out the penetrating cold. The brick became a slick, cold uncomfortable thing in our bed. It was then that I would creep into mother's room and seek the warmth of her body. She would knowingly take me in her arms and snuggle me close to her as I went blissfully back to sleep.

The next morning when I awoke, she would no doubt have been gone for several hours to prepare a large breakfast for the family and whatever farm hands or stray boys that had sought the shelter of our roof. It has always been a mystery to me how mother was able to rear six children and sandwich some twenty odd years of teaching in between, but this she did. During this particular era of which I speak, she was teaching at Bowerbank, a small one teacher school located about a mile-and-a-half from what was then the San Joaquin Power Plant and about a four-mile drive from our place. If the Model A Ford happened to be out of order, daddy would hitch old Skip to a little two-wheel cart that was used for such emergencies. Instead of feeling inconvenienced, mother seemed to derive a certain amount of pleasure from these trips. She never seemed to notice the cold, and she always looked happy as she sat there perched primly upon the seat of the cart, reining Skip expertly around the chuckholes and through the sagebrush. We always took the cut through Section Five, and mother seldom failed to remind us that it was our father who had carved these roads from the sagebrush. I was not too impressed with the condition of the roads, and I sometimes wondered if it would not have been quicker to follow the main route, but I did not say so since I, too, enjoyed these adventures through the sagebrush. Occasionally we were rewarded by the sight of a fresh outburst of mushrooms by the roadside, or perhaps a humorous looking jackrabbit darting across our path. The wheels of the cart, as they passed lightly over the sandy loam, made a pleasing sound combined with the steady beat of old Skip's hoofs. Mother took a particular pride in the fact that Skip had once been a racehorse. I don't think that he had forgotten it either, for he delivered us to school in record time considering the hazards of the journey.

After our horse was unharnessed and placed in a stall with that of the Antongiovanni family, who arrived in a similar conveyance, Tilton and I made a dash for the schoolhouse to warm ourselves around the huge old coal-burning stove. This stove, though very impressive in size, was somewhat of a disappointment in the amount of heat produced.

The average enrollment at Bowerbank ran around twelve to fifteen students and didn't fluctuate very much. It consisted largely of the Antongiovanni family, a few students from the families who worked at the power plant, and Tilton and me. Mother had many and varied interests in life, but the three most prominent were gardening, preserving food, and teaching. Whatever she pursued, she did so with her whole heart and soul and her teaching was no exception. I can see her still, working diligently with the older children as they studied the constitution, or worked on arithmetic or grammar while I listened and pretended to be studying my own lessons.

Lunchtime was the high point of the day, and somehow mother had inveigled the school board into acquiring an electric range. This was indeed an innovation at this time. The children were encouraged to bring things to cook in order that they might have a hot lunch. My favorite lunch was Campbell's vegetable soup—I was always fascinated by the little alphabet macaroni it contained. We often had baked potatoes and once I even remember having roast duck.

Friday afternoons were reserved for art, and we all looked forward to this with great anticipation. Some of the projects carried out with the older children were quite unique. I have no idea where mother learned to do them, but somehow she did. One project in particular I remember quite vividly. We used oval discs of plywood to which we attached coils of rope then covered with raffia. We painted the rope peacock blue and gilded it with gold paint. A piece of violent orange silk was placed in the bottom of the tray upon which we mounted peacock feathers and covered with glass. I thought they were the most beautiful things I had ever seen!

There was one boy in particular who was an outstanding student in every way. His name was Guido Antongiovanni. Guido was the eldest of the Antongiovanni children, and he made baskets of broom straw and raffia that he fashioned in the Indian motif. Mother often boasted that his baskets were so well made that they would hold water, and indeed they were works of art. Later he learned to cover overnight bags with ostrich leather. Guido graduated with high honors from our little school, and we were all greatly saddened to hear shortly afterward that he had died of pneumonia.

My father used to say, "In spring, it is as natural for a woman to get out and dig in the garden as it is for a hen to scratch." Being married to mother no doubt had a decided influence upon his point of view. As spring approached, each child at school was allowed to make a small garden. We joyfully worked the ground, tied strings to stakes to make straight rows and planted our seeds with anticipation. Somehow as time went by, mother's plot seemed to progress better than the rest, but we all enjoyed our gardens, and if nothing else succeeded, our radishes did. We had them in great abundance. I recall the many trees that mother planted during this time. To live in a place and not plant trees was to her a sacrilege. Sometimes she was tempted to plant them beyond the reach of a hose, and I can remember carrying water in buckets to the parched alkali soil in which they struggled.

Mother claimed to have started the first school orchestra in Kern County where each student in the school participated. I think this might have been a slight exaggeration, but this was her claim. Nevertheless, I was astonished when she announced that I might have any instrument I wished to play in the orchestra. Times were always hard as I was growing up, and to have this much money spent on such a luxury for me alone was incomprehensible. I said in utter disbelief, "Do you mean that I could even have a saxophone?" This was during the jazz era, and I had seen movies where the saxophone player seemed to be the life of every party. He would get right out on the dance floor with his instrument and mingle with the dancers as he reared his head upward and downward in a sort of Indian war dance fashion. All the while he was going through these gyrations, he would presumably be making beautiful music, but due to the fact it was during the era of the silent film, I will never know. I thought that being able to play a saxophone would be a sure key to social success, and now mother was telling me that I might own one of these magnificent instruments! It was beyond belief, but I was to learn that this was characteristic of life with mother.

Just when I had been brought to think things were so bad that we surely would starve to death, mother would come forth with some new earthshaking investment. Like the time she bought the Ampico electric player piano! Of course, we were all delighted, and mother convinced us that it was a real buy because it had cost two thousand dollars new, and we were getting it, only slightly used, for twelve hundred. In our enthusiasm, nobody stopped to think that times were exceedingly hard and that twelve hundred dollars was a small fortune, probably more than a year's salary derived from school teaching. We were all ecstatic over this huge new wonderful instrument that produced music like magic by the touch of a finger.

By now a little of the luster had worn off my saxophone. I found that I could barely play the scale standing perfectly still, let alone moving about like the playboy in the film. But the Ampico was the answer to everything, no tedious hours of practicing, no pumping, just lots of good music and plenty loud. In addition to this, it had words printed on the roller and many an hour I spent as I sat and wailed mournfully to "Flow Gently Sweet Afton." This was my favorite, and I still remember all the words to "Moonbeam Kiss Her for Me."

Mother came from a musical family. Both her mother and her father had fine voices. They had formed a family singing group, which was the core of their home and social life. Consequently, she probably wasted thousands of dollars on her unmusical offspring. Each of us was exposed to piano lessons. Cecil had a trumpet, Darrel an alto horn, Frances a piccolo, I the sax, and we all took turns beating on the fine set of drums that followed the purchases of the baby grand electric piano. The trouble was that she failed to recognize the hereditary tendencies from daddy's side of the family. Not that daddy didn't love music! I can see him yet in church earnestly endeavoring to sing, with his chin lowered and one eyebrow sincerely arched as he struggled in vain to keep in harmony with the others.

Like daddy, we all tried, and Darrel even wrote an original composition for the piano entitled, "To A Hen." I'm afraid that the title however was the only part that was original. Cecil used to march in the Kern County High School band, proudly clad in the flashy new uniforms of the day, which sported twenty-eight-inch bellbottom trousers, and felt sport hats pulled rakishly down over one eye. He marched with the band for miles on end, enjoying the many privileges and the glory of being a part of this institution, yet he never played a single note. But then Cecil always was a good actor and nobody but the family ever knew.

By the time I was in the eighth grade, mother decided to send me to town to get some of the sticks out of my hair. Here I lived with mother's sister, Aunt Dee, and her kindly husband George Taylor. By now the saxophone, that beautiful golden instrument, had become a millstone about my neck. However, I still felt a strong sense of obligation toward this extravagant venture, and so I enrolled in the school orchestra. We lived some distance from the school, and it was necessary for me to carry this instrument for many blocks. By the time I approached the school, it weighed at least a thousand pounds, and I had to stop for frequent rests. Each time I stopped, I would shift it from one arm to the other. I developed an ache in my right arm, and I still attribute my bursitis of more recent years to the carrying of that loathsome instrument. Upon reaching school, I went directly to orchestra where I learned to play the first bar of "Tenting Tonight." After the first bar, I was lost and I simply sat there for the rest of the time with the reed poised in my mouth. Bob Kitchen and Joe Himmovitz sat directly in front of me with their saxophones, and they carried on for the saxophone section.

The following year I entered Kern County High School, where I dutifully lugged my saxophone to the music building and tried out for the school orchestra. Bob Kitchen and Joe Himmovitz preceded me, and I commenced to have some misgivings as the bandleader was shouting at them in an unkind manner and pounding them on the shoulder rather hard in time to the music. When it came my turn, he neither pounded me on the shoulder nor shouted, he just rolled his eyes in a very unpleasant way and said I was hopeless. I was deeply hurt, tears welled in my eyes, and I fled choking with hurt and anger. During the next four years, whenever I encountered this man, I would raise my chin high in the air and cut him dead. As I think back now, I doubt very much if he remembered me, but at least it made me feel better at the time. This ended my musical career and about the only good that ever came out of it was years later when I sold the instrument for forty dollars.

Frances then was mother's only ray of hope as far as musical accomplishment was concerned. Fran was always the smartest one in the family, and she had shown plucking management to get herself and her piccolo in the all-boy band at Kern County High School. Furthermore she was making some progress on the piano. She could play, "Falling Waters," which was in itself quite a piece, and every time mother wished to display the accomplishments of her "musical family," it was Frances who upheld the family pride with, "Falling Waters." I can't remember her ever playing anything else, and when she practiced, it was always "Falling Waters" but, as I said, it was "quite a Piece," and we were very proud of her.

In retrospect, although I was shocked by the expenditure at the time, I believe now that the purchase of the baby grand player piano was the best investment mother ever made, musically at least. Many were the happy hours we spent as we danced, sang and beat on the drums in time to the good loud music of the Ampico electric.

Tilton:

I believe I had one of the happiest childhoods that any human could ever have. My mother made this possible almost single-handedly. Now, as I look back, it all took place with the absence of material things except on rare occasions. Included here are a few of these special times that I remember most vividly.

I can recall so well hounding my folks for a pocket watch. Though I was not yet of school age, and could not tell time, I believed a watch packed a lot of importance. Finally, the great day arrived. Mother gave me a dollar to buy a pocket watch. In 1924 that was all they cost, tax and all. I went to the Miller and Lux General Store located in Buttonwillow. Upon looking into the showcase, a new problem arose that I had not considered. There must have been a half-dozen different brands and styles of dollar watches. The clerk asked me which one I wanted. As a young boy raised with horses that heard the men talk about fast horses, I placed a lot of importance on speed, so it was only natural to want the fastest watch they had, and that was my answer.

Another time at an age when I was learning to break horses, I wanted a certain set of books that dealt with this subject. This was in 1933, right in the middle of the depression. I wrote to the company and priced the desired books. They were quite high priced, and I knew there was no money like that in the entire family. I guess the depression was telling on the book company also, as I kept hearing from them about every thirty days, and each letter quoted a cheaper price. Finally I thought they were cheap enough for me to approach mother with the proposition that involved ten dollars. She weighed the matter very carefully and said, "We will see." During the next sixty days she would talk to me about it. After making sure it was not a passing fancy, I was allowed to purchase the books for which I was most grateful. They turned out to be very helpful to me. I did not fully appreciate what a sacrifice that ten dollars must have been at that time, until later as a grown man. I still have this set of books and cannot look at them without thinking what a mother will do to make one of her children's lives a happier one.

To show further how liberal mother was with her children, I remember that while we were in high school our agriculture department at Wasco

High School put on a rodeo. My folks owned a lot of cattle and horses at that time, so I volunteered to furnish all the stock, panels, bucking barrels, etc. for the show. I never asked anyone, just started hauling things away. I even took the steers my parents were trying to fatten. Not once was I questioned about all this, or even did I think I was supposed to tell anyone.

Of all the hardships outside of losing members of the family, the worst I can remember my mother being hurt was when she stepped on and killed a little pet quail called "Chee-Chee Bird." The little fellow would follow her anywhere and moved as fast as lightning. To this day, her voice saddens when you mention him. Mother was an unusual force in affecting the ways of her children.

On the morning of April 16th, 1966, mother was unable to speak. A pleading expression on her mother's face told Frances that she had suffered a stroke. It was obvious that mother knew her time had come. In spite of all the medication it took to keep her out of pain, our mother's mind had remained perfectly clear. Fannie was transported by ambulance to Cottage Hospital in Bakersfield, where the daughter who was most like her spent what was, no doubt, the longest day of her life as she tried to comfort her stricken mother.

Finally, at midnight, with Frances's hand upon her pulse, Fannie Tracy succumbed to her old adversary a month after her 90th birthday. She was laid to rest in Bakersfield at Union Cemetery beside her beloved husband and two of her sons, Charles Wellington and William Darrel Tracy.

What a legacy of memories she left to her family: from the barefoot girl who rode sidesaddle through the sagebrush and over the plains, the mischief-maker who rebelled against an education, and to the young lady who later struggled so hard to achieve one; from the dutiful daughter who washed her father's foot rags to the wife of Will Tracy, the rearer of six children, and the mother who suffered the loss of an eight-year-old son. Twenty years a dedicated teacher, raiser of ostriches, grower of camellias, survivor of a great depression, woman who battled to save her home, and widow of a devoted husband of thirty-seven years—little wonder that Fannie Tracy felt compelled to record the story of her life!

From left, Cecil, Frances, Darrel, Fannie, Martha, and Tilton in Frances and Skip's home, 1962

My Brother Cecil

Firstborn

Cecil Foster Tracy was born to Fannie Rowlee Tracy and William Tracy on January 14, 1905. He was the eldest of six children, and he was followed by three brothers and two sisters: Darrel, Frances, "Teddy," Martha, and Tilton, in that order.

He was born at home, delivered by his maternal grandmother, Martha Ann Rowlee, the beloved little pioneer lady who brought so many babies into the world.

Since I was not born until Cecil was ten, I do not remember much about his childhood, but I do remember a great deal of what happened after that, for although Cecil was not my favorite brother, he certainly was the most colorful! I do recall that he and Darrel shared many pleasures growing up on the ranch together; fishing and hunting wild game were but two of them. Should a flock of sand hill cranes fly over the ranch, the boys might be missing for two or three days, and they thought nothing of crawling on their bellies for hours in order to sneak up on a flock of wild ducks.

Bronc riding was another of their shared activities. As a little girl, I listened with awe when they returned from a rodeo and related their adventures. Sometimes they would win as much as a hundred dollars in prize money and, indeed, that was a great deal of money for a couple of kids to bring home in those days.

Cecil riding bucking horse. He was once named Kern County's "Champion Bronc Rider of the Year." Most of Cecil's photographs were lost when his house burned in 1974. This photo must have been returned after the fire by his friend Bertha Coe.

"A Living Statue"

Cecil inherited a sense of poetry and romanticism from our father William (who enjoyed reciting from memory all eight cantos of *Lady of the Lake*), and was endowed with the artistic sensibilities of our multi-talented mother Fannie. As a range cowboy for cattle barons Miller and Lux, awarded Best All-Around Cowboy at age 27, and a poised equestrian, Cecil the showman would become the Old West personified. A visionary and impassioned chaser of rainbows, Cecil found his pot of gold, his crowning glory, during the year 1949, commemorating the Centennial of the California Gold Rush. The most enduring image of his artistry lies in the spectacular "Living Statue."

The city of Bakersfield sponsored a celebration to commemorate the Gold Rush. It became an annual celebration called "Frontier Days." People turned out in pioneer costumes for this popular event; men were encouraged to allow their beards to grow, and those who did not were apt to find themselves doused in a watering tank or imprisoned in an iron cage. It was all in good fun, and culminated in a lengthy parade. It was for this parade that Cecil conceived the idea of a "living statue." Dressed in typical western cowboy garb, complete with chaps, cowboy hat, boots and spurs, he gilded himself as well as his favorite horse, Captain. Horse and rider mounted a float, where they posed motionless during the entire parade, Cecil holding his hat high above his head. Captain remained quiet and composed, shifting his weight only occasionally, with Cecil stopping to rest his arm only once or twice as they passed through street intersections. The float was so striking and spectacular that Cecil soon began receiving invitations to ride in other parades. He rode in a San Francisco parade, one in Sacramento, and finally in Santa Barbara.

The following is a quotation from a Santa Barbara paper: "Cecil Tracy, Buttonwillow rancher, won the grand sweepstakes prize today at the Santa Barbara Fiesta. Yesterday, Tracy and his horse, both of them gilded for the occasion, won the first prize for being the most unique, unusual and most beautiful in the Santa Barbara Fiesta parade."

Yes, Cecil had truly created something unique, unusual, and beautiful, but as the spectators gazed in wonder at his magnificent creation, they could not realize the planning, patience, and perseverance that went into its conception.

Captain was one of our favorite horses. He was a foal of Rosie Pony, who was part Shetland. Unlike his mother, he was of a stocky build, and pleasingly rounded. He was also a very intelligent horse, and quite capable of opening certain gates. Several weeks before a parade, Cecil would place wooden planks over an irrigation ditch. There he would instruct Captain to stand as he mounted him. Horse and rider would then stay perfectly still for a period of time. These training sessions were repeated and extended into longer intervals until the day of the parade.

After the job of transporting his horse and all of the gear, Cecil would smear Captain from one end to the other with axle grease. This step was very important, for had the grease not been applied, the gold paint would have closed the horse's pores and caused his death. After a gilded bridle and saddle were placed on Captain, Cecil then applied the grease and the gilt to himself and his clothing. Imagine the inconvenience of all this; yet, it was nothing compared to the cleanup that took place after the parade! First, Cecil cleaned himself, and changed his clothes. Then he began on Captain, who received a thorough scraping. After this tedious and messy operation, he was washed several times with various soaps and detergents. Finally, Captain himself would complete the job by rolling in the dirt and shaking himself freely.

Cecil truly was a showman. Had he been born in Hollywood instead of Buttonwillow, there is little doubt but that he would have been an actor in those early cowboy movies. How he would have loved that! But Cecil lived his life as grandly as he could, considering the circumstances in which life had placed him.

Cecil and his horse, Captain, would maintain the pose you see here, never moving a muscle until the parade halted. Then Cecil would lower his arm for a quick rest and allow Captain to shift his weight.

Farewell, Old Friend

In 1954 Captain was 34 years old and lay hopelessly ill in an area just west of the old barn. It became Cecil's lot to put a merciful end to the suffering of his old friend and companion of so many years. What pain must have tortured the heart of this sentimental man; what memories must have flooded the mind: of the gilded statue's first conception in 1949, when he and Captain rode in Bakersfield's Frontier Day's parade; of the glory days when he and his horse glistened in the sun as they thrilled throngs of spectators from Sacramento to the Golden Gate of San Francisco to the Santa Barbara shores for the Fiesta Day Parade. Surely that had to have been the finest hour for this dreamer of dreams. Now it was time to bid farewell to his comrade, his beloved Captain. Only a silver-mounted pistol would do to perform the merciful task.

But they had an understanding, these two. Cecil had a kinship with Captain, as he did with all creatures, and he set to digging a grave for his old friend. Just west of JB's home, a grassy mound surrounded by a small lake and a eucalyptus grove marks Captain's final resting place. Cecil landscaped the area to provide a suitable tribute to this remarkable animal. Today, Captain is memorialized along with other family members in the serenity of today's Tracy Ranch Memorial Garden.

JB Selvidge, 1999

Perhaps the culmination of a vision of Cecil, the Tracy Ranch Memorial Garden today mirrors not only the majestic eucalyptus, oaks, and palms but the ranch heritage as well. In 1960 Cecil excavated the pond, creating the island that is believed to be the gravesite of his horse, Captain. JB witnessed the fateful moment and approximate location (then a pasture) when Cecil was forced to mercifully put a bullet in dying Captain's head. Jack Frey recalls Cecil digging Captain's grave by hand a day later. Larry Frey speculates that the island and surrounding pond was Captain's final tribute. The rest of us agree.

Wild Animals

As a youngster, another activity included the capturing of wild animals—an interest that continued into Cecil's adult life. A pet coyote by the name of Toby comes to mind. Toby, although tamed and kept on a leash, still had an appetite for chicken. One day, he broke loose from his leash and ran into a flock of our mother's baby chicks. As fast as Toby would grab a baby chick, Cecil would release it from his mouth, and although Toby would willingly give up the chick, he would quickly grab another—until Cecil finally realized that he had better capture Toby instead of rescuing the chicks.

A youthful Cecil Tracy, "bronco buster" in his custom-made chaps

Snakes were another source of fascination to the boys. It was not unusual to come upon a twenty-gallon drum containing a half dozen or more rattlesnakes, all shaking their tails at once. About the time Cecil and Darrel were in their last two years of high school, mother leased an apartment in Bakersfield for them. For several weeks they kept a small sidewinder rattlesnake in a fruit jar, and their pet king snake had the run of their apartment.

One day our mother visited the apartment with a friend, and they were sitting alone in the living room having a good chat when, suddenly, mother's friend stopped talking. With her mouth slightly ajar, she was sitting as if transfixed and seemed to be peering at an area above mother's head. Sensing that something was amiss, mother turned to locate the source of her friend's attention. There, to her surprise, was the king snake hanging by the very tip of his tail from the chandelier and reaching for mother's head as a means of getting down. Someone had put him up there and then forgotten to take him down!

Another day the boys invited a teacher and a few friends to stop by the apartment to witness what would happen when a rattlesnake encountered a king snake. The sidewinder was released from his fruit jar, and both snakes were turned loose in the street. As the boys well knew would happen, the king snake quickly grabbed the rattler by the head and slowly began to swallow him. *Very* slowly, that is, for this gruesome process took place over a period of several hours.

After the show, the witnesses departed and the boys placed the king snake and the half-consumed rattler just inside the screened front door of the apartment. As luck would have it, this was the exact time Aunt Dee chose to pay a visit. She was deathly afraid of snakes and, as a rule, would phone the boys when she planned to visit. "Put those damned snakes away," she would tell them. Of course, this was the one time she had neglected to call.

In the meantime two salesmen who were selling lots is Cambria Pines had approached Aunt Dee. Not knowing how to get rid of them, she decided to take them down to the boys' apartment. With the salesmen in tow, she arrived at the front-screened door. There before her eyes lay the gruesome spectacle—two snakes, one half-devoured! She screamed and almost knocked the salesmen off their feet. Letting go with a stream of profanities, she ran to the opposite side of the courtyard. By now I'm sure the salesmen were wondering just what they had gotten themselves into. And when their fellow apartment dwellers heard Aunt Dee's terrified screams, one can only speculate as to what they thought of these two wild country boys who kept snakes in their apartment.

The Prisoner's Song

"The Prisoner's Song!" I had almost forgotten about "The Prisoner's Song!" This was Cecil's favorite piece. He simply *loved* that song—we all did! Through hours of practice and sheer determination, he had learned to bang it out on the Ampico baby grand piano. It was the only piece he ever learned to play. In deep and sonorous tone, his voice accompanied the piano and we all gathered 'round to sing that sad and mournful song. Cecil played this song so often that my sister Frances and I still remember the words:

The Prisoner's Song[1]

Oh! Please meet me tonight in the moon-light
Please meet me tonight all a-lone
For I have a sad stoor-ee to tell you
It's a stoor-ee that's never been told.

Oh! I wish I had someone to love me
Someone to-oo call me their own
Oh! I wish I had someone to live with
'Cause I'm ty-red of living alone.

I'll be carried to a new jail tomorrrrow
Leaving my poor darling alone
With the cold prison bars all around me,
And my hay-ed on a pillow of stone.

Now if I had the wings of an an-gel
Over these prison walls I would fly
And I'd fly to the arms of my poor darlin'
And there I'd be will-ing to die.

Now I have a grand ship on the o-cean
All mounted in sil-ver and gold
And before my poor darlin' would suffer
Oh! That ship would be anchored and sold!

Recently, Frances and I were highly amused to read about "The Prisoner's Song" in *From Bluebirds to Big Screens,* written by David E. Urner, a respected appliance dealer in Bakersfield for more than sixty years. From Mr. Urner's book:

> There is one song of the twenties that produced more record sales than any other we ever carried in the store. I could never account for its immense popularity—a popularity that seemed to have no geographical boundaries. But there is no doubt about it, the song that wove its spell over people everywhere was the mournful but very sing-able The Prisoner's Song, music and lyrics written by Guy Massey and published in 1924.
>
> A footnote about the song referred to it as the Rosary of Hill-Billy Songs, whatever that means. It further revealed that the writer: "after years of disappointment and disillusion, died at an Army Hospital happy in the knowledge the whole world was singing his song—and he had attained fame no [other] had ever had in the entire history of Music."

Mr. Urner, in describing his showroom, also mentions the Ampico grand piano that mother had purchased from his store in the early twenties:

> This room provided a fairly plush setting—carpeted, drapes on the walls, a room large enough to display the astonishing Ampico reproducing grand piano, the Brunswick Panatrope, a sofa and comfortable chairs.

Another Side

The carefree side of Cecil's nature has now been revealed, but he cannot be so lightly dismissed, for there were many facets to the personality of this complex and creative individual. First and foremost, he was a devoted son who revered his parents. He had a strong sense of family and, more than once, he subordinated his interests to the greater good of the family as a whole.

Our sister Frances describes him as having a "deep appreciation of nature and all living things. At an early age his mind became one with the many species of the bird and animal world around him. He had a love of the land where he grew up and all things connected with it, thereby laying a foundation upon which those who came after have built and preserved for posterity."

Long before conservation was a popular cause, Cecil established a bird sanctuary around the home place, and he participated for many years in the annual California condor count.

"Save the Ranch!"

In 1929, while I was in the eighth grade at Emerson School, mother moved to Bakersfield to establish a residence for the older children attending Bakersfield High School. She soon found herself with too much time on her hands and took a teaching job on the outskirts of town, returning with the children to the ranch on the weekends.

During mother's absence, some smooth-talking stockbrokers approached our father, or, as mother later called them, "crooks." She agreed to the investment of $5,000, but she would not put her signature on any notes. Before she knew what was happening, our father had not only invested the $5,000, but had mortgaged the entire ranch to the extent of $100,000. This was a great deal of money in 1929, and the ranch was now in extreme jeopardy. Because of the community property laws and the fact that she had not signed any of the notes, mother decided to fight the case. She engaged attorneys Dorsey and Campbell, and a court case ensued. They easily won the case, but now there was $6,000 to be paid in attorneys' fees, and the ranch, which had been a profitable business, was now in shambles.

Father's unwise investments plunged the ranch into a decline that, along with the Great Depression, made it almost impossible to survive. It was at this point that our dear father found himself broken in spirit, and mother became the head of the family.

Cecil and Darrel were about halfway through high school when mother took them aside and made them aware of the acute financial dilemma that existed. She asked if they wished to let the ranch go, or would they prefer to make a Herculean effort to try and save it. Cecil and Darrel agreed—they wished to try to save the ranch. They both quit school and took jobs at the San Joaquin Light and Power Plant, which was about three or four miles south of our ranch. Here they each worked for $4 a day. Cecil worked for two years and Darrel for one. They turned their entire salaries over to mother before eventually returning to school and completing their high school educations. During this time, Mother taught school for an entire year with all her salary going toward attorney fees. Meanwhile, Frances stayed out of school for the year and took care of the household at the ranch.

Will Tracy was a man who, above all else, was known for his honesty and integrity; it is difficult to understand that he could have made the mistake that he did. Perhaps it was his own integrity that made him incapable of detecting fraud in others. At any rate, Fannie forgave him. A quote from her book reflects her attitude: "This whole matter greatly depressed Mr. Tracy, as he felt to blame. While he really was, I never felt that way. Instead, I felt it was a time when we should pull together."

Work at the Power Plant

When Cecil and Darrel went to work at the San Joaquin Light and Power Company plant, they were working with men who were seasoned employees of greater age and experience. They were looked upon as two green young kids still wet behind the ears, and indeed they were. Cecil and Darrel accepted this; in fact, they not only accepted it, they played it to the hilt until it became a game of who was kidding whom!

They ate at the boarding house with the other men. One morning when it was mentioned that they were to have pancakes for breakfast, Darrel said, "Pancakes! What's them?"

The men had a sort of contest as to who could lift a heavy iron boiler lid the greatest number of times. Cecil discreetly observed this contest, and he knew exactly how many times the strongest man was able to lift it. In the meantime, when there was no one to watch, Cecil practiced lifting the weight until he could exceed the limit of the best contestant. Needless to say, there came a day when, much to the surprise of the older men, this green young "kid" could out-lift them all.

Difficult Years

After our father suffered the stroke in 1931 and could no longer function as head of the family, mother turned to Cecil. She also remained very active, and she and Cecil managed as best they could to survive the many desperate depression years that followed. They lived off the land as farmers used to do. Mother loved to tend her garden, and she was seldom without a large flock of chickens or turkeys to care for, while Cecil pastured a few head of cattle. At one time he ventured to purchase a rather large herd but, due to some crooked negotiators, Cecil was about to lose them and, possibly, the ranch as well. Mother went to see her attorney, Henry Mack. "Fannie," he eventually told her, "I've done everything I know how to do short of telephoning the President of the United States."

"Well then," Fannie said, "let's phone the President!" A call was then made to President Franklin Roosevelt. Here is a quotation regarding the situation from mother's book:

> Though it was a shot in the dark, I refused to give up hope. I waited until 5:00 P.M. that day, but no answer came to our telephone call, so I went home. The next morning about nine, Mr. Ray from the Mortgage Company drove up in our yard with a telegram from President Roosevelt. He was a big man who found it difficult to get out of his little car. As I came close to him he said, "We are not taking your cattle today, nor forty-five other herds in Kern County." Apparently, the President had intervened. At least we were temporarily saved. When you realize the numbers that were in the same boat as I, you get an idea of how difficult it was to save anything. They had tagged the cattle of Tracy Ranch for sale. Next would have been the land.

No wonder, then, that the family members, always staunch Republicans, now became enthusiastic Democrats. Upon the victory of Roosevelt's second term, Cecil and mother celebrated by building a fire so large that it blistered some of the paint in the living room. It would have been easy for Cecil and mother to give up the struggle during these many difficult years and let the ranch go. In addition to the hard times, there were other lawsuits to be settled, and although oil leases had made it possible to hang on, it was next to impossible to secure loans to do anything. Two or three sections of land had already been sold to pay off debts, but Cecil and mother were determined to hang on to the rest. Deep in their hearts they knew that there would come a day when this somewhat spotty alkali soil and undeveloped land could be brought into production. Regardless of Cecil's faults, he always had the integrity of the ranch at heart, plus the vision to see what might one day be accomplished.

University?

In 1935 I had just graduated from Bakersfield Junior College. My sister Frances had been out of school for a few years. It was now, right in the middle of these depressed times, that we came to our mother with the request to attend a university in order to complete our education. We knew that mother would somehow find a way—she always did.

Due to no fault of her own, mother had lost her teaching position, so she prevailed upon Cecil, and in order to help finance our education, he secured a job in the oil fields where he remained for the next two years. Without his help, it is doubtful that Frances and I could have graduated from a major university.

Cecil Tracy in his youth runs for Supervisor of Kern County in 1930—he did not win the race

My Brother Darrel

When I think of my brother Darrel, specific adjectives come to mind. He was devout, patient, considerate, kind, and conscientious. Above all, he was a man of strong faith who practiced his religion. If any Tracy of my generation gets to heaven, it will be by hanging onto Darrel Tracy's coattails!

Young Darrel and ducks

Darrel and Cecil "ham it up" Tracy style

Darrel, the second son of Fannie and Will Tracy, was born on December 3, 1906, and he was named William Darrel after our father. He and Cecil were as different as two brothers could be; it is said that in high school, while Cecil chased after the girls, it was the girls who chased after Darrel. However, growing up together on the ranch, the brothers were very close. To my knowledge, the photo on the left is the only photo that survived the fire when Cecil's home burned.

Together, they found much to amuse themselves. As youngsters they rode bareback, collected bird eggs, fished for catfish, explored coyote dens, and captured snakes. As they grew older they avidly hunted ducks and geese. As young adults, both were excellent riders and participated in local rodeos.

In 1930, after the home place and eighty acres of land had been sold to a duck hunter, mother purchased the local grocery store in Buttonwillow and managed to establish Darrel there as postmaster. He and Cecil ran the store together until around 1933, when the duck hunter defaulted on his payments and the home place and the eighty acres were reclaimed. Cecil then returned home to take over the operation of the ranch and, for several years, Darrel ran the store while maintaining his job as postmaster. But those were difficult depression years, and he was eventually forced to sell the store.

Milys Crawford

Wildwood School: Cecil, top row, third from left; Darrel, top right; Teddy, bottom left

Darrel was thirty-one in 1934 when he married Marie Fine, a teacher from Oklahoma. He and Marie had two daughters, Kathryn Darrelen born October 17, 1939, and Margaret Marilyn born November 7, 1941. Marie was a dedicated teacher who devoted more than sixty years to her profession.

Darrel marries schoolteacher Marie Fine, 1934

The Darrel Tracy Family, circa 1946

The Postmaster

Uncle Sam is very particular about how his mail is delivered, and so is the general public. Postal regulations are explicit and must be strictly adhered to, for people get very upset if anything goes wrong with their mail. For more than thirty years Darrel Tracy satisfied those demands by not only providing good service but also conducting himself with warmth and courtesy. No matter his or her status, each individual who entered the post office was greeted with a smile and a friendly "hello."

Darrel had always felt a strong emotional tie to the ranch and was interested in its welfare. After he retired from the post office in 1963, he returned to work on the ranch. During that period he and Cecil restored the old tank house that, by then, had become a symbol of our childhood.

In 1964 Darrel was stricken with a sudden heart attack that led to his death. Services were held in Bakersfield, and there was not enough space to provide for the throngs of citizens who came to say farewell. The outpouring of public affection awarded him was very gratifying to the family. It was touching to see so many of the Hispanic community, including grandmothers with mantilla-covered heads, come to honor the memory of their friendly postmaster. No doubt some were remembering how he used to take time to play Santa on Christmas morning and how he personally delivered the packages that had arrived too late for the mail carrier.

The post office was built on Front Street, circa 1936

Postmaster Darrel Tracy plants a Buttonwillow tree at the Buttonwillow Post Office

Dedication of the Buttonwillow Tree, February 24, 1952.
At the old Miller and Lux Headquarters, this tree served as a post office where cowhands and farmers placed messages in tin cans tied to the tree. Cecil Tracy (on horseback) attends its dedication as an historical landmark. Other attendees, left to right, were Johnny Parsons, Charles Parsons, "Chromie" Johns, Darrel Tracy, and first mayor Joe Taziola.

From left, Jimmy (J.T.) and Tilton Tracy, Bill Peltier, Larry and Nancy Frey, Bill Tracy, and Jack (with Jeff) Thompson at the Buttonwillow Tree dedication in 1952

Darrel on HD-6 "crawler" after his retirement from the post office

My Father Darrel Tracy

By Marilyn Tracy

Many of those in the family now reading this book never knew my father, William Darrel Tracy. Aunt Martha has prevailed upon me to contribute some of my memories of her brother so that others might know more about him.

I loved my dad. The sun rose and set around him. The greatest punishment to endure was to somehow, actually—or in my mind—disappoint him. I would rather have died a thousand deaths. All this he inspired without much demanding. Somehow, all the childhood lessons, all the necessary disciplines of child rearing, he taught and accomplished in such a manner that I felt compelled to obey so as not to fail his expectations. I never remember him withholding his love or compassion—even when a reprimand was required. I was loved unconditionally.

Growing up for me was being with my dad—anywhere and everywhere. I was his little shadow. He taught me how to use tools properly, paint properly, saw, hammer, shoot, hunt, fish, ride horses, work cattle, look for bird nests, spot Indian artifacts, pick mushrooms, hike in the woods, listen to the geese go honking by in a quiet pre-dawn sky, mark my birds, tie my rope, saddle a horse, kill a chicken, clean a duck, be tough and never say "quit." That's the short list.

There was no end to his patience in teaching practical everyday things or things about nature and the outdoors. And those things never seemed like a "lecture"—he made them seem like wonderful things to learn or great adventures and worthwhile accomplishments. He loved the outdoors and he loved his ranch and his family. That love spilled out naturally in the way he lived his life and it was contagious. Yet, in his gentle way, he required discipline. I remember "practicing" opening and closing the refrigerator door and the back door quietly. There was no running in and out of the house slamming doors.

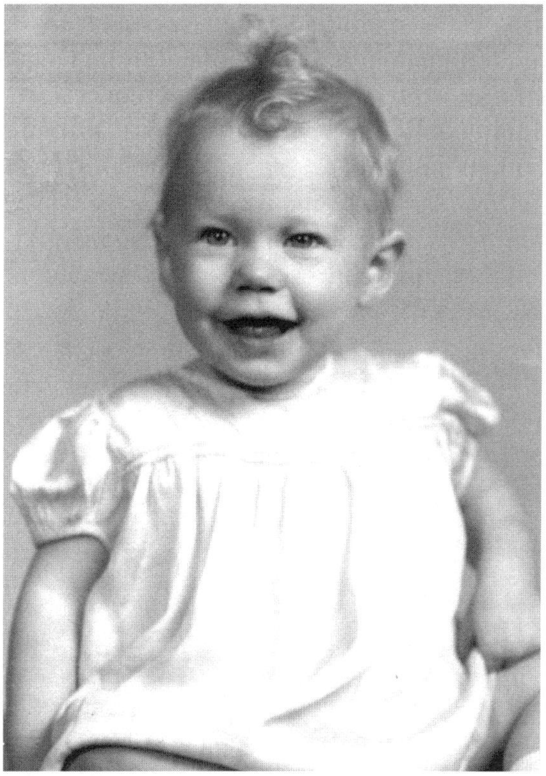

Sisters Darrelen (left) and Marilyn Tracy

A work ethic was established early. When I was about seven or eight I was assigned a chore that I cheerfully accomplished before I went away to play. Before long daddy found me and asked about my chore. I told him that I had finished it. He firmly told me that one chore might be finished but that doesn't mean all the work is done. You should then look to see what else might need to be done. And then, if you can't see any work to do, ask what needs to be done. Then, if there still isn't anything you can do to help, you may go play.

A "no" was a "no" in our home. There was no questioning, no complaining, no arguing, and most of all no whining "why?" Nor was there any, "I'm bored." With that statement, you would be put to work very quickly. Among adults, we children were to be seen and not heard. One spoke when spoken to. Good manners were a must. And you always minded and had respect for your elders, especially Grandma Tracy!

However, along with that—and vivid in my mind—is the memory of standing on a chair at the kitchen sink at night—just him and me—using a sewing needle to puncture and "blow" various bird's eggs we had collected on our adventures. He taught me how to spy nests and the tricks the birds would use to keep us away from these nests. And we never took all the eggs. If a nest contained only one egg, we left it alone and looked for another nest. He was a bird enthusiast, and he knew all the indigenous varieties, their habits, and their habitats.

Once our egg collecting took us to Buena Vista Lake. At that time it was a vast marsh of tulles and grasses. We took a rowboat in search of pelican eggs, which we eventually found. It was hard work, rowing in and out of the rushes (one could easily get lost in them), and I remember the cacophony from such a number and variety of water birds! At sunset a huge cloud of mosquitoes engulfed us. It was pretty frightening. Because we couldn't row fast enough to get away from them, there was no escape. A similar scene in the movie *The African Queen* reminds me of that experience.

One time daddy spied a great horned owl's nest in a tall tree by "Waynie's" place, and he sent me up the tree to collect an owl egg. Reaching the nest and carefully peering over, I discovered a mess of carrion and maggots. It was quite disgusting, but we got our egg.

Another time we were "checking on things" in Gooselake and found a badger's den. The badger came out fighting, but after awhile daddy realized it was a pregnant female, so we left her alone. She was probably protecting some babies she had already delivered, because boy, was she mad!

When we went to bed at night, daddy would come in and tell us a "tale." He was a great storyteller and, coupled with our own imaginations, his stories were better than television. The stories were always about his boyhood experiences on the ranch: the bull that pinned him with its horns against the wooden corral as a child, and Grandma Tracy rescuing him as he calmly picked cockleburs from between the bull's horns; the rotten egg fight with Uncle Cecil; riding ostriches; swimming in the tank house; milking the cows; hunting for chicken eggs; the glimpses of his father who allowed no smoking, drinking or cussing on the ranch, often followed by his own admonition to us, "Girls, don't ever let me chew and drip tobacco juice down my shirt!" Of course he never did chew tobacco, smoke or drink. Smoking is the only thing I remember him being adamant that my sister and I not do. He asked us to promise, and for that reason I never smoked.

Cecil, Tilton, Darrel, Fannie, circa 1962

I have also been told that daddy was quite a good bronc rider as a young man. He could rope, too. Sometimes I would ride behind his saddle when he helped with the roping at Uncle Tilton's corrals. I remember his warnings about getting a thumb caught in the rope while making a dally around the horn.

At some point dad undertook to memorize Sir Walter Scott's *Lady of the Lake*. Anyone in his company for very long was required to listen to his recital of the poem, usually while driving somewhere. I remember it the most at 4:30 a.m. driving out to the duck blind. It became a bit of a trial to endure, but to this day a line from the poem that does not leave me is,

Woe worth the chase,
Woe worth the day,
That costs thy life
My gallant gray.

After daddy died, I felt those prophetic lines applied especially to him.

Another "must" to endure during any travels was stopping to read every single historical marker along the way. He loved historical facts and information, not to mention artifacts. He and Uncle Cecil had a penchant for accumulating anything old.

Daddy was a duck hunter. I can't really remember how old I was when I started hunting ducks with him. It just seems that I always did, even before I could shoot. I was his bird dog. Since I didn't have hip boots, he would piggyback me out to the blind. As I got older, we still performed this same ritual. I would hold my shotgun across his chest, and he would trudge with me on his back through the sucking mud to the blind. Looking back, I don't know how he performed that feat! And there were still decoys to carry and set out. Now I can appreciate the effort and strength this took; performed so willingly just so I could shoot with him. We hunted a pond just east of what is now the Buttonwillow truck stop. Coming home from hunting in the pickup with our limits of ducks, he would reach over with a big grin on his face, slap my knee and say, "Well, you don't get 'em unless you go after 'em!"

I learned to shoot with Uncle Cecil's 410. Daddy always stressed gun safety; even as little children we were not allowed to point our toy cap pistols at anyone. He would take me out to practice on tin cans at the "dry lakes." I later loved shooting rabbits from the back of the pickup at night. At that time the rabbits were a real threat to our crops. After taking an NRA course, probably around my sixteenth birthday, he gave me my first gun—a Remington 20-gauge pump, which I still shoot.

Bruce Crawford, Gene Guenther and Paul Freed were his best duck hunting buddies. It was always a treat when they were out to hunt. Paul Freed mounted the ducks for daddy that I now have on my walls.

Darrel, Marilyn, and ducks

Mamie and Bud lived in back of our house, and from the time I was a baby until I went away to college, Mamie took care of the house and us. She called my father the "Flying Saucer" because he was always on the go. Daddy always showed the utmost respect for Mamie and Bud. He went to considerable effort and trouble to pave under the clothesline with concrete and brick so Mamie wouldn't have to walk in dirt or mud while hanging out the clothes. Her handprints and name are still there in the concrete—now at my kitchen door. The first cake I ever baked was for daddy's birthday when I was perhaps six or seven. Mamie and I made it and put on the top: "I CAN BAKE A CAKE."

Once I got a little "smart" with Bud, and daddy got quite upset with me, which was unusual. He made me apologize and I was so embarrassed. The first time I saw my dad cry was at Bud's funeral. He had great reverence and respect for his elders and the "old timers"—Henry Alvidrez and Arnold Rojas come immediately to mind. But daddy had great respect for almost everyone, regardless of his or her station or color. He always treated people with kindness. He was very understanding and compassionate of those less fortunate. In many aspects he was too good a man.

Daddy thought that everyone should be up and "at 'em" at the crack of dawn—if not before! His was the "rise and shine" routine: turn on lots of lights, slam doors, rattle pots in the kitchen, and recite:

The lark is up to meet the sun.
The bee is on the wing.
The ant his labor has begun.
The woods with music ring.

"Time to rise and shine. Half the day's gone and no work's done!"

Then there was the morning fitness regime of jumping jacks, squats and push-ups. However, there was also the twenty-minute-nap routine on the living room floor when he came home from the post office at lunchtime!

Daddy suffered from the family's "Tracy Headaches." Mother always suspected they were triggered by milk because daddy loved a big glass of cold milk.

He loved to have his back scratched, which Darrelen and I usually did. He enjoyed it so much! "Girls," he would say, " a dollar an hour." Once we called his bluff and took turns scratching his back for an hour to get our dollar.

Daddy would come home at night and stomp the dust from his boots as he came up the sidewalk to the back door. I knew by that sound that my daddy was home, and I would run to greet him. After his death I missed that sound so much.

Next to daddy, I loved Uncle Tilton and working cattle on the ranch. I remember taking an unprecedented day off from school to help move some of Leonard Neal's cattle that were coming in by railroad from Arizona. We met the cattle train in Bakersfield and rode out with it to the Buttonwillow corrals. Once the train was underway, daddy and I climbed on top and walked along the wooden boards on top of the cars as the train rumbled along. We were careful not to trip or catch a boot on the protruding screws that had been worked loose by the constant vibration of the train. In retrospect I realize it was a dangerous thing to do, but in many respects daddy had a daredevil side, and we did quite a few reckless things that mother never knew about.

Another time, when the Brushfield ranch was just being developed, we took a horseback ride to "check on things." We rode through brush and cockleburs that were over our heads (hence the name of the ranch). A full-fledged dust storm came up, lifting the fresh topsoil from the land that had been cleared. We were engulfed in dust and couldn't see our way. We tied kerchiefs over our faces and gave the horses their rein. They found the way back to the ranch.

Daddy could do anything—construction, concrete and brickwork, wiring, plumbing, repair work, etc. He was a jack-of-all-trades. When we were kids, daddy rigged up a bucking barrel in the back yard. It was a lot of fun, but it disappeared within a short time. My mother, I suspect. He also put up a huge wheel (from an old harvest machine, I think), mounted vertically so it would spin around. We would belt ourselves onto the spokes by the waist and hang on while being turned all the way upside down. Great fun in those days!

Darrel with the wheel in the front yard

Father's most outstanding characteristic, one I particularly hope to convey with these stories, is our understanding that we were always loved and cared for. None of "us kids" was ever made to feel we were an imposition or a bother. While other parents look for ways to get away from their kids, daddy always had time for us. He would come looking for me or even wait until I was home from school so we could go do something together. How many fathers take their little girls to a Postmasters' Convention in San Francisco? When we were away at college, we would occasionally receive canvas bank bags tied with a postal tag and marked, "Tracy Ranch Gold." This meant jerky and maybe some pomegranates from Dad!

Daddy was Postmaster of Buttonwillow from 1930 until his retirement in 1963. That position grew out of being in the store business with Uncle Cecil. During hard times on the ranch, Uncle Cecil and my father left the ranch in an effort to support themselves and save the ranch some expense. They opened a store in Buttonwillow that, in those days, meant that they did everything: groceries, butchering, and handling the mail. I guess daddy got stuck with the mail end of it. Eventually, they were able to sell the store and Uncle Cecil returned to the ranch. But by that time daddy had become postmaster, and he also feared the ranch could not support both him and Uncle Cecil. So daddy and Henry Alvidrez built the post office building in Buttonwillow, but daddy always talked about returning to the ranch.

Daddy had an especially close relationship with Uncle Cecil. In many ways they were opposites, but they so loved this ranch and this land, and their common bond made them inseparable. I remember going out to Uncle Cecil's and seeing them pore over maps, books, and plans in the library. They were always collecting farm "antiques" and looking for birds. The two of them must have felt like kids again, going out to check on a flock of geese or a den of coyotes. I can imagine they had wide grins on their faces while on a mission and up to some kind of "mischief!"

I could go on and on about my childhood experiences with dad. But I don't want this to become "my" story. I relate these personal vignettes in an effort to convey to others the person and man that was my father. He had his faults, certainly, but his sterling character and fine qualities far outweighed them. He was a kind, compassionate man. He was religious; he was honest and true to a fault; he was loving. He was the best father in the world.

Daddy died of a massive heart attack on a hot July 24th in 1964 while working at the Gooselake on the ranch he loved. He was fifty-seven.

A Daughter's Tribute

Martha Frey:

In 1989, after the death of her mother, Marilyn returned to the area to make her home on the ranch. Childhood recollections of a devoted father prompted her to choose Gooselake for her home site, the very place where she and her father once hunted ducks and searched for Indian relics.

Marilyn's unique home is a tribute to her father's memory. In it she installed the original mailboxes from Darrel's first post office, and above them the former POST OFFICE sign. The dining room features a case displaying the collection of birds eggs they gathered together. Throughout her home mounted ducks adorn the walls. Mallard, sprig and teal bring back memories of the father who once carried her piggyback to the duck blind. She lives happily here, content to look out over her pond and review the memories of a child who once hunted with her father on the Gooselake.

Martha and Marilyn celebrate the Christmas christening of her new home at Gooselake

The Gooselake Pond

By Martha Frey

A Poem for Marilyn

Now giant tractors rip the soil
Where teams of horses once did toil,
Their mighty engines day and night
Prepare the ground for fields of white.

And where the rattler once did coil
With rabbit burrow and virgin soil,
Crop dusters fly above the earth
To give the land a second birth.

Still, bullfrogs croak and mallards call
By pond or stream and trees so tall.
And men still come with shell and gun
To seek their prey upon first sun.

While in the pond the catfish bite
And swim below the surface bright,
But seldom does the "ky-ote" call
He's rarely heard as shadows fall.

Still, owl and blackbird take their flight
Where Indians trod before their plight.
And people search for artifacts
Where those before have left their tracks.

Just as water seeks its source,
Man determines to change its course.
The people build a GRAND CANAL
To replace pump, artesian well.

Now water flows from Shasta Lake
To West L.A. and Golden Gate.
And honey bee and almond tree
Take place where sagebrush used to be.

And rows of cotton propagate
Where fields of mushroom grew of late.
Time has stemmed the mosquito's flight,
Who plied his needle day and night.

And man still tries to prove his worth
To make this place on Mother Earth
A place to live in harmony
A place almost mosquito free!

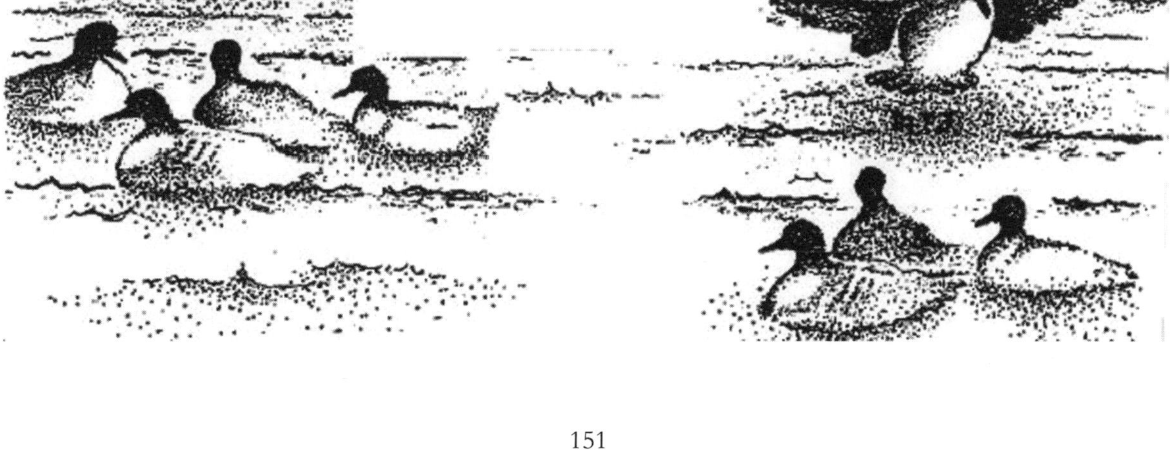

Marilyn's Poem

(Marilyn Tracy, in a poem to her family shortly
after moving into the house she built at Gooselake.)

From the beginning and as I grew
there were the words:
"To Thine Own Self be True."
When faced with a decision in my life
and in the midst
Of considerable strife
While wrestling the demons
of fear and doubt,
That one faint verse kept ringing out.
So many factors I had to weigh:
"Do this, do that, what if?" and
"What is the right way?"
But always in my heart I knew
I loved the most this place
Where I grew—
Hunting ducks, riding horses,
learning things
That my dear Dad knew.
So, the thought of "home" was nothing new,
but a place of my own on the ranch
Could that really come true?
Several years the battle inside raged
while my heart and thoughts paced
Within my comfortable cage.
Then, at last, I finally knew,
and those words came
Echoing through.
Grandfather Tracy said,
and he knew
"Above all else, to thine own self be true!"
So I gathered my courage, hope and love,
held my breath…
And flew like a dove.
The days that followed were dark and involved,
full of trepidation and
Problems to be solved.
There were days and nights
spent with specters
And tear, but
Always my family
and friends
Drew me near.

Now as if by some magic
(but thanks to Larry and Mike)
All has vanished that once was panic
And everything I held so fond
has come true
With this house on the Pond.
Thanks I must give,
so much
To so many:
My Mom and Dad, grandparents, too,
My aunts and uncles,
and cousins aplenty.
All live in my heart and in my soul
This is "God's Country,"
as my Dad often told.
So with this poem and gift I hope you see
That they come with gratitude
and love… from me.

Marilyn's house on the Gooselake pond

Top: Still an avid hunter, Marilyn is accompanied here in 1997 by friend Bill Kuch and a deer that she bagged in Wyoming. Bottom: Marilyn in 1998, on the slopes of Colorado.

My Sister Frances

Fannie revealed a tendency toward alliteration when she named her two daughters Frances Fay and Martha May. Our maternal grandmother, Martha Ann Rowlee, delivered Frances at home on June 27, 1909.

After two active boys (especially Cecil!), this daughter turned out to be an easy child. Mother claimed that she could take Frances with her anyplace, and that she was always well behaved.

Frances often made school dresses for me, and when I needed a haircut it was Frances who applied the scissors. I well remember the long tedious hours I spent perched on a stool while Frances hacked away at my hair. During the twenties it was popular for girls to wear a "boyish bob," so I had a boyish bob. Another time we studied pictures of a "sweetheart" haircut and I had one of those. The picture portrayed a girl with bangs cut in two decided arcs above the eyebrows, with a sharp point toward the nose. That haircut was *not* a success! Even I could tell that I looked just like an owl!

From the time I was a small child, Frances played "Easter Bunny" to me. She enjoyed coloring eggs and hiding them for my pleasure. I must have been almost ready for high school by the time we finally discontinued this childhood pursuit. By then, of course, I knew there was no Easter Bunny, and she knew that I knew, but the charade was so enjoyable that we hated to relinquish it.

Tag-Along

From Frances' passage in "Reflections of Mother," *The Tracy Saga:*

> My two older brothers were constant teases, and it seems I spent all my childhood trying to tag along after them and that they were constantly trying to get rid of me. In my own youthful estimation I reached maturity the day I was able to reach my little hand up to the cross beam on the slanting walls inside the tank house. This act enabled me to pull myself up so I could step onto the first landing, thence up the ladder and up to the water tank, clear up on top of the tank house into the sky where the boys were! At last I could look out over the house, barns, and fields. It was a wonderful place to dream and think, or look at the sunsets and, even more important, the boys could no longer ever totally escape me.
>
> Dad was very proud of this pretty tank house. It still stands today, though leaning some from age. Around it several of the concessions at our big Red Cross picnic were located. Cecil, by means fair or foul, had come by a lot of soda pop and cartons of gum at the close of that picnic. He had them cached in a little old trunk hidden out in the blacksmith shop. Apparently he began to miss a few items, for I found a lock on it one day.

The lock didn't baffle me long, but the big old gopher snake I found coiled inside did. I had been raised with snakes and older brothers, and I knew that a gopher snake's bite was harmless. But, time after time, I'd open that trunk thinking that I'd get up enough nerve to let him attempt to strike while I grabbed the gum. But as my hand would cautiously approach near him, that black tongue whipping in and out at me would always win, and Cecil's goodies were at least very safe from one little bandit.

Frances Tracy, age 3, on Rosie pony led by big brother Darrel

Even though I had two brothers to keep me in hot water, life at times became lonely for a little girl way out in the desert. Then a very wonderful thing was going to happen. I had been chosen to be flower girl at an impending wedding. I had never been to a wedding, and to think that I was to be part of it was overwhelming. The wedding was some weeks away, however, and in the interim I had been shipped off to the hills back of McKittrick for a visit with my aunt and uncle. The dress I was to wear became the concern of my aunt. I don't recall how, but some yardage appeared from someplace. Perhaps my uncle bought it at Reward, a little town beyond McKittrick. Anyway it was a limp cotton material of a drab color but, luckily, my aunt Hazel was artistic and handy with a needle, so she started the "wedding dress." The pretty wide band of hand smocking around the waist fascinated and pleased me so much I became full of anticipation for the big event. As the time went by, however, I began to feel apprehensive about getting down out of the hills and, sure enough, my fears were confirmed when, a week before the wedding, my uncle announced there would be no way to get me back to Buttonwillow in time for the event. The dress was laid aside and the hem was never hung, for what need would a small barefoot girl like me have for such a dress in the simple ways of country life. It was a wonderful play dress though, and now when I take out an old snapshot of a little girl in braids and crossed bare feet sitting at the piano, the sunlight that comes through the window reflects little strands of thread hanging from a hemline. Yes, just a make-believe dress was all it was ever meant to be.

My mother, had she been near, would never have let this happen because to her nothing was ever impossible. In the lives of all of us, we could always count on her for encouragement in any project that, to most adult minds, would seem impractical or impossible.

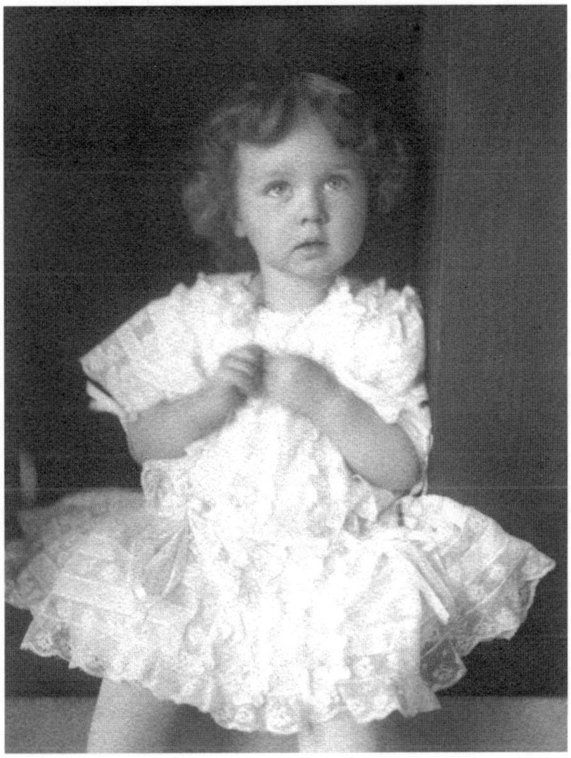

Little Fran, age 4, 1913

Fran's Early Memories

By Frances Tracy Selvidge, 1999:

Birthing

My birthing, attended by grandmother over ninety years ago west of Bakersfield in sagebrush country, coincides with that of my mother's birds hatching their young troop of ostriches.

Later, when our new Ford replaced horse and buggy, there was more room, time and speed to travel to what would be Our Town before returning home and back to bed.

The ostriches and I grew, and soon the ostrich feather factory came into being in the back lot of my grandmother's house on Chester Lane. My education began in the first grade at Lowell School on the site of the present St. Francis Church. Miss Ballagh was my teacher.

A Chronicle of Memories

Dusty Rosedale Road was always posted with signs warning: "The world is coming to an end," "Jesus is coming," "Get with God," etc., etc. Dates for these events were predicted. I always noticed the "little red church" at the corner of Rosedale Street and Allen Road (later moved to our Kern County Museum). At one time the Ashe family lived in it. Their twin girls and I became high school friends years later. Their father was the long-time manager of Hughes Drug Store.

My first movie was *Snow White and the Seven Dwarfs* at the Grogg Theatre on Chester Avenue, and I remember attending the Nickelodeon (5¢) in the shadow of the old Clock Tower at 17th and Chester, on the corner next to the first Beale Library. I went with my parents to see the play *Abie's Irish Rose* at the Bakersfield Opera House (now the Nile Theatre) with its gilded opera boxes close to the stage. (My parents were in attendance when Sarah Bernhardt appeared there after she became an amputee.)

My husband, Skip, reminisced that he came from Los Angeles to fill in as projectionist. He complained that the streetcars passing on 19th Street jarred the theatre enough to jiggle the needle off the track of the phonograph, putting the music off-sync with the picture.

My first taste of politics was standing by my mother as a child, watching election returns being projected on the building across the street from the upstairs window of the Bakersfield *Morning Echo* on Chester Avenue.

I recall my excitement, as a little girl, inspecting the aftermath of the Hockeimer Department Store fire (later Brock's)—the sight of the singed mannequins in the broken-out show windows, the broken glass, the smell of it all—these will forever live in my memory. Who could forget the Cupid's Palace on 19th street, with its cute chairs and tables, ice cream sundaes, candies, and the "Cupie Dolls" so popular at the time?

I remember Redlick's Department Store on 18th and Chester, with its wood floors and trolley baskets overhead, and the old courthouse building with the elevator attendant who operated a big wheel and transported us to our floors with a jolt.

In the early twenties the California Theatre (the site of a bank today) provided wonderful entertainment, with fresh vaudeville acts every week, a master of ceremonies in the person of "Vince Silk," comedy, sing-a-longs, and a silent movie—all accompanied by the five-piece Bayless Family Orchestra. We couldn't ask for more than that! To wrap it up, we would often finish the evening with the bunch at the Chinese Eatery down on 18th Street! Sometimes we preferred dancing all evening at the Coconut Grove, above Judd's Dress Shop, or when it was over—the Woolworth's store that is recognized today as an award-winning Art Deco building.

My First Job

Some relatives of Uncle George's went every summer to work in the fruit cannery at Campbell near San Jose. This sounded like an adventure to me and a chance to get some work. I was fifteen and my cousin Dorothy from Los Angeles, the same age, was coming for the summer. She had helped me with the dishwashing when we boarded the power line crew.

My cousin in San Jose assured me he could get us a job at the cannery where he had some influence, and that we could live in the girls' dormitory at the park accommodating the cannery families. I was now in possession of a 1924 Ford Coupe so had my own transportation, and off we went. We arrived after doing some sightseeing of the California Missions along the way.

Upon reporting for work the first day, we discovered that my cousin had represented his new recruits as experienced fruit workers! Good gosh! We knew ***nothing!*** We were put in our spaces at the bottom of the last of the many long tables. With a quick look at the workers on each side, we knew which knife to pick up for tackling the apricots being dumped in our box. They were much harder to pit than peaches!

By observation, Dorothy began picking up the "how" and "why" of cutting them, and I slowly followed suit. Fortunately, no one was looking over our shoulders, but at the end of the day our punched tally cards told the truth. We stuck it out and gradually were moved to the next table, and our wages were increased until we were at the head of the numerous tables. We began getting winks and extra punches slipped onto our cards. With overtime, we were finally making top dollar—$25 a week!

We began to get ideas about going to San Francisco on a weekend, but my Ford, like others of its time, was having front-end "shimmy" problems. No way was I going to take off in it to San Francisco! But the mechanic at our storage garage assured me he could cure the problem and, confident in his repair, we went on a clothes-buying spree in San Jose, before taking off for the city.

All went fine in San Francisco until we started for home down Market Street and I hit the streetcar tracks. Wow! Did we ever start to "shimmy" then! I drove several blocks before the shimmying ceased. Altogether, it proved to be one of the most embarrassing episodes of my teenage life.

The Treasure Hunt

Cecil and I discovered a one-room deserted cabin in the sagebrush beyond the north end of Wildwood Road. Inside the one door was a murder threat, supposedly scrawled in blood. A search of the surrounding area for some mayhem, however, revealed nothing. A ditch crossing some hundred feet beyond the cabin was the only way to reach level ground.

Aha! A perfect setting for the party at the end of a treasure hunt Cecil was planning for our high school friends in Bakersfield. Excited over our plans, our brother Darrel pitched in to help. He and Cecil loaded a wagon with a piano, a wood-heating stove, and some benches—all we needed for decor. The stove was one of those with an extension on the front to catch the overflow of ashes from the burning wood.

The treasure hunt started after dark, with each carload having the first written clue. Many miles were traversed that night, and many fence posts were counted. Stars were consulted for directions, and we had to locate clues hidden in lonely trees; one clue was buried in a bucket of flour. Eventually everyone ended up at the lonely cabin.

These were prohibition days, and those who imbibed (men only) carried their own flask. Blanche Ford, a talented pianist, was elected to furnish dance music and our celebration began! Everyone loved to dance in those days. Everyone had a partner. My brother selected Howard Laster for me.

Busted at Sagebrush Flats

We were all essentially well behaved, but we were sure stirring up a storm around midnight when, suddenly, there was a loud banging on the door. "This is the Sheriff! Open up!!"

Blanche at the piano cried, "Oh, my God, my reputation!" and scuttled to the farthest corner of the cabin; Howard dashed to the ashes at the stove front where he shook out his bottle of whiskey as fast as he could.

By this time the two "so-called deputies" were cold and tired. They had been waiting in the ditch outside for a signal from Cecil who confessed to the sting a short time later. It turned out that one of the deputies was Doug Miller, a character who worked for the ranch between his drinking sprees. He had done quite a lot of duty as a deputy sheriff in his heyday and knew how to carry off the scam. Things eventually quieted down. We all agreed it had been quite a party!

Teenaged Frances voted most popular girl at Bakersfield High School, 1927

Teenagers Pack Trip, circa 1927

A group of teenaged boys and girls decided on a two-week adventure—a pack trip into the High Sierras culminating in a hike to the top of Mt. Whitney. The experiences we had on the pack trip helped to weave strength and versatility into the fabric of our young lives and gave us memories never to be forgotten.

I learned that a pack trip takes a lot of work and much preparation, and I shall relate some of the humorous and near-tragic moments. We were a bunch of green-horn kids starting out on the trail under the following circumstances:

There were eleven of us traveling in two groups with Chris Stockton's Grandpa, C. Stockton, as our chaperone and guide. Each of us *did* know how to ride the saddle horses that the ranch could provide for our party of six. However, the two mares had only the knowledge that my brother Cecil was able to give them in a quickie course on packing heavy kayak loads. Those making the trip included Cecil, Pauline Cassidy, Stephanie Massey (later Stockton), Jimmy Arp, Ralph Smith, Leon Selby, Chris Stockton, "Grandpa" C. Stockton, and me.

Money was scarce, so Cecil made the kayaks out of wood and covered them with stiff cowhides that had been softened. He cut them to fit the boxes and, in order to make a sturdy pack container, he laced them up the corners with narrow leather lacing. By leaving the hair on the hides, they were smooth and easy to handle.

The Send-Off

With two novice mares loaded with all our supplies and tied in tandem with five other mounts, Cecil left the ranch with lead rope in hand. His mount was a bronc still in training, and the rest of us were to meet him when he got to the Jack Ranch Pack Station in the low

Sierra Mountains. This required a pre-arranged stay at a ranch in the foothills.

That night during the stopover, one of the mares apparently decided that enough was enough, and when the heavy weight was again loaded onto her in the morning, but not yet tied down, she had her say. Off she went around the corral while bucking, kicking, denting pots and pans, spilling food, and coating everything with the castor oil that was in the load! Fortunately, the kayaks were still intact.

Helter-Skelter

When Cecil rode into Jack Ranch, we heard about the problems he had experienced that morning. However, everyone got packed and organized to take off the next morning. Our first full day on the trail found us up and anxious to get started from the pack station. Our breakfast of pancakes and bacon, served from an array of cookware dented as a result of Cecil's Holocaust, was delicious and welcome. (We would replace many culinary items along the way with odds and ends discarded at vacant campsites.) We had gone a little heavy on pancake batter, so this furnished extra pancakes to be thrown in the pack—possibly for a bread substitute along the trail. There was also an ample supply of canned sardines. Fortified with a supply of jerky in our pockets, we were ready to break camp. The others hit the trail for Frog Meadows; Cecil and I were slower because we were leading the pack mares.

Cecil was in the lead just ahead of me with his bronc when the lead rope inadvertently got under his horse's tail. Mr. Bronc did not like that, and he took off down the hill towards twin trees with Cecil still on top and hanging onto the mare's lead rope. I closed my eyes and waited for the resulting crash as the kayaks on each side of the mare hit the trees! "Helter-skelter" describes the scene mildly! The pack boxes were ruined. Back to the pack station we went, where we rented new kayaks and reclaimed our scattered supplies. We arrived with the rest of the bunch—much later than expected, and how!

Close Friends

Pauline Cassidy was my pal at school, and she was on this trip. Our gear was simple. A small canvas bag, homemade, held it all: an extra pair of Levi's, long underwear, shirt, socks, comb and lipstick, all tied back of our saddles. We had practiced at home getting in and out of our one-man, not-too-warm sleeping bag. The last one in could reach the zipper and follow her body curves to the top. A tight fit, and neither of us could turn without the other. With our Levi's folded for a pillow, we decided "can do" on the trip. Those were the days before air mattresses, thermal blankets, vinyl, freeze-dried foods and wristwatches. We gathered pine boughs for mattresses and used canvas for vinyl. We followed the sun, moon and stars for time.

After breakfast the following morning, Pauline and I were proud of how neatly we had packed our kitchen gear, but when its weight was nestled onto that mare's back before tie-down, away she went again! Once again knives, forks, pans and dishes flew in all directions as she galloped down the meadow. Woe is me! Fortunately, this was her last fling and she settled down to be a good girl the rest of the trip.

Refreshed

At midday our second day out, we came to a fork in the river. It appeared to be a likely spot to take a bath and, with swimwear on, we cautiously approached a calm spot. I touched the water and knew that I was not going to plunge in! By sitting on a rock at the edge, my feet grew slowly numb as they entered the water. By the time I was numb up to the knees I gathered my nerve, took a deep breath, and plunged down and up with lightening speed. Never so cold in my life! Thank goodness for towels.

Mad Cow

The trail ahead, a steep one, led us over a mountain and into a beautiful little valley to our next campsite. There was evidence of cattle grazing in the area and, as we approached our site, my kooky cousin (Jimmy Arp), true-to-form, got a wild idea. He had spotted a lone milk-cow without her calf (which was probably in a small thicket of trees farther down). "I'm going to have fresh milk for dinner!" Jimmy said as he retrieved a bucket from the pack, mounted his horse, uncoiled his lariat, and took off. We had dismissed him from our minds and were continuing to make camp when we heard cries of distress from the thicket. A couple of us took off on our horses towards the cries, and there we found Cousin Jimmy, treed by a wild cow. He was hanging onto one end of a rope, the cow's neck was at the other. Upon seeing that help had arrived, Jimmy cut his precious lariat twenty feet short of the cow and escaped from the thicket. We wrangled out the cow and calf, rope trailing behind. Needless to say we had no fresh milk for dinner; instead we had one would-be milker minus a bucket and twenty feet of rope.

For Goodness Snakes!!

Our two groups would meet each night at a different meadow. On one occasion after we had stopped to rest for the day, everyone went off on foot to explore the area. I chose to fix dinner for everyone that evening, so I stayed in camp.

I was hammering some jerky in preparation for jerky stew when Cecil walked into camp with his gun and a great long rattlesnake he had shot. He said, "Fran, if you will cook him, I'll skin him."

"Okay," I told him.

He cut the snake into finger-sized morsels; then into the deep-fry they went. We both agreed not to say a word at the dinner table as to what they were eating. No one could have resisted that platter of golden brown "whatever it was" food.

As the platter was passed around the table and they helped themselves, there were a few queries as to its identity. A few guesses, but that was all. Unfortunately, somehow during the afternoon kooky cousin Jimmy had learned what it was, and he began snickering and behaving as if he was concealing a big joke. Then, bursting into laughter, he revealed our secret.

A horrified Stephanie Massey jumped up from the table on the run and disappeared behind a tree, not yet ready for reptile! Most of the others were so hungry they didn't follow suit. When Stephanie was asked about the episode many years later she said, "It became the best story I had to tell on myself the rest of my life!" (Stephanie Massey Stockton became a fine artist who, in the year 2001, was still painting at 98 years of age.)

Mush! Mush!

Our two weeks were up, and we had just made it down from climbing Mt. Whitney to our base camp where our youngest member had stayed with the horses. This was one time that he had a chance to cook for the rest of us and, while scavenging within the remaining food supply, he came across a partially full, five-pound bag of corn meal. Deciding on mush, he started with one kettle of water and the remainder of the bag. As it bubbled toward the top of the pot, he retrieved another kettle and added more water. The results were the same. By the time we got back to camp, this kid had filled every cup, bowl and pot with corn meal mush!

Our Showman

Near the end of the trip, our food shortage was desperate. Our trip had taken almost two weeks more than originally planned. I think we were at Long Meadows when we came across several people who were just coming in. There was a little store here, and Cecil spotted a wild horse in a nearby pasture. After sizing the horse up, Cecil decided that, if he could catch him, he could ride him. In order to get enough money for food for us, Cecil passed his hat around the crowd, telling them he would ride the horse. Riding the horse in fine cowboy style, he gave them a good show for their money, and he bought enough food to get us home.

We were a scruffy looking bunch when we finally arrived home! Now we girls could see our entire faces instead of just our noses in a compact; bangs were a fad, and ours had grown down past our eyebrows while our eyes peeked out from very deep tans. We luxuriated in hot baths with lathered soap.

Martha remembers overhearing me telling mother later, "Mom, do you know that cold pancakes wrapped around sardines are perfectly *delicious?"*

Three Hitchhikers

In earlier times our Uncle George was active in the "Native Sons of the Golden West." They always met in the old "W.O.W"—the Woodmen of the World lodge building on the corner of 18th and Eye streets.

While Martha and I were attending UCLA in 1935, the Native Sons were planning a large picnic at Fort Tejon, located in the foothills south of Bakersfield. We had not seen our family for some time and, since they were planning to attend the picnic, we thought it would be fun to surprise them by meeting them there. Minus a car and unable to spare the money for transportation, I considered the idea of hitchhiking to Fort Tejon.

We had become close friends with Jessie Spencer, a girl in our dormitory. High-spirited and game for adventure, Jessie joined us in the plan. It would be about seventy miles, but we assumed we could persuade a family member to return us to the dorm after the picnic.

The morning dawned bright and clear as we struck out. As we inhaled the fresh mountain air we were filled with excitement. What devils we were! What great fun, what a lark! It would be something to tell our children and grandchildren!

We were a bit choosy concerning the cars we hailed, but we had no trouble getting rides and, before long, we arrived at the Fort Tejon picnic area. But alas, there was no one in sight! We waited for a couple of hours, but still no picnickers arrived. Though I was apprehensive about our return trip, I suggested we not be choosy about selecting a car for the ride home—we should just take the first one that came along. It turned out to be a rather dilapidated old sedan, driven by a middle-aged man.

I got into the front seat and, scowling at me, the other two piled into the back. As we proceeded at a slow pace, my new partner and I struck up a conversation. He seemed to be a pleasant sort, and when I told him that my sister and I had been raised in the Buttonwillow area, he perked up. "Well," he said, "when I was a young man I worked on a ranch in that area."

When I mentioned "Tracy Ranch," he looked at me in surprise. "Well then," he said, "you must be Frances, the little girl I used to joggle on my knee when you were tiny!" It turned out he had worked for our father for some time when I was a child. I was the only girl in the family at that time, so he presumed this had to be me.

Our next "hitch" almost took us to our destination. It was with a nice couple; however, we were chastised by the lady about the dangers of hitchhiking.

An exchange of letters with our family later revealed that the picnic had been changed to the following week.

Rescued in Pixley

I should mention that during our lives mother took in many boys to put "on the right path." One such young man in his late teens or early twenties was mechanically inclined and a very good chauffeur. His name was Pat Rheiner and we kids liked him, but it had been years since we had seen or heard of him.

When I was finishing high school, mother bought a cabin at Huntington Lake. For the next thirteen years we had fun times and some hard times at the cabin. However, it was always a chore to close it for the winter season. One such year Cecil and I had to go "close the cabin." We took off in Cecil's second-hand Chrysler roadster and made the trip up with only a little carburetor trouble, but on the way back it was a different story.

With me driving and Cecil sitting on the front fender manipulating the gas feed to the engine, we made it clear down the grade. Right in the middle of the town of Pixley, however, something broke underneath and we were stranded.

We unloaded the blankets and pillows we had brought from the cabin and made a bed right beside Highway 99 in the center of Pixley. Cecil had 25¢ in his pocket. The next morning I told Cecil to take the quarter, see if he could make a 15¢ call to Bakersfield or Buttonwillow, and use the rest for a cup of coffee. He did this, then he went scouting the town.

Well, lo and behold, while walking through that *very* small town, Cecil came upon a mechanic's small shop, and who would be the owner but our Pat! It turned out that we had a broken axel, and Pat was our savior. With a little borrowed cash, a little to eat, and a little of Pat's Irish luck, we were on our way!

Easter, 2000, ninety-one-year-old Frances recalls the Tank House:

I celebrated Easter in the new millennium with many of the family at the ranch. We ate outdoors in the lovely spring weather, and the old tank house was in my direct view. While the younger ones were at play and the older ones visited, I sat quietly and thought of what the building had meant to me.

There she stands in her aged glory—one of my father's first dreams come true. Freshly painted white, she stood out in the desert landscape of those homestead days one hundred years ago. Now she is shaded from above by towering eucalyptus trees—trees planted by my mother as a bride. Due to its overflowing tank, one tree grew to championship proportions.

Our tank house was not just an ordinary tank house like those found on stilts to water animals. Instead, it is an edifice thoughtfully planned out beforehand to accommodate many activities inside her boarded exterior. How proud my dad must have been when the last nail had been driven, the pipes to the new horse trough were ready, and the windmill had been installed high above the three-story building! All waited for that ever-loving evening breeze from the north; then the age-old litany would begin: the windmill would start to turn and the familiar chorus would be heard, each phase of the process contributing its own solo. The whispering of the wind against the blades above, the groan of the pump below as it pulled and retracted, and then the trickle of water at the trough would complete the melodic cycle. It must have been music to my father's ears as, over and over again, the cycle was repeated. When the water in the tank reached the top, down it flowed through the pipe to splash upon the thirsty ground.

Top of the World

In time the tank house would serve as a source of entertainment to the Tracy children. The boys grew and began to explore their environment. They reveled in escaping from a tag-along sister as she grew in their wake, not yet tall enough to reach that first rung of the ladder leading to the catwalk on the second floor. Once you were there, a ladder beneath the tank would entice you up and out of a side door onto a balcony under the windmill. At this level you were able to enter the cool tank room.

The owl habitat was there, and you could look for eggs and perhaps find a baby owl or two. Or take a dip in the tank, if the water was deep enough. Or you could go back down and sit on the little balcony above the first floor, where my brothers had once experimented by jumping to the ground with the aid of an umbrella. Fortunately, no bones were broken!

At last, if you were brave enough, you could summon your courage and climb that last exterior ladder, the one that led to the very top of the world. There you could sit beneath the windmill and dream as you watched the birds fly by.

Assisting Father

By the time I was a small child, my father was running his cattle under the "91" brand. When it was time to butcher, the bottom floor of the tank house served the purpose. At butchering time, when the boys might be adventuring where I could not follow, I would watch and learn "the anatomy of a cow" from dad. The skinning process fascinated me, and it was not long before I would be on the other end of a sharp knife, joyously cutting the fascia that divided the hide from the beef. I was so proud when I could do this without cutting the hide, thus decreasing its sales value.

A Near Tragedy

A near tragedy was averted one time when my brother Darrel fell from the second floor of the tank house. He was saved from hitting the floor full force when his foot was pierced by a meat hook suspending him in the air by one heel. He was able to take some weight off his foot with his hands, which barely touched the floor, until his anxious family rescued him.

Coming of Age

This tag-a-long bided her time until the day came when she, too, could reach the ladder and explore all the wonderful places the boys could go. Soon there would be a younger brother for a playmate and, still later, a sister, too. Each of us would remember with vivid clarity the very day we were able to grasp that first rung of the ladder leading to the world of magic above. It was our coming of age.

Changing Times

After a portion of the ranch was sold in 1928, along with the family home, the tank house was converted into sleeping quarters for visiting duck hunters. That's when the outside staircase was installed on the western exposure, as it remains today.

When the people who purchased the ranch defaulted on their payments in 1930, mother repossessed it. She surveyed the bottom floor of the tank house and said, "This will be a perfect place for the incubator, and we can continue with the hatchery that we started in Buttonwillow." So, with a cot in the corner of the old tank house, I took over the business of hatching baby chicks with 2,500 eggs as a starter.

Killer Bees

For years, bees had made their home in the walls of the tank house, and we would harvest their honey. But in 1985 the first "Killer Bees" to be found in Kern County were discovered there, too. The old tank house and its notorious occupants caused quite a stir in the local newspaper!

Salute to an Old Friend

In this year of the millennium I have passed my ninety-first birthday and, like my father's cattle, oddly enough, one might say that I too am wearing the "91."

I cannot depart this long life without saying "goodbye" and "thank you" to the old tank house for the part it played in the entertainment of our youthful lives.

May God bless, Frances

UCLA or Bust

Had it not been for my sister, I may never have attended a university. I knew that my family was in no position to send me away to school. In 1935 I had graduated from junior college, but Bakersfield did not then include a graduate school. At that point Frances decided that she should complete her own education. In spite of the situation, Frances approached our mother and, as usual, "Fannie would find a way." No one valued an education more than she, who had fought so hard for her own!

At UCLA we located housing at a dormitory near "The Village" called "Doheny Hall." It was an attractive two-story edifice housing eighty-five girls and, for $30 a month, we were provided with a nicely furnished room. Twice a week our beds were freshly changed and clean towels provided. Sheer luxury! We shared kitchen privileges with eight other girls. In pairs we took turns cooking four nights a week.

When we arrived on campus we passed by a sorority house where Frances and I observed a mother and daughter clinging to one another in a tearful farewell embrace. Our reaction was simultaneous, and we looked at one another in total disgust. How could anyone fortunate enough to attend a university be boo-hooing about it!

Frances majored in Physical Education and I majored in art. By the second semester I was thrilled to find in my mailbox an invitation to join Delta Epsilon, UCLA's honorary society for art students. I had made the grade!

Frances became active in Phi Beta, an honorary sorority promoting music, drama and dance. It was in this creative atmosphere that Frances was introduced to pantomime. By this time UCLA's physical education department had become famous for performing a dance recital each year. Students created the choreography, designed and made their own costumes, and danced to classical music. One time they were invited to perform at the Hollywood Bowl accompanied by a 90-piece philharmonic orchestra. This rich background of production and stage direction would later be of benefit to Frances—and the town of Buttonwillow—when we became involved in the Chamber of Commerce shows.

My Brother Tilton

First Memory

My earliest recollection of my youngest brother, Tilton, is of standing by his crib and watching mother change his diaper. In my hands I clutched a sack of jellybeans. These brightly colored candies were a precious possession, for in our home candy was a rare treat. Suddenly a golden stream of liquid shot into the air and arched directly into my bag of jellybeans! In a state of shock, I looked at mother. Without missing a beat, she said, "Oh, just go rinse them off and they will be perfectly all right." It was with great relief that I followed her suggestion.

"To Me My Horse!"

Albert Tilton Tracy was born at home on the ranch on July 16, 1919, just after World War I. Named after the Tilton family (Fannie's sister Nellie's husband and children), he was the last of Fannie and Will Tracy's brood, and perhaps the best of the lot. But there were times as he was growing up when his sister Martha did not think so.

Almost before Tilton could talk, he was riding a horse. In fact, his first sentence may have been, "*To me my horse*!" By the time he was three or four years old, my father kept a horse bridled for him. Strangers often made their way into our yard, usually to ask permission to view the ostriches. When Tilton would spot one of these visitors, he would scramble up on his horse and go tearing across the yard at breakneck speed. People watched in astonishment and exclaimed to one another at the sight of this tiny tot, who looked like a mosquito on horseback, speeding before their eyes. This, of course, was exactly the reaction that the little rascal had hoped to achieve. At the time, I was six or seven years of age, and old enough to be aware of his tricks. Therefore, it was with some disdain that I observed the antics of this little show-off.

The Happy Little Boy

Housekeeping was never high on mother's list of priorities, and by the time Tilton was born our home was bereft of order. Meals were often sketchy, dishes unwashed, beds unmade, the laundry neglected. This did not seem to bother anyone else as much as it did me—least of all Tilton. His memories are reflected in mother's book when he says: *"I believe that I had one of the happiest childhoods that any child could ever have had."* Tilton grew up like a weed, untended and carefree. While our parents were very liberal with their children, mother did occasionally deliver lectures upon the virtues of good behavior. Tilton escaped any such tutelage. What he did not learn by good example, he apparently did not need to know, for in my opinion he turned out to be the best of the lot. Tilton was one of those individuals who seem to possess a special capacity for the enjoyment of life.

Kiddo

"To me my horse!" Tilton on a blue roan

One of Tilton's favorite horses was named "Kiddo." Kiddo was a blue roan, foal of Rosie Pony and brother to Captain. His unusual coloring was made up of a mixture of blue, black and white hair. He had a black mane and tail. We children were all fond of Kiddo, for like Captain, he was an intelligent horse and also had a smooth gate that provided an easy ride.

Runaway Horses

When Tilton was about five, and I perhaps nine, we sometimes sought relief from the summer heat by riding our horses bareback to a water well about a mile east of our home. (We were never allowed to ride with saddles after our brother Teddy was almost dragged to death when his foot was caught in a stirrup.) Tilton bridled Kiddo, and I, old June. At our young age, bridling a horse was a difficult task. We had learned, however, to grab them by the nose, thrust the bit into the mouth, encase the ears in the halter, then toss the reins over the head. The next hurdle was to get through the barbed wire gate that separated us from the field with the well. It took the combined strength and ingenuity of the two of us to let down the gate and close it again. The horses waited patiently until this fete was accomplished.

Since we were not yet able to mount our horses from the ground, we had to find some means to get up to their bare backs. We usually did so by maneuvering the horses close to a wooden gate or horse trough, from which we could climb onto their backs.

Once the gate was closed and our horses were mounted, they lost no time in taking charge. From that moment, they were in complete control! With noses thrust forward, ears laid back, and tails flying, they shot forth like a couple of bullets. With only a handful of mane to cling to, it was all that Tilton and I could do to hang on. They charged through brush, jumped over ditches and anything else that lay in their path. On these occasions, Tilton always wore a purple bathing suit, and I can still see his little purple bottom bouncing up and down on his horse. It seemed as if we covered that one mile in one minute.

When we arrived at the well, the horses stopped abruptly and allowed us to tie them. We refreshed our spirits as well as our bodies in the ice-cold water that gushed from a pipe above a shallow sandy-bottomed hole. When we were ready to leave, we mounted our horses and braced ourselves for the same wild ride, this time in reverse. As we reached the troublesome gate, the horses again stopped abruptly then waited patiently for us to open it. Now docile as a pair of innocent lambs, these two rascals followed their "masters" into the barn like a couple of obedient children.

The Little Imp

By the time I was in the eighth grade, our home and eighty acres of our ranch had been sold to one of the wealthy duck hunters. With this money mother purchased a grocery store near the Union Pump Station, about a half-mile east of Buttonwillow. Her plan was for Cecil and Darrel to run the store. We made our home in back of the store in one of mother's "oil field houses."

Since mother now devoted her time to working in the store, little or no time was spent housekeeping. Occasionally I tried to put the house in order, and sometimes I tried to put Tilton in order! Mother seemed oblivious to the shortcomings of her youngest son. It was one of my cleaning sprees that led to an altercation between Tilton and me.

One morning, after washing dishes, sweeping and cleaning, I asked Tilton to empty some wastebaskets. He responded by making one of his faces at me. Thrusting out his chin, he made a disgusting sound by sucking air through his front teeth. This was just one time too many! Since Tilton was small for his age and I was four years older, I was able to catch him. I grabbed him by the wrists, swung him off the ground and began whirling him 'round and 'round until he was flying through the air. When we had reached a fair momentum, I released him and sent him tumbling into a field of Russian thistle.

Now I knew that I was in for some trouble, so I dashed to the store for mother's protection. It was not long before Tilton made his entrance, waving a hoe menacingly above his head. He was out to kill me! Thanks to mother's intervention, I'm still here to tell the story.

Little Blacksmith

As the "Little Imp" continued to grow, there were times when even his sibling sister had to acknowledge certain attributes of his character. He made friends easily; other boys just seemed to like to follow him around. Perhaps it was his never-ending enthusiasm.

In early days our father had installed a blacksmith shop on the ranch. The shop contained a forge that was used for making horseshoes and repairing farm equipment. Tilton must have spent a great deal of time at that forge, for I clearly recall that when he returned to the house his face would be thoroughly blackened with soot. Below his mouth, however, there always remained a clean area caused by his habit of biting his tongue as he concentrated upon his work. In the wake of this little blacksmith there often trailed one or two small boys. It was easy to see that these young fellows had great respect and admiration for their leader with the unusually patterned face.

Measles

In 1931, when Tilton was twelve, he caught the measles. Since father had never been exposed to this childhood disease, he too came down with them. This caused him, at the age of sixty-five, to suffer a severe stroke. Although he did not become crippled, he was never quite the same again. He was left with a slight paralysis of his left side and some impairment of his mind. It became necessary for mother to shoulder the burden of providing for her family. In the middle of the depression, with a lawsuit pending and huge debts to pay, it is no wonder that she had little time to devote to Tilton or to keeping house. Still, Tilton always knew he was loved.

The Little Man

In 1935, when I was in my second year of junior college, I met and fell in love with Bakersfield Renegade football star Jack Frey, my future husband. I was thrilled when, as a symbol of his affection, he asked me to wear his gold football and chain.

One day while playing soccer in PE, I lost both the football and the gold chain. After a desperate search on hands and knees I recovered the football, but not the chain. I was distraught, for I hated to tell Jack that I had lost it, and I had no idea where I could find money to replace this valuable item.

At that time I was living at Aunt Dee's home in Bakersfield. Tilton, a freshman at Bakersfield High School, was riding the school bus from Buttonwillow; consequently, I did not see him often. However, I encountered him that day, as he was about to board the bus. For some reason, I told him of my dilemma. To my astonishment, this little brother reached into his pocket and pulled out a roll of bills. He peeled off the fifteen dollars necessary to replace the gold chain and handed it to me! He had just sold his 4-H calf at the fair.

A transformation had taken place; the little imp was turning into a little man whom I would admire above all others for the rest of my life.

Passage from "Reflections of Mother," from *The Tracy Saga* written by Tilton Tracy:

> My father became somewhat of an invalid in my boyhood, so all the responsibility of making a living fell on mother's shoulders. Due to these circumstances she had little time to worry about us children. We just kind of grew up. I enjoyed this freedom very much and wondered why so many of my friends could not join me in all the ventures I undertook. There was one such boy, a son of a doctor in Bakersfield. In 1932 we decided to go on a pack trip to the High Sierras, though we were only thirteen years old. We went to the store where this family traded and charged all our groceries to the doctor, put the shoes on our saddle and pack horses, and took off for six weeks. Being new to the pack business, we had overlooked a few items that were needed. We finally came upon a camp of air force officers in the backcountry that could not get over two kids way back there by themselves. They furnished us with the needed food items and equipment, but none of it meant so much as a split bamboo fishing rod that was in the booty. We were no more around the corner when we had to decide who was to own the rod. We did not have a coin to flip, so we spit on a rock

and flipped it. I was the lucky boy that day. The next day, while fishing in the gorge of the Kern River, I slipped on a mossy rock, lost my rod, fish, and almost my life in that churning river. I thought so much of the rod, I let myself go through the rapids the second time in hopes of recovering it, but to no avail.

A Summer of Boyhood Bliss

By Victor Brown, 1998:

It was the summer of '33, as I recall, and we were both thirteen years old. Tilton Tracy and I had become fast friends the previous summer when we were part of a group of six youngsters under the guidance of Jim Arp and his wife, making a party of eight. The lessons learned on that seven-week trip gave us a foundation that made the following adventure possible.

A Rescue Mission

Drought conditions existed in the southern Sierras that summer, and the head game warden for the district had asked for help in packing into the Salmon Creek headquarters on a trout rescue program. Since Tilton and I each owned a saddle horse and our families had plenty of other animals for packhorses, my father volunteered us for duty. Our assignment was to meet at Stockdale, shoe our horses and buy proper attire (such as cowboy boots, leather chaps, hats and Levi's) that we would need for a rough trip. We were able to secure about all our needs at Paul Horning's 19th Street store, which carried all sorts of western equestrian gear from horse shoeing tools to pack saddles and kayaks. (Kayaks are the boxes into which food, utensils, etc. are packed and hung on the packsaddles, one on each side of the packhorse.)

Once all the necessary items had been obtained it was time to see that each horse was properly shod. We had three horses. Tilton rode in on Lindy, his personal horse. Mine was Silver, and we packed Silver's mother, Bessie. I don't recall how we knew what size shoes to get, but Tilton was raised with livestock and we were able to choose correctly, with a good supply of extra shoes to replace those we expected to lose over the rocky terrain we would cross.

Now it was time for a couple of kids to do something stupid. Since Tilton had done some horseshoeing on the Tracy Ranch, he began the task on whichever horses needed shoes. Not having much to do, I wandered into the house and stumbled onto a bottle of my father's sweet white wine. Feeling pretty grown-up, we began taking sips as the work progressed. In spite of the wine's effect, Tilton was able to drive each nail perfectly and for years bragged about the fact that he "never lost a nail," meaning that none of the horseshoe nails went into the quick, where it would have disabled the horse.

So, with our clothes, tools, tackle and other necessities packed in the kayaks and the kayaks themselves secured to the packsaddle and covered with our bedding and a tarpaulin, we left early the next morning on our way to Kernville, some sixty miles from Stockdale. Little did we know that before we returned home we would have ridden nearly 250 miles over a six-week period!

That first day was long, hot, and tiring. Fortunately, at dusk we reached the Polkinghorn ranch at Mill Creek, about halfway to Kernville. There we were able to put our horses in a corral, give them plenty of hay and water, and were treated to a nice meal at the ranch house before bedding down in the barn.

Dawn found us saying goodbye to the Polkinghorns, and with full stomachs and canteens, we headed for Kernville by way of Bodfish and Isabella. Arriving that evening, we learned that the game warden had been called away to help out on a forest fire. So, the next morning we rode up to our cabin on the river at Salmon Creek, where my grandparents were living. There we had good food and a corral with hay and water for the horses.

Frustration

For what seemed like weeks, Tilton and I would hitchhike down to Kernville trying to contact Warden Welch, who seemed always to be away, somewhere. These were tough days, because this was a remote area in that time and cars were few and far between. We were becoming frustrated.

At the height of that frustration, we decided to go to Mt. Whitney. On our last trip down to Kernville we rode our horses and took old Bess, our packhorse. As usual, the warden wasn't available so we went to Bert James' grocery store where my dad had a charge account, and filled the kayaks with everything we thought we'd need: condensed milk, condiments, sides of bacon, Bisquick, corn meal, dried beans, onions, hooks, flies, some salmon eggs plus several boxes of shells for our ancient single shot .22 rifle. Probably some other stuff too, like hard candy, but I can't remember. Anyway, by the time we left Kernville to return to Salmon Creek, we felt fully supplied and equipped, and totally grown-up!

Adventure and Crisis

Two days later we said "so long" to my grandparents and headed up the road past Fairview, past the intake where the road ended in those days, and on up the Kern River trail. We rode to Brush Creek, where we left the river and headed east up the Brush Creek Trail. Climbing about a thousand feet, we hit the Rincon Trail, which skirted high above the east rim of the Kern River gorge and returned to the river near "the forks," where the Little Kern joins the main Kern and where there were some lush meadows.

Our first crisis came when we noticed our fishing rods had slipped from the top of the pack and dropped somewhere along the trail. We backtracked for a few miles, but were unable to find the rods. Now our problem was compounded because it was getting dark and neither of us had ever been over the Rincon. We missed the point where the trail branched off to the river and hours later total darkness found us high upon a mountain and totally lost. We found a spot where we could tie the horses and bed down with enough dirt and pine needles banked up against a large tree to create a bed-size level spot.

At daylight we awoke, tightened cinches and started back down the mountain. We found the trail down to the river and by midday had forded the river, selected a camping spot, unpacked old Bess, and after watering them staked the three horses out in spots where they would have plenty to eat.

Boyhood Paradise

The next few days provided one of the greatest vacations of our lives. We swam in the river and hiked around, exploring the area. We found a nice clump of willows and cut ourselves willow rods, tied on our reels, made line guides from bailing wire and caught scores of beautiful Kern River Gilbert rainbow trout. We set up targets and "sighted in" our antique .22 rifle in preparation for when we came across a brush rabbit. Naturally, we swam and sunbathed in the buff. Also, there was plenty of time to experiment with various menus and cooking techniques. We could have written a cookbook on what you can do with bacon, cornmeal, Bisquick, trout, rabbit, and (of course) pink beans.

After nearly a week, we decided to move on. Our next stop was Kern Flats, which had excellent pasture and was a popular stopover for people packing into the High Sierras. Good luck caught us on our way to "the flats." Along the trail we met a friend of my dad's, Frank Munzer. When he learned we had lost our fly rods, he graciously gave us his as he was on his way out. So for the rest of our trip we had one nice outfit with which we took turns catching trout.

WHAT A LIFE!!!! We camped at the lower end of Kern Flats, and a day or two later discovered there were three men staying in a nice camping compound at the other end of "the flats." They had a wall tent, rock fire pit, and even a hitching rail for their horses. They invited us for dinner a couple of times, and since they were planning to leave in a few days, gave us some canned rations they wouldn't be needing. Again we spent the better part of a week enjoying the same pastimes as at the Forks, then decided to move up to Kern Lakes, another place with superb pasture and unlimited possibilities for exploring.

A Bath in Hell's Hole

We had an interesting side trip up to Hell's Hole and across the river to Jordan Hot Springs, where we enjoyed a hot bath in the bathtub someone had packed in years before. Although Little Kern Lake lies just a few hundred feet from the river, it has no outlet other than seepage, which created plenty of grass for our stock. It is fed only by small Coyote Creek, which is fishless, but this small lake contained a strain of landlocked rainbows with a beautiful coloration all their own. The water was warmer than the river, so better for swimming. The main Kern Lake is on the river, and was created when an earthquake caused the short gorge to fill with house-sized boulders, making an effective dam. A great area for exploration and fishing!

The Really High Country

After a few days we decided to move on to the really high country. So it was up the river trail to Camp Lewis, a ranger station with a small store with some emergency items. This was in the southeast corner of Sequoia Park. We checked in with the ranger (after hiding the .22), crossed the bridge over the river and began the climb up Golden Trout Creek toward Little Whitney and Big Whitney meadows, spending time at each. Golden Trout Creek was amazing. It was teeming with small goldens in the five-to-seven-inch range. Millions of them. Starved. We could put two or three snelled flies on our primitive leaders and pull two or three trout out in a single cast. We ate trout. And when we weren't eating them, we were drying them with lots of salt and pepper into trout jerky for trail snacks.

At Big Whitney Meadow we quickly learned the inverse ratio between altitude and temperature. We nearly froze in our bed, and in the morning there was an inch or two of ice on the water in our buckets and dishpan. We were above the timberline, so fuel was scarce. Fortunately, there had been cattle in the meadow and we soon learned the best use of dried cow dung.

Kings of the Mountain

Our next move was over the well-named Siberian Pass to Rock Creek, where we only stayed overnight due to a shortage of good pasture for the horses. Then on to Crabtree Meadow, where we set up camp for our ride to Whitney. Fishing was good in the creek, but we were unable to catch the larger goldens in the small lakes on the trail up Mt. Whitney.

We rode the horses as far as they could travel, then, using rocks as hitching posts, tethered them and continued on foot to the little rock house on top. We signed the book, threw some rocks off the east side just for the hell of it, then returned to the patiently waiting horses and rode back down to our Crabtree Meadow camp.

It was time to move on, so we took the John Muir Trail toward the South American Lakes, considered to be the origin of the Kern River. However, when we reached Wallace Creek we decided to take the trail down to Junction Meadow on the Kern, so named because here is where the main tributaries join to form the upper Kern. Good pasture, spectacular fishing and scenery—the Kern River from Junction to Funston meadows flows through a gorgeous, nearly perfectly straight north-south canyon in the Sierras, carved out by some large prehistoric glacier. Its sides are rounded and nicely forested.

Home Again…

We were now on our way home, so our stopovers were few and of short duration. Unfortunately my memory fails me as to our exact route home. All I can report is that we made it. Oddly enough, I do recall our trip home from Frog Meadow the year before when Tilton and I had the job of herding all of our stock home from that trip. That was a real challenge, to get some twelve horses safely down the mountains and on to Buttonwillow!

I can truly say that the two most memorable summers of my life were compressed into those weeks I spent with Tilton Tracy, already A MAN AT THE EARLY AGE OF THIRTEEN!

End of Vic Brown's story

The Quarterback and the Rotten Egg

During his high school years, Tilton became friendly with a fellow student named Bob Montgomery. Bob was a quarterback on the Bakersfield Drillers. He was a star player, and was well known for his passing ability. For a few summers he left a comfortable home in Bakersfield to cast his lot with Tilton on the ranch. The accommodations were not of the highest quality—there were times when the boys were lucky to find a bowl of beans to sustain them. But Bob seemed to enjoy these summers with Tilton, for although the living may have been a little rough, the experiences were memorable.

Cecil and the boys sometimes amused themselves with playful warfare. One dark summer night while they were engaged in one of these games, Bob gained access to a batch of rotten turkey eggs. In total darkness, the fellows dashed about the yard in some form of "hide and go seek." Although Bob could not see his victim, he heard Cecil running beyond the tank house. Picking up one of the rotten eggs, he made a wild guess as to Cecil's location and heaved it in that direction. As luck would have it, the egg caught Cecil square in the face! Cecil, too amused to conceal Bob's good luck, revealed himself to the boys with rotten egg dripping from his face. This may have been the most accurate pass of Bob Montgomery's illustrious football career!

A Colorful Episode

In the summer of 1935, I had just graduated from junior college and was dating Jack Frey. I spent part of that summer on the ranch, doing the cooking. Tilton was there, and so was Bob Montgomery. To relieve the monotony of a dull hot summer, we sometimes played tricks on one another.

One day, looking through the kitchen cupboard, I discovered a bottle each of red and green cake coloring. Now, if you have ever had an experience with cake coloring you will know that it is almost impossible to remove from skin or fabric. This gave me an idea. The next morning I made it a point to awaken just before dawn. Armed with two throat swabs and the two bottles of cake coloring, I stole quietly out to the front-screened porch where the boys were sleeping. Because of the hot weather, they were uncovered and clad only in their shorts. I could barely make out the outline of their bodies as I held the two bottles in my left and, and administered the swabs with the other.

Starting with their legs, I began to decorate their bodies with the alternating colors. When I got to their faces, it was all I could do to contain my laughter, for as they felt the tickle of the swabs on their lips and eyebrows they twitched and twisted in a most amusing manner. But, sleeping soundly as boys of that age usually do, they did not awaken. My artistry now complete, I made a stealthy retreat to my bed, and anxiously awaited the morning.

I was pleased with the results of my deed when Tilton and Bob made their appearance at the breakfast table. Although their faces were shiny from a thorough cleansing, there were still traces of red and green color. They had scrubbed their lips until they were raw. (Later Bob told me that they had used Dutch Cleanser—a scouring powder—to try to remove the evidence of my nocturnal visit.)

As the years go by, I seldom run into Bob but what he reminds me of this escapade. He says that he was awakened by the sound of Tilton's laughter, and then had his laugh in turn. At first, Tilton thought that the coloring was the result of perspiration and a cheaply colored mattress, but when they looked at the mattress they found it was brown! It was then, of course, that they realized what had happened to them.

Many years later Bob was to feel the grief of losing his very best friend and I a dear brother in a tragic accident. But many are the laughs we have shared as we recalled the colorful episode of the summer of '35.

High School in Wasco

By 1938 Tilton's proclivity for leadership began to manifest itself. Then in his junior year at Wasco High School, he was running for class president. Guess who was running against him! Tilton won the race against none other than Kathryn Pollard, and before long he would also win her heart. Kit was automatically appointed vice president, and this team would eventually become a team for life. In time Kit would inform her four sons that, although it was Tilton who got his picture in the yearbook as class president, it was she who did all the work. While Tilton practiced after school with the football team, it was she and her cabinet who carried out the duties of class president! During his senior year in 1939, Tilton was elected president of the student body.

As he courted Kathryn in his stripped-down Ford, he once again helped to organize the annual 4-H Club Rodeo for the school, even providing livestock from the ranch. Like his two older brothers, Tilton participated in rodeos, riding broncos as well as bulls, and he was a good left-hand roper. As our father said, "You would laugh to see him swing a sixty-foot riata from the left side of his horse, for his right hand was on his left side."

MARTHA COMES OF AGE

Emerson School

I would think most educators today would frown upon the system of segregation employed at Emerson Elementary School when I attended there in 1928 and 1929. Seventh and eighth grade students were divided into five categories, and the method was simple. The smartest kids went into First Division, while the dumbest went into Fifth Division with the problem kids. I was surprised to learn just recently that a similar system is employed as students enter Bakersfield High School. It is the age-old educational dilemma: there is little doubt that teachers can teach more effectively when children are grouped by ability, but it makes it hard on the "Fifth Division" kids.

Alice Rankin

When I entered Mrs. Newell's room at Emerson School it was easy to see that Alice Rankin was the best-looking girl in class. She and I had landed in Third Division, halfway between the smartest ones and the dumbest ones. I felt lucky to be there.

Alice Rankin's father was Walker Rankin; the Rankins were pioneer ranchers in the mountainous area east of Bakersfield, and the area still bears the name "Walker's Basin." Perhaps, because Mother Nature had failed to deliver a son to this tall and stalwart rancher, she put forth all of her efforts by creating two beautiful daughters for him. And just to make it more interesting, she made Alice blonde and her sister Rachel brunette.

Getting children off to school has always been a hardship for families living on far-flung ranches. Although the Rankins maintained a home in Bakersfield, I can remember a time or two when, on a Monday morning, Alice would arrive at school late and looking a bit sleepy. I surmised that she had spent the weekend at the Basin and was weary from the early morning trek down that tortuous mountain road.

Later, when Alice and I attended junior college, there was still no question as to who was the most attractive girl in school. Now tall and handsome, she stood out above the others. But it was not for her physical attributes alone that she was recognized. Unpretentious and unaffected, with her simple good taste, this cattleman's daughter was soon to gain the respect and admiration of her fellow classmates. In 1935 she was voted "Most Popular" at Bakersfield Junior College.

Still later, Alice and I attended UCLA together. She majored in physical education, as did my sister Frances, and I majored in art. In time Alice became a physical education instructor at Kern County Union High School. During World War II she married Bob Beard, at the time a Lieutenant Colonel in the U.S. Air Force. Eventually they settled on a 25,000-acre cattle ranch near El Paso, Texas. With his partner Joe Mendiburu, Bob managed their "Double U" Cattle Company where he and Alice brought up their four children: Rob, John, Walker "Wick" and Mary Lou "Coosie." But we were destined to rekindle our friendship for, in 1995, due to Bob's failing health, they returned to Walker's Basin. When he died in 1996, their son Rob took charge of the "Double U."

Huntington Lake

The Cabin

Martha and Frances:

Right out of the blue, the impulsive Fannie bought a cabin at Huntington Lake! The year was 1928, a year before the great stock market crash; Frances had just graduated from high school, and I from the seventh grade. Frances and her friend Pauline Cassidy had heard that Fresno State College held summer school at Huntington Lake and that it was a fun place to go. Frances approached mother and, as usual, "Fannie found a way."

Father had made his fatal investments; clearly, there was no way we could have afforded the luxury of a summer cabin. But for Frances and me, like the purchase of the Ampico electric piano, this turned out to be one of the best investments Fannie ever made.

Huntington Lake Cabin

In those days it took five hours to reach Huntington Lake and, after a slow climb up the steep toll road from Big Creek, mother, accompanied by Pauline with Frances at the wheel, arrived at the lake. When they arrived they learned there were no dormitory tents left to rent, so mother decided to look for a cabin to lease for the summer. She searched around the fringes of the school campus for a rental possibility and eventually found a cabin with a "for sale" sign.

Leaving the girls to wait in the car, Fannie knocked on the door and when a woman answered, the two of them disappeared inside the cabin. A short time later she returned to the car and said, "Come on in girls and look at the house. I hope you like it. I just bought it!" That was mother—even then, during "depression" times for our family.

We later learned what had taken place inside the cabin. After a cursory examination Fannie, taken with its rustic appearance, told the woman that she was interested in purchasing it, and a price was agreed upon. When the subject of a down payment arose, Fannie knew that she did not have sufficient funds in the bank to make any kind of a deposit; however, the woman seemed not to know how much of a deposit would be required. "Well," mother said, "I have $5 in my purse. Would that be enough to hold it?" The woman said that she guessed that it would, and that is how we happened to acquire the cabin at Huntington Lake!

Since there was space for another cabin on the lot, mother sent the ranch carpenter up to build a small secondary cabin. By leasing one cabin and living in the other, Frances and I garnered enough money to buy groceries and attend classes. Many were the happy, carefree summers my sister and I spent at Huntington Lake!

Martha in riding pants at the Huntington Lake cabin

The Football Player

One day at Bakersfield Junior College, I was walking down the sidewalk with a girlfriend when I noticed a couple of fellows approaching from the opposite direction. One of them caught my eye immediately. He was very blonde, almost a towhead; short but with a stocky build. The sleeves of his crisp white shirt were rolled up to just below the elbow and, like most young men of that era, he was wearing cords. The most striking thing about him was that he looked so clean and innocent, like a little boy who had just been freshly scrubbed.

I asked my companion if she knew who the very blonde boy was. "Oh yes," she said. " That's Jack Frey. Don't you know him? He is one of our football players!"

Mr. Walling and the Lakeshore Resort Hotel

Mr. Walling cut quite a swath around Lakeshore. He was a big man and he smoked a big cigar. He wore tan riding pants that ballooned out at each side and a matching gabardine shirt. English riding boots completed the picture. He owned the grocery store, the soda fountain, the tent houses for the college kids, and the local tavern. In fact, he owned just about everything around Lakeshore except the post office.

But best of all he owned the beautiful large Lakeshore Resort Hotel which was a mecca for college students. Two nights a week we donned our best Levi's and, with cuffs rolled up two notches around our boots, we headed for the dance at his resort hotel. The huge lobby awaited our arrival. Furniture had been cleared away, and an enormous fire snapped and crackled while reflecting flames on the highly polished hardwood floor. We would back up to the mammoth fireplace to warm ourselves until the rivets in our Levi's told us it was time to move away.

When the hotel orchestra struck up we began to dance, and we danced the night away. We did the "Fresno Hop" almost 'till the break of day.

Summer School

The campus was a delightful place to attend school. Classes were held in outdoor pavilions, and as we listened to the lectures of our professors we watched chipmunks scamper up the trees and inhaled the scent of mountain pines. The crisp, cool atmosphere encouraged a relaxed and informal relationship between faculty and students and, together, we played croquet, hiked, swam, boated and went to concerts. There was a little outdoor theater and, every Friday night, a school dance.

In the summer of 1934, simply because it fit into my schedule between art classes, I signed up for a small, two-unit course entitled "Taxonomy and Field Botany."

Martha Tracy, high school graduation photo

Jack Frey, high school graduation photo

On the first day I attended the botany class, I noticed a good-looking young fellow with very blonde hair. He looked surprisingly like the football player I had seen at Bakersfield Junior College. Could they be the same? One day after class I made it a point to linger. When I asked the professor if the blonde young man was from Bakersfield, the professor said he was. I told him that he was one of our football players.

Later, a botany class field trip to visit the Giant Sequoias was organized, and a notice was posted on the bulletin board. Jack Frey, among others, had offered to take some students in his car. Guess who signed up! On the way to the Sequoias I discreetly chose the back seat of the car. At that time cars still had running boards, and when we arrived Jack and I sat on the running board of his Ford and became somewhat acquainted. On the return trip I occupied the *front* seat.

Not long after this, Jack invited me to accompany him on a trip across the lake in his uncle's motorboat. The reason? To gather plant specimens for our class, of course! On the appointed day we met at the dock. From afar a small group of bystanders watched as Mr. Walling launched his brand new speedboat. It was a beauty! Its mahogany-covered exterior glistened in the sun. We soon heard the roar of its powerful engine, and a great spray of water arose as it took off.

In the meantime, my clean and innocent-looking date had wrapped a rope around the starter of his uncle's small outboard motor. Pfst! Nothing happened. He tried again, pfst! The engine failed to start. After watching him repeat this process several times, I noticed that the face beneath the blonde hair was beginning to redden. I could hardly believe my ears when cuss words came from the lips of my newfound friend! Eventually, when the motor started, we both sighed with relief.

That day Jack told me that his mother knew me, but I did not make the connection. I assured him I did not know his mother. But the next time we met he informed me that his mother had been one of my teachers at Emerson School. Then I realized I had been in his home for the orchestra party! Since most of my high school years were spent at Aunt Dee's, we had been living within a few blocks of one another. I learned that Jack was spending the summer with his mother's sister, Aunt Maxine, and her husband, Frank Falenzer. They, too, owned a cabin at the lake!

Left: Jack in the boat. Above: Jack falling in love.

Courtship

Somewhat shy and young for my age, I never had an inordinate number of boyfriends, but the summer I met Jack was a good one, for I had two fellows from Fresno State on the string and was basking in my popularity. I suspect Jack knew this, for I later learned that he had visited one of the Lakeshore dances and had peeked through the window to see me dancing. Perhaps this enhanced our relationship, for Jack didn't know how to dance and, in fact, had never even *been* to a dance.

Soon my summer of popularity drew to a close and I was back at Bakersfield Junior College. Jack Frey was tucked away in my subconscious, and I was still thinking about the fellows from Fresno State. A few weeks had passed when, one day in study hall, I saw Jack approaching and my heart began to pound!

Jack invited me to attend a movie with him at the beautiful new Fox Theater. While we were watching *The Barrett's of Wimpole Street,* I shyly placed my hand in his, and that was it! Goodbye, Fresno State!

Jack had several customers for whom he cut lawns, and he took great pride in doing a good job. Consequently, he always had a pocketful of money. No girl was ever courted more handsomely than I. When we went to the movies, we always sat in the loges; I received a box of candy on special days; car doors were always opened on my behalf; and he was careful to walk on the proper side of the sidewalk when we were together—he did this the rest of his life. And I noticed that he kept the engine of his little Ford as clean as he kept its exterior. I had never seen an engine look like that on the ranch! All of these things made a favorable impression on me, and soon I was head over heels in love.

Jack and I both needed to complete our education. It was a long distance from Davis School of Agriculture to UCLA, so we wrote love letters and looked forward to vacations when we could be together.

Jack at the ranch, 1935

Martha in front of Doheny Hall, UCLA

Martha, UCLA graduation photo, 1937

Jack, UC Davis graduation photo, 1937

The Art Teacher

Since my degree allowed me to teach only one subject, art, I was overjoyed when I secured a position in Fellows, a small town situated in the oilfields about ten miles west of Taft, California. All of the female teachers were required to live in a teacherage and, for the next two years, I shared living quarters with fifteen other women. At the same time, Jack took a job with his Uncle Frank who was superintendent of an oil lease near Coalinga. At the teacherage, our housing was free, and a pleasant Mrs. Henderson served us three excellent meals a day on white linen tablecloths. We paid around $30 a month for the best food I had ever eaten. I have since wondered how she accomplished this feat without even the aid of an automatic dishwasher!

At the time, I was quite sure that I was the world's greatest art teacher. But, looking back, I am not so certain. One day as two little boys passed by my open window, I heard myself referred to as "Old Lady Tracy." I was all of twenty-one and, no doubt, looked more like a teenager. Nevertheless, Fellows was a rich district, and I gladly collected my generous monthly salary of $160, twelve months a year.

Left: Martha, student teacher. Above: teacher group, "Old Lady Tracy," front row, third from right.

Jack and I had been dating for more than four years, and we were anxious to be married. After teaching for a year-and-a-half, we were married on December 23, 1938, in Bakersfield during my Christmas vacation. Our wedding took place at the beautiful old Baptist Church on Truxtun Avenue. It has since been converted into a restaurant, but it is well preserved and still beautiful.

Martha and Jack Frey's wedding photo

Jack, Martha, and bridesmaid, Frances

It was a small wedding and Frances and mother managed a small reception for Jack and me at Aunt Dee's home. I wore a lovely ivory-colored wedding dress, and Frances was my only attendant. Since we didn't feel we could afford a photographer, Jack's thoughtful Aunt Maxine took the only pictures we have. I continued to board at the teacherage while finishing out the school year and, in the spring, gave up my good teaching job to join Jack in Coalinga.

Martha

Jack

The Jack Freys

The Oil Fields

At the Universal Oil Lease in the Coalinga oil fields, Jack's Aunt Maxine and his Uncle Frank Falenzer ruled supreme. They took us under their wing and were more than generous to this young couple. Uncle Frank had an old three-room bunkhouse converted into a home for us. The rooms were each 12' x 12' and a tiny bathroom was added at one end of the building. One room became a kitchen, one a living room and the third, of course, was a bedroom. Since the building rested on a windy hill, two cables passed over the roof and were anchored to the ground on either side to keep it from blowing away.

To my satisfaction, the interior was painted white, and I was allowed to select the solid blue linoleum I liked for the kitchen. Brightly colored curtains soon adorned the kitchen windows, and Blue Willow dishes were displayed on the plate rails. True, the wooden batting sometimes covered seams in undesirable places, but this was insignificant to me. My nesting instincts must already have been apparent when I was teaching—while other teachers were investing in cars and fine clothes, I had been buying furniture. Consequently, we had a beautiful maple dining room set for our new home, and I had purchased an ancient iron bedstead for $2.50. Under Aunt Maxine's careful guidance, we opened an account at Sears, Roebuck and Company, and thus we completed the furnishing of our first home. We were now in business; I was ecstatic!

Jack welding

Jack and Model A

ON GUARD . . .
The
AMERICAN HOME
15¢
January 1943
★

THE **JACK FREYS** of Oilfields, California, converted a bunk house into a home: no mean accomplishment, but hard work can do wonders when backed up by determination. They streamlined their kitchen making this symbol of utility a color feast for the eye. Bright materials, hanging shelves, and home-painted pictures add up to the desired result—a home to be lived in and admired. Even did their own plumbing!

The Freys' kitchen and living room

Nelle F. Mosgrove

Nelle F. Mosgrove

The Jack Frey home, Oilfields, California—once a typical bunk house

Earl Sargent, a former football teammate who was Jack's friend, was hired as a coach at Coalinga Junior College. His wife, Esther, had been Dean of Women at Bakersfield Junior College, and they quickly became our best friends. Earl sat at our little kitchen table many nights, patiently attempting to teach the three of us how to play bridge. Jack did not take to the game, but it has afforded me much pleasure over the years.

Babies

I was already pregnant when I arrived on the lease. Aunt Maxine hemmed six dozen flannelette diapers; none of those newfangled gauzy diapers for this baby! She had also informed Jack that women in this family did not wash clothes while they were still nursing. Therefore, Jack was assigned to the outdoor washing machine. (And you can bet that I would nurse my babies as long as my milk held out!)

"Hasn't Hatched Yet," a cartoon by Martha Frey

Esther Sargent lined a bassinet for our expected baby, and we anxiously awaited "its" arrival. (No sonograms then.) On the day the baby was due, Jack kept poking his head through the door to see if there was any action. Three weeks later, on February 11, 1940, I delivered a baby girl whom we named Nancy Jane. She was long overdue, I was thirty pounds overweight, and she weighed nine pounds, nine ounces. Having exceeded her estimated time of arrival by three weeks, I sometimes think she was given a jump-start that she has been unwilling to relinquish!

By April of 1941, I was pregnant again. Having experienced some of the joys of motherhood, I was not quite so ecstatic this time. But I was thrilled when I delivered a baby boy on February 3, 1942. He weighed eight pounds, and we named him Lawrence William after his two fine grandfathers.

Martha, three weeks overdue with Nancy

Irene (Conkey) Frey, Jack's mother

If all mothers-in-law were as wise and good as mine, there would be no mother-in-law jokes. When Jack and I began dating, I was welcomed into the Frey home without hesitation. Since Jack was in his second year of junior college and had never gone out with a girl, perhaps his parents were getting worried.

Irene Frey was completely different from my mother, the unpredictable Fannie. As a teacher, her classes were well organized. In her home, the rooms were clean, dishes washed, beds made, laundry done, and meals planned. This sense of order appealed to my nature. On a Sunday afternoon one could count on a good pot roast dinner where Irene presided at her table with warmth and humor.

Irene Conkey Frey, b. 2/13/89 – D. 5/4/52

Lawrence John Frey, b. 9/1/86 – D. 8/31/55

For the next four-and-a-half years, until Jack and my marriage on December 21, 1938, most of my spare time was spent in the hospitality of the Frey home.

Little had I known, back in 1928 when I entered my history class at Emerson School, that one day the stern-looking teacher would become my mother-in-law! Her demeanor then warned mischievous boys not to mess with Mrs. Frey! And I could not have imagined that one day she would be feeding my three-year-old daughter, Nancy, with a fork through the rungs of a chair while Nancy pretended to be a bird in a cage!

Under Irene's tutelage I had learned to really study for the first time in my life. If you had a test on Friday, your paper was corrected and returned to you the following Monday, and you looked up any questions you had missed.

Irene also taught orchestra each morning before school. It was there, with my now-detested E flat alto saxophone, that I learned to play the first two lines of "Tenting To-night." At the end of school year the orchestra was invited to a party in the Frey home. There I met my future sister-in-law, Effie, a friendly girl with a Dutch bob who would later become a part of my life. But her brother Jack had skinned out. He was not about to stick around for that orchestra party! Consequently, I did not know that there was a boy in her family. In fact, although he attended Emerson school while his mother was teaching there, no one ever knew that he was her son. Intuitively, Jack's mother knew that he preferred it that way and, with her usual good judgment, she respected his feelings.

Jack and Effie Frey

Jack's father, Lawrence J. Frey

I remember my father-in-law as a kind and gentle man who took pleasure in stuffing the holiday turkey for his family each Christmas and Thanksgiving. Lawrence was a second generation American, the eleventh of twelve children born to Joseph Frey. Joseph had been born in Alsace Lorraine to a family who considered themselves French. When he was just a boy, Joseph's parents sent him to the United States to avoid his being drafted into the German army. They later sent his three younger brothers to join him, and the brothers settled in Nevada where they became prominent cattle ranchers. Lawrence was born there on September 1, 1886.

Jack's father graduated from the University of Nevada as a petroleum engineer. It was at the university that he met and married Irene Maude Conkey. Finding no openings in his field, Lawrence and Irene settled in Oregon to try their hand at farming. Their daughter Effie Laura Frey was born on December 10, 1912, and Jack was born on June 21, 1914.

The Oregon climate did not agree with Lawrence's bronchial condition, however, and when Jack was eighteen months of age they moved to California. They settled in the desolate plains of Lost Hills where Lawrence went to work as a roustabout in the oil fields. Irene began her teaching career and eventually became principal of the elementary school. When Jack was in the seventh grade, the family moved to Bakersfield and purchased a home at 2526 Sunset Avenue. Irene spent the next twenty years teaching elementary school, and Lawrence became a driller for Standard Oil Co.

These four brothers and pioneer cattlemen (Jack's great uncles) migrated from Alsace Lorraine to Nevada

Residence and ranch of Joseph Frey at Reno, Washoe County, Nevada. Located in 1879.

Ranch of Joseph Frey at Franktown, Washoe County, Nevada. Located in 1860.

I have a vivid recollection of watching my father-in-law pack his old black lunch pail for work, preparing to spend a freezing cold and windy night climbing a drilling rig. This was not the life he would have chosen, but he did it without complaint.

Making ice cream are Lawrence, rear left, Jack, rear right, Nancy and Larry

Lawrence eventually took up a homesteading claim in the desert area of Lancaster, California, where he built a rather rough cabin. On his days off he would visit his farm where he had some apples trees and grew a few acres of alfalfa. There, too, he cared for several hives of bees and processed their honey. This seemed to fulfill his farming instincts.

In 1950, when he and Irene had reached the age of retirement, they gave up their home on Sunset Avenue and moved to the place in Lancaster where Lawrence soon constructed a more conventional little house for his wife. I always thought my mother-in-law was a very good sport to make this move because I knew she hated to give up her nice home and all the conveniences of living in the city.

On May 4th, 1952, an acquaintance I barely knew appeared on my doorstep. He informed me that Jack's parents had been involved in a serious automobile accident. He said that Irene had been killed instantly but that Lawrence was still alive. We rushed to Lancaster to learn that their car had collided with another at an intersection, and Lawrence had been critically injured. Services for Irene were held in Lancaster on May 7th, 1952, my 38th birthday. It was with great sorrow that her family said farewell. She had shown me just how generous and caring a mother-in-law could be. Lawrence made a slow and painful recovery only to die of cancer three years later on August 31, 1955, one day short of his seventieth birthday. He is at rest in Union Cemetery in Bakersfield. These two people were devoted grandparents. Their wisdom and generosity had a profound influence upon all of our lives.

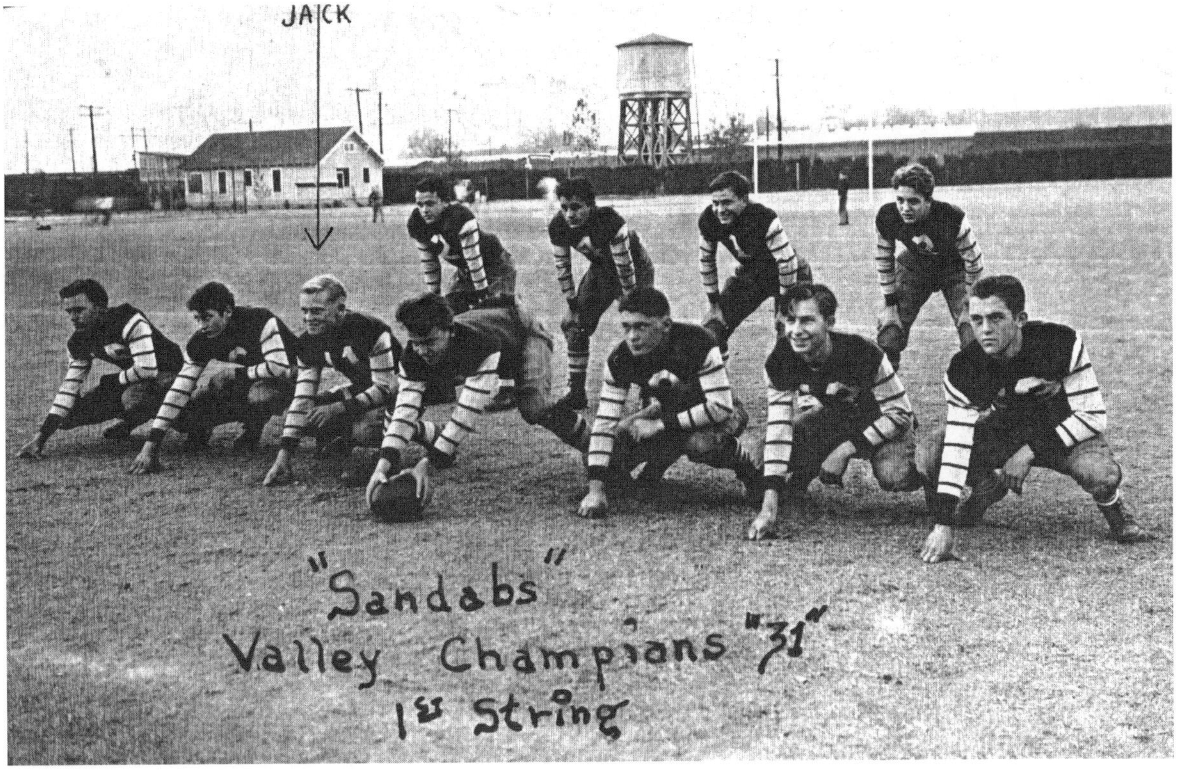

Sandabs, 1931 (Jack Frey, front row, third from left)

Basketball, 1932 (Jack Frey, front row, third from left)

Bakersfield Renegades (Jack Frey, fourth from left)

Zeta Xi Fraternity, Davis School of Agriculture, 1936 (Jack Frey, seated, third from right)

Nancy and Grandma Frey

Nancy and Grandma Conkey

Nancy with Aunt Effie (Frey) Parker

Fannie Summons Her Children

Sometime during the 1940 Christmas season mother gathered her children and their families around. She informed us that, should anyone wish to return to the ranch and endeavor to develop it into a successful enterprise, we would be welcome to do so. Father was now an invalid; the war in Europe was looming close; and mother was worried that the war would leave her with no male to tend the ranch. Looking ahead, she calculated that if she could get all her sons and sons-in-law to move back onto the ranch, odds were the government would probably leave her with at least one or two men to supervise production of vital farm products.

A few years earlier our parents had given each of their children a parcel of land, and mother had installed a salvaged oil field house on every parcel—no doubt the beginning of a long-range plan to entice her children back to the ranch. Now she was hinting that, if we could establish a viable farming operation, the rest of the land holdings would eventually be turned over to the enterprise.

After many long years of hardship and struggle, mother and Cecil had been able to salvage twelve hundred acres of our father's original ranch. As I stated earlier, some parcels had been sold to wealthy hunters from Los Angeles for duck clubs; also, one of our father's favorite pieces of property had been sold to pay off a year-old grocery bill. Through sheer determination and the benefit of an occasional oil lease, they had managed to free the ranch of debt; however, it was in a terribly run-down condition. Vintage fences were allowing cattle to break into neighbors' fields to forage for better pasture, and the remaining farm equipment was obsolete and/or worn out. Our father had farmed a few patches of land with teams of horses and he had raised draft horses for market, but there was little level land, only one or two water wells and, worst of all, the mosquitoes were about to carry the place away!

Still, there were twelve hundred acres of undeveloped land, and my husband was more than eager to meet the challenge. Jack was an ambitious man, and he had farming in his blood. Why else would he have attended an agricultural college? No "women's libber" I, for I had been raised to believe that a man had to be given the right to work in his chosen field. Jack began spending more and more of his days off helping out on the ranch; meanwhile, I remained at the oil fields where I was caring for our first baby and would soon be pregnant with our second.

In 1942, Jack spent the entire summer in the valley, helping to operate the Sill Brothers' potato shed. It was grueling work and, by the end of the summer, he had lost ten pounds and looked like a college boy again. After the birth of our son in February, Jack returned to the oil fields just long enough to claim his wife and babies before setting out on a "glorious" new life as a farmer. With a sinking heart, I packed myself and our two small children to return to the God-forsaken land of my childhood.

Back to Buttonwillow, Full Circle

In the heat of August 1942, I gave up my little three-room palace atop the hill in the oil fields of Coalinga, and we returned to the ranch. At that time, 7th Standard Road was nothing more than a rabbit path filled with ruts and covered with alkali, and a tornado of dust trailed our 1940 Ford as we bounced our way toward the home place where we would spend the next two years living with mom and Cecil.

Jack and I had been married three years; Nancy was two and Larry was seven-months. I had been sublimely happy in my little home. Having been brought up on a ranch during hard times, I appreciated the orderliness of my Coalinga life and the fact that there were *no mosquitoes* in the oil fields! But this euphoria was not to last, for persistent forces were about to overtake me.

These were to be months that I would not soon forget. Jack was one step closer to realizing his dream of becoming a farmer but, knowing that without an income we would soon use up our meager savings, he took a job with Standard Oil. He was hired to help dig ditches for the oil lines, and before long his boss tapped him on the shoulder. "Son," he said, "you don't have to go at it quite so hard!" After two weeks, Jack took a job closer to home, and he spent the next three months working for the Pacific Gas and Electric Company in Buttonwillow.

As I said earlier, I now realize it had always been mother's objective to lure her brood back to the ranch. Unbeknownst to me, she had been encouraging Jack. The ranch was at a standstill, and she recognized the potential in this industrious young man. We were welcomed home and given the best bedroom in the house.

At the ranch, the antiquated washing machine hadn't worked in God knows how long. Forget Aunt Maxine; I was now on my own. It was easy for a mother of two small children to see that a functioning washing machine was high on the list of priorities. Because we were now involved in World War II, there was a shortage of many items, including washing machines. We were fortunate to acquire a "Bendix," one of the last automatic washing machines still available in Bakersfield, and Bendix became my best friend. Jack crawled under the house and poured a thick, three-foot-square block of cement, and the Bendix was anchored to the floor to keep it from dancing around the house. Mother was so enthralled with this magic that she began washing old pillows, feathers and all! (In 1974 when Cecil's house burned to the ground, the cement block remained, a monument to old Bendix.)

With winter approaching we became very aware that there was no heat in the house. Although we had wood-burning stoves, we had no wood. Jack installed a propane floor furnace that served to warm the kitchen and one additional room. When we were not sleeping, we confined ourselves to this heated area.

As she had done all her life, mother awoke early and cooked breakfast for the men before enjoying the freedom to work in her garden for the rest of the day. From that time on, I took over the household tasks—the laundry, meal preparation, and general housekeeping chores. Each morning I arose to a kitchen sink filled with dirty dishes, two little ones to feed and bathe, and a meal to prepare by noon. After nursing Larry, I put three diapers on him and tossed him into the playpen. (No Pampers then.) Since there were a couple of cows, we had some dairy products, and I made butter and baked bread, six loaves at a time. We had a trickle of meat from the frozen food locker in Buttonwillow, a few vegetables from mother's garden, and a grocery bill up around $20 a month. Jack and I paid half while Cecil and mom split the other half. We were certainly not living high off the hog!

By winter most rooms in the house were so cold that I wore Jack's fleece-lined jacket while making the beds and cleaning that old house from stem to stern. Nancy was still wetting her bed, but she had learned to change her own panties during the day. When I put Larry into his crib, he was zipped into a sheet that allowed only his head to emerge, and this kept the covers in place all night as he slept.

My one diversion each week took place on Friday when I put the children down for their naps, shampooed my hair, and went to Buttonwillow to have my hair set for 50¢. Mother kept an eye on the children while I did the weekly grocery shopping.

During this period, we often wondered about the circumstances in which we had placed ourselves. We were now living in Mosquitoville, USA, and our children were constantly covered with welts from their bites. I can remember running from the back door of our house to the door of our car, only to find my legs covered with the damnable insects.

I especially remember one evening not long after we arrived on the ranch. The four of us were sitting on two cots in a crude little workroom at the back of the house. Cecil and mom had been contemplating purchasing a small herd of cattle. "You buy 'em, mom," Cecil told her, "and I'll herd 'em." When Jack's eyes met mine, a moment of truth flashed between us. I knew we were both wondering why we had left our snug little home in Coalinga to live in someone else's house under these strange circumstances. But somehow Jack and I felt exceptionally close during this trying period of our lives. We must have felt like two people clinging to a life raft. But, of course, there were lighter moments.

Our bedroom was adjacent to the only functioning bathroom in the house. Consequently, it was easy to hear the flushing of the toilet—and even the unwinding of the toilet tissue as it spun off the roll. (During previous visits to the ranch, we had found this useful item to be sadly missing. Now, under my efficient management, it was kept well stocked.) Cecil was apparently enjoying this newfound luxury, for as we lay in our bed we would sometimes count the number of spins. "Zing, zing, zing," went the toilet tissue. "Zing, zing, zing." Oh happy day! "Zing, zing, zing!"

The Day I Shot Toby

During this time I experienced one of the more dreadful scenarios of my life. As usual, Cecil had a pet coyote about the place. This coyote was named Toby. One Sunday when Jack's

parents paid us a visit at the ranch, Toby broke loose from his leash and attacked a pig. Jack's father happened to be nearby and, because the coyote had the pig by the throat, he took a board and killed the predator. Or so he thought.

The next day I was alone on the ranch with the children when I heard a disturbance among the chickens. I looked out the window and there, to my amazement, was Toby. He had risen from the dead. As I approached the poor creature, he came toward me in a cowed manner, but with his tail wagging in a friendly way. I recoiled as I observed the hideous spectacle before me. One eye was hanging loose from its socket, and his bloody coat was a mass of cuts and bruises. Poor Toby! What was I to do? I could not allow him to suffer in this horrible condition.

I went to the closet where Jack kept a loaded revolver. Taking the gun, I approached the innocent victim. He lay down and rolled over on his back as if to play with me. With trembling hand I placed the revolver at the side of his head and pulled the trigger. His body stiffened and he lay quite still. "Thank heaven," I thought, " the dastardly deed is done!"

I went back into the house and put the pistol away, but when I looked out the window I was shocked to see that Toby was still alive and in worse shape than ever. His mouth was opening and closing as if gasping for air. This went on for some time, but I could not summon the courage to go out and face the creature a second time. I was almost sick to my stomach, but worse was yet to come. Toby got up on his feet and staggered out of sight!

After what seemed like an eternity, I heard a truck drive into the yard. I did not know the truck driver or why he was there, but I was never so glad to see anyone in my life. I could have hugged him! I related my tale of woe and asked if he would kindly take the pistol, locate the poor animal, and put him out of his misery. He found Toby in the shade of the blacksmith shop and put a bullet through his heart, which is what I should have done in the first place. When Jack returned that evening, I fell into his arms, for now it was I, not the coyote, who needed comforting.

Cecil and the Dead Coyote

When Jack and I and our two children left a secure life in the oil fields of Coalinga to try our hand at farming, I knew that life would be rough. But I had no idea that my husband would soon be sleeping in a bed thirty feet above the ground in a eucalyptus tree! One of our first ventures was to purchase a flock of 500 turkeys to be raised for market. The project was going fairly well, until we ran into a coyote problem.

Needless to say, turkeys and coyotes are a bad combination, and we had both. Cecil devised a plan for a bed high up in a eucalyptus tree where he could guard the flock of turkeys in the pen below. When Jack was to take his turn as night watchman, he asked Cecil how he could tell when a coyote was approaching. Cecil told him not to worry, the turkeys would let him know. And the turkeys did. One night Jack awakened to a very loud racket from the turkey pen. There, in the well-lighted pen was Mister Coyote. With considerable excitement, Jack grabbed his shotgun and made short work of the stealthy visitor.

A short time later when it was Cecil's turn to guard the flock, he answered the call for help with a shotgun blast that brought down the coyote. He then propped the dead coyote up into a standing position a few feet from the pen. As daylight approached Tilton and Jack arrived to load the turkeys for market. Since no one had noticed the coyote, Cecil pointed it out in a hushed but excited voice. Tilton quickly grabbed his deer rifle, and taking careful aim, shot at the coyote. Since the coyote did not move, he fired a second shot, but still the coyote showed no sign of moving. After the third shot, Tilton caught on to the joke. Upon examining the three bullet holes, they found that all were within an inch or two of the coyote's heart. Tilton had proved to be a good marksman, and he enjoyed the joke along with the rest of us.

Our Own Home, Fannie's Bargain Houses

In the spring of 1943, the children and I were visiting Jack's parents in Bakersfield for a few days when we suddenly received a call informing us that we were not to return home because the ranch was under water from the floods! The next three months were spent in the relative comfort of their home, and while I was there I dreamed of our new place.

During the depression, oil companies sold off the bunkhouses that had formerly housed their employees. These houses sold for as little as $25 and $30 apiece. Although it cost about $100 to move one, it was still a bargain. Mother was quick to take advantage of this opportunity and she had purchased several of them. It is difficult to imagine how Fannie obtained funds to purchase houses during those depression days. Perhaps she had made an oil lease—oil leases had saved our lives more than once.

At any rate, she bought several of these old houses. A few were put down on lots in Buttonwillow for rentals; two of them were allocated to Darrel and Cecil. Thinking always of the future, one house was allocated to each of her other three children. These houses were placed in various locations on the ranch as an incentive for future occupancy. Frances's house was placed on Section 31, just north of the home place; Frances never occupied hers. Tilton and I landed on Section 4 on Tracy Avenue near 7th Standard Road. Since Cecil would logically inherit the home place, no house was placed on the ranch for him.

At that time I could not have cared less where our house was located, for I had no intention of ever living in it; nor did I ever dream that I would one day marry a farmer.

Doug Miller

Doug Miller was one of several drifters who would show up on the ranch from time to time to work for a small wage and room and board. Doug was an interesting character. His quick wit and sense of humor were exceeded only by his ability to tell a good story. Under different circumstances he might have been a good cartoonist. As it was, he became a self-made carpenter and painter. He had remodeled the old family home for mom and Cecil, and he had done a pretty good job.

You guessed it! Not long after we returned to the ranch, I learned where Doug was working and wrote to ask if he would be willing to transform the oil field house into a home for us. We did not hear from him for some time, but one day without warning he showed up on the ranch ready to go to work! He agreed to work for $2.50 a day plus a few groceries. Doug was a hard worker, but he was also fond of hard liquor. Twice during the remodeling it was necessary to retrieve him from the "Road Camp."

This Old House

Like our house in Coalinga, this old house had at one time been a men's bunkhouse. Whereas the house in Coalinga had consisted of three bare rooms, this one had four plus an eight-foot porch that ran the length of the house on either side. Originally, neither house had any plumbing or wiring. This one had a stout tin roof. There was already a lumber shortage when we moved to the ranch, and Jack had wisely purchased lumber for our house as soon as we arrived. Frances and I drove to Los Angeles where we visited a wrecking yard and purchased all the windows and doors that were needed.

Doug knocked out the center partition of the house to form a fair-sized living room. The south room became the kitchen; the two porches were enclosed to make space for three bedrooms, a laundry porch, and (again) a tiny bathroom. I requested that the living room be encircled with wainscoting.

The house was nearing completion as I anxiously waited in Bakersfield for the floodwaters to recede. Sometimes in the evenings, after I had put the children to bed, I would get out the wallpaper that I had so carefully selected, unroll about a yard of each pattern, and place it where I could just sit and gaze at the cheerful designs. There was a fresh-looking blue-and-white stripe for the living room, a dainty rose pattern for Nancy's room, a gray-and-white stripe for Larry's, and a large pattern of gorgeous red roses for our bedroom. This was my favorite; its stark white background and those bold red roses with bright green foliage fairly made my heart sing.

When I learned that the paper was finally being hung, I drove out to see how the job was progressing. Filled with anticipation, I entered the living room. There was the beautiful shiny white wainscoting and, above it, the treasured blue-and-white stripes covered an assortment of lumps and bumps. In Larry's room, the alternating stripes were not always alternating. Our bedroom held the only wallpaper job that was passable; here, the vivid rose pattern was so lively that my eye was too distracted to detect the flaws. Nancy's room had not yet been papered. With tears in my eyes, I surveyed the disaster.

Doug Miller works on our "new" home, circa 1943

Poor Doug had worked long and hard to create a home for us out of this disreputable old bunkhouse, sometimes arising at three in the morning and working until sunset. Surely he had earned his $2.50 per day. His nerves were now so badly frayed for lack of drink that he could not properly complete the job. We thanked him for his work, paid him the money due, and sent him on his migratory path.

One of the Worst Days of My Life

Although neither the linoleum nor the carpeting had been laid, I was desperate to move into our new home. Jack could not find the time to help me, so one day I took it upon myself to pack the car with our clothes and transfer them to our new house. At least this would be a start.

With neither a desert cooler nor a fan to relieve us from the summer heat, I chose a particularly hot summer day to carry out my mission. I located the pipe destined to be the clothes rod, but it had not yet been cut to fit the closet. Finding a worn hacksaw, I attacked the pipe but, after much vigorous sawing, I had barely scratched the surface. Determined, I resumed my sawing. I sawed and sawed and sawed; still, I could see I was getting nowhere. But I would not give up. It would have been difficult to tell whether the sweat on my brow exceeded the tears on my cheeks.

There comes a moment in the life of each human soul when his or her courage reaches rock bottom. Surely this had to be mine. Nearing a state of exhaustion, I gave a final thrust to the saw and the stubborn pipe gave way. I hung the clothes in place but made no attempt to tackle a second closet!

Jack soon poured the concrete block at the corner of the house for the Bendix, and he wired the house for electricity. His mother generously provided money to replace the wallpaper, and we were finally ready to move in.

Rag Carpeting

When Grandmother Rowlee visited our home in Coalinga, she had offered to sew carpet rags for our "someday home." I had no idea when that "someday" home would materialize, but, because grandma was almost blind, this was one of the few activities she could still perform. I cut thin strips from worn clothing that she tacked together with needle and thread. I would thread several needles for her at one time and put them within her reach. As the strips were joined, she rolled them into balls until we eventually had enough to carpet a house. When we arrived on the ranch, it seemed that the "someday house" was arriving sooner than I had anticipated. We sent the carpet balls to Fresno where the Kerr Rug Company wove them into carpeting on their broad looms.

A Home at Last!

Although the carpeting had not yet been laid, I could wait no longer. We moved into our new home. When the carpeting for the living room was unrolled, it was eight inches short, so Jack borrowed a carpet stretcher and in the evenings after hard days at work, he pounded away at that old carpet stretcher until his knees were black and blue. In time, he was able to stretch it the eight inches, and the tautness caused by the stretching made it lie pleasingly flat.

And it was BEAUTIFUL! The multicolor bands of color seemed to dance before our eyes. My grandmother's hands, never idle, had indeed created a useful and cherished gift for our new home.

Cousin Bill Tracy with our son Larry

The beautiful dining room furniture was set in place. Blue-and-white gingham curtains soon hung at the kitchen windows, and our little family was home at last! Here, for the next twenty-five years, we raised our family. Here, we celebrated holidays with Jack's loving family or partied with our friends in smoke-filled rooms. To this day the doorways still bear scars where the hubs of tricycles cut notches. (As I write this, the old house stands idle; one corner—where the Bendix stood—holds up while all else around it sags.) In 1967 we were able to build our dream house, and this is where we spent the next thirty years. Our new home provided us with much pleasure but, looking back, I realize a palace could not have meant as much to me as those two homes created from disreputable old bunkhouses.

Old House, 1952, with Larry in front

Love and Marriage

Nancy Jane Frey marries Covell Brown in Bakersfield, California, 1961

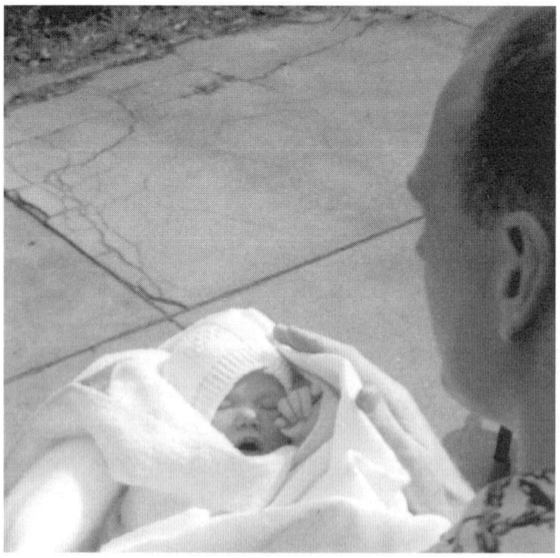

Tracy Annette Brown is born to Nancy and Covell Brown in Boston, Massachusetts, March 15, 1962

Carla Anne (Seidel) Brown, daughter of Nancy and Covell Brown, obtains PhD in psychology from Harvard University (shown here with future husband Ed De Fraine)

Tracy Annette Brown weds Paul Michael Koll at the Tracy Ranch, May 4, 1991

Eleven-month-old Ian Michael Koll, son of Tracy and Paul Koll, born December 18, 1992

Carla Anne (Seidel) Brown marries Ed De Fraine in Nashville, Tennessee, October 9, 1993

Evan Anders De Fraine, son of Carla and Ed De Fraine, born February 1, 1996, in Atlanta, Georgia

Aislin Isabella De Fraine, daughter of Carla and Ed De Fraine, born 2001

Nancy (Frey) marries Randall "Randy" Earnest Trahan, September 9, 1995

Nancy's daughters, Tracy Koll and Carla De Fraine, attendants

Claudia Irene Lee marries Lawrence William Frey in Bakersfield, California, February 28, 1965

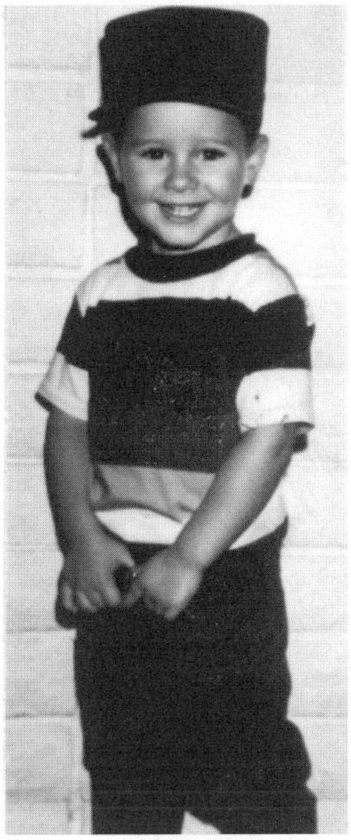

Michael Tilton Frey, son of Claudia and Larry Frey, born September 21, 1965

Jennifer Ann Goins marries Michael Tilton Frey in Fresno, California, May 30, 1987

Christmas, 1999: Justin Michael Frey, top, born May 25, 1990, and Nicholas Daniel Frey, born January 12, 1993, sons of Jennifer and Michael Tilton Frey

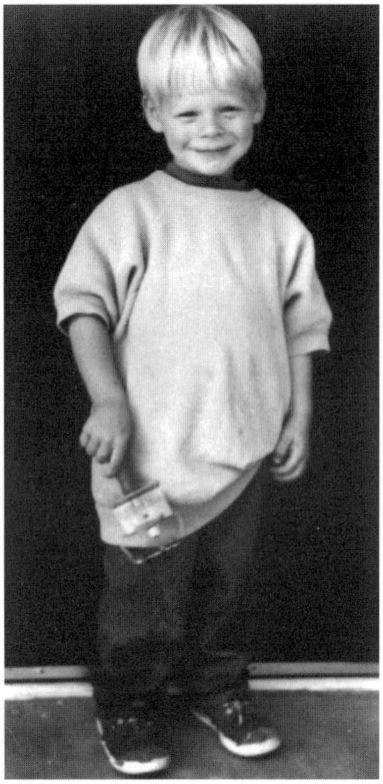

Jack Randolph Frey, son of Claudia and Larry Frey, born August 7, 1968

Kelie Wilson marries Jack Randolph Frey at Tracy Ranch, August 27, 1994

Wade Wilson Frey, son of Kelie and Jack Randolph Frey, born January 28, 1997

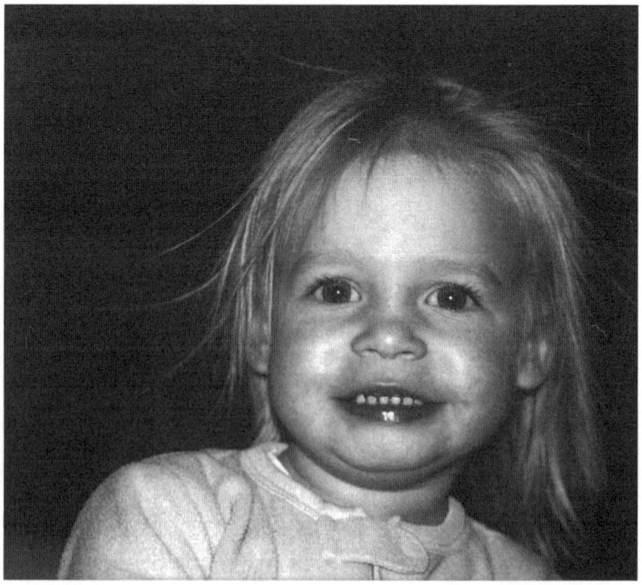

Madeline Kinsey Frey, daughter of Kelie and Jack Randolph Frey, born February 24, 1999, in Bakersfield

STEPHEN'S SAFARI

By Martha Frey

When I was but a little boy
Perhaps of four or maybe three
We set out in our pickup,
My grandpa, my Daddy and me.
I said to them: "I want to go
some AMINALS for to see."
My grandpa said: "All right my boy
Just sit right here upon my knee,
And we'll set out and look
Some AMINALS for to see."
We drove but for a little while
When we came upon a cow
Lying in the shade
Beneath a willow tree.
My grandpa said, "Stevie, is that
An AMINAL there beneath the tree?"
I looked right at the cow and studied carefully.
Then I said, "No, Grandpa, that is no
AMINAL, there beneath the tree."
We then drove on about a mile
Until quite suddenly
We came upon a horse
Not far beyond a tree.
My grandpa said, "Stevie, is that
An AMINAL, there beyond that tree?"
I looked right at the horse, and studied carefully.
Then I said, "No, grandpa, that is no
AMINAL there beyond the tree.
We drove on for quite a while
Until eventually
We came upon a field of sheep
Grazing peacefully.
Grandpa said, Stevie, are those things AMINALS,
So far as you can see?

I looked right at the sheep,
And again, I studied carefully.
Then I said, "YES, Grandpa, ***those are aminals***
As sure as sure can be!"
But I did not tell my grandpa
As I sat there on his knee

That these were not the kinds of AMINALS
That I'd set out to see.
These were not the kinds of AMINALS
I'd seen at home on our TV!

Stephen Lawrence Frey, son of Claudia
and Larry Frey, born August 4, 1972

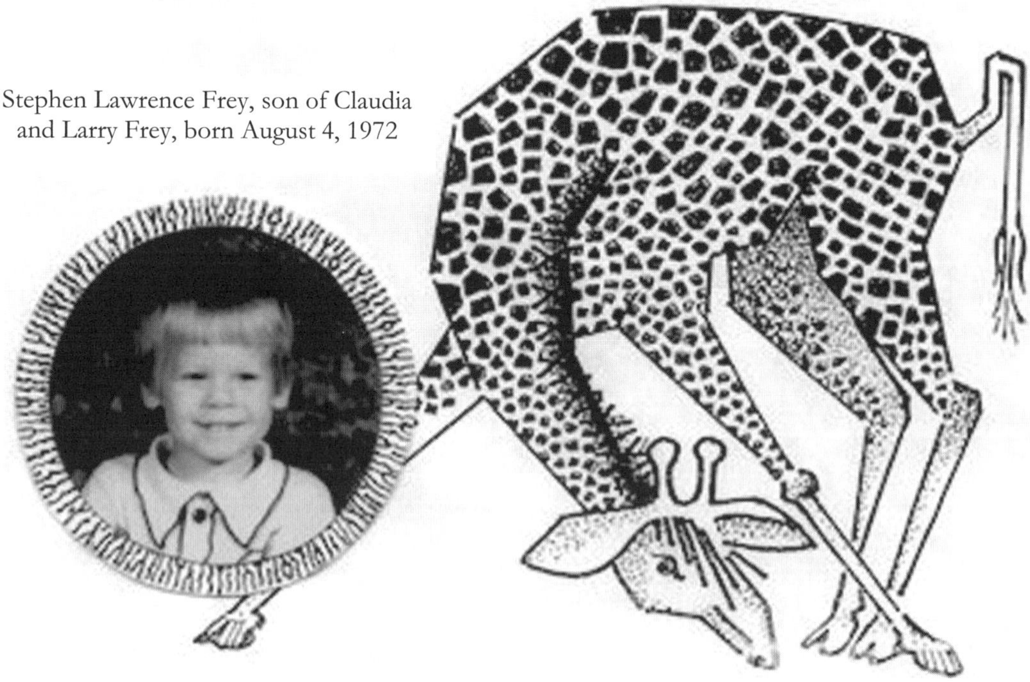

Tennis players left to right: Mike, Steve, and Jack Randolph Frey, circa 1976. Mike wants to know how his grandfather Frey got to be "boss of the ranch;" Steve imitates "The Fonze;" Jack likes Oreo cookies.

Left to right: Steve, Mike, and Jack Randolph Frey, 1998

Melissa Midzor marries Stephen Lawrence Frey on October 1, 2000

Rebecca Lake marries Larry Frey November 18, 2001, in Las Vegas, Nevada. They met at the Celebration of Western Culture in 1987.

FRANCES AND SKIP

While we were attending UCLA, Frances began dating Elden James Selvidge, better known as "Skip." Skip, who had worked at the Fox Studios for several years, would become her husband and the father of their three children: John Bentley (JB), born on January 27, 1941; James Wesley, born on August 8, 1942; and Tracy Ann, born on October 24, 1945. Forever practical, Frances planned her simple wedding near the campus on the day that I graduated, making it easier for the family to attend both ceremonies. She would receive her teaching degree a year later.

Frances, 1934

Frances and Skip's wedding, June 11, 1937

Frances and Skip eventually built a home in the hills above West Hollywood near Mulholland Drive. Their two sons were born in Los Angeles; Tracy Ann was not born until 1945, after Skip gave up his job as a projectionist and, with Fannie's encouragement, had joined forces with the rest of us already on the ranch. In addition to caring for her three children, Frances taught in the local elementary schools for a few years. But, like our versatile mother, her talents would take her in many directions. Over a period of time she would teach square dancing, the preservation and arrangement of flowers, and wreath and candle making. She would become an avid gardener and eventually assume the care of the hundreds of camellias that Fannie had propagated. Later, in 1952, Frances and I would find the perfect outlet for our creative activities, as you will discover in a later chapter, "Good Times."

In 1999 the family gathered to celebrate Frances's 90th birthday. Still alert, she was pursuing her numerous hobbies with interest.

The Elden Selvidge Family

JB, Tracy, and Wesley

Frances and Skip enjoy a Mediterranean cruise

Frances

Elden "Skip"

The Wesley Selvidge Family

Janet Louise Funk marries James Wesley Selvidge in Wasco, California, on June 20, 1963

Janet and Wesley add two children to the family tree: Andrea Susanne born on January 20, 1971, and Jason Wesley, born November 8, 1972. Here Andrea celebrates a birthday while Jason looks on.

A now grown-up Andrea Selvidge in front of her home in Bakersfield. Andrea is the first member of the family to become an attorney. She graduated from the University of Southern California in 1992 and the San Diego School of Law in 1996.

Jason Selvidge stands with father Wesley in a field of carrots. Jason graduated from the University of Denver in 1994. After completing a five-year apprentice on the ranch, Jason became a full-fledged partner in 1999.

The Gary Gill Family

Tracy Ann Selvidge marries Gary Gill at the Tracy Ranch homesite on September 15, 1984. They pose here with their son, Casey, in 1999.

Ten-month-old Casey Garrett Gill, son of Tracy Ann and Gary Gill, was born on December 30, 1985

Casey Gill, Bakersfield High School Band member, 2000

KIT AND TILTON

During the early 1930s while Tilton and some of his classmates were still bedeviling the inexperienced teacher at Wildwood School, a young girl named Kathryn Nelson Pollard was growing up in Westminster, California, south of Los Angeles. She was born on August 1, 1921. This slender wisp of a girl was fair with blue eyes and a tendency to freckle. Her curly blonde hair was touched with just a glint of red that she undoubtedly inherited from her father, Robert Lee Pollard, known as Lee, who operated a dairy farm in the area. Lee was married to Willa Elizabeth Caperton, who would come to be known to the Tracy family as "Polly."

Kathryn was the youngest of three children. A brother, John Winston, and a sister, Elizabeth, had preceded her. In 1933 when Kathryn was twelve and Elizabeth, fourteen, they were attending school in Garden Grove. That was the year of the Long Beach earthquake and, during the quake as the children were exiting the school, a chunk of ornamental cement fell from a doorway and struck Elizabeth in the head. She was killed instantly. This was Kathryn's first brush with tragedy, but it would not be her last.

According to a close cousin, Kathryn seemed to go into a period of quiet withdrawal following the accident. Her parent's concern was one of the reasons they left Southern California and relocated in the Semitropic area near Wasco. Lee Pollard switched from dairyman to cattleman, and for several years he bought and sold beef cattle. Polly worked at the Semitropic School, where she helped prepare lunches for the children. Kathryn attended Wasco High School, and it was here that her life became enmeshed with that of Albert Tilton Tracy. (Just as Darrel and Cecil had done, Tilton stayed out of school for a couple of years to help during hard times on the ranch. He probably did not meet Kathryn until his junior year in 1938.)

Tilton and Kit in high school

By the time Kathryn and Tilton graduated from Wasco High in 1939, there was no doubt concerning their love for each other. They were anxious to begin their lives as man and wife, but one obstacle lay in their path. Kathryn's mother was strongly opposed to their relationship. Because of her own unhappy marriage, she had become embittered toward her husband's way of life. Horses and cows were now an anathema to her, as were bridles, saddles, and anything else representing a cattleman's way of life. Her daughter was now in love with a "cowboy." It didn't matter that the "cowboy" had been a leader at school, popular with teachers and students alike.

It didn't matter that he came from a respectable family—Polly would do all in her power to dissuade her daughter from this marriage. She could not have dreamed that the marriage would provide her daughter with supreme happiness; nor could any of us have foreseen that it would also endure such deep sorrow.

As so often happens, Polly's opposition served only to enhance the relationship, and Kathryn Pollard and Tilton Tracy were married in Bakersfield on March 16, 1941. Two close friends, Janet Potter and Frank Gafner, witnessed the ceremony. Only at the last moment did one of Polly's friends persuade her to attend. It could not have been a joyous occasion.

In defense of Kathryn's mother, it is only fair to remember that by this time she had not only suffered the loss of a daughter in sudden and violent death, but she had also lost her only son. John Winston Pollard was killed, along with his co-pilot, when their army bomber crashed on December 23rd, 1938, the very day that Jack and I were married. This double tragedy, combined with her own marriage that was about to fail, no doubt contributed to her unhappy state of mind. Nevertheless, after her divorce, Polly would make her home with the young couple and live with them for several years.

For someone opposed to western culture, Polly had landed right in the middle of it! She did her share by helping with the chores of the busy household, but, like the mosquitoes that pestered us, she never missed an opportunity to apply her little needle. In spite of her attitude, Kathryn and Tilton always treated her with courtesy and respect.

Although the United States was not yet officially involved in World War II, our country certainly seemed to be gearing up for it. To provide for his young bride, Tilton took a job with Consolidated Aircraft, an airplane manufacturing company in San Diego, and they spent the first year of their marriage there. While Tilton was riveting wings on airplanes that were being turned out by the thousands, Kit was "turning out" their first son. He made his appearance on October 16, 1942, and was named Robert William after his two grandfathers. Kit later confided that this was the happiest year of her life.

A Home for Kit

Tilton and Kathryn decided to accept mother's offer and move back to the Tracy ranch. They settled originally in the Morgan Place, northeast of the ranch. At this time mother dubbed Kathryn "Kit" and most of us called her by that endearment. By 1945 Kit was anxious for a home of her own and, even on a salary of $100 per month, Tilton wanted to have his little family on the ranch. Mother had designated his house to be put down on Section 4, not far from Jack's and mine. Tilton hired Doug Miller, just as we had done, to put the house in shape for his family. In the meantime, "inflation" had set in. Instead of the $2.50 per day that we had paid him, Doug now commanded $5. Nevertheless, we would all be grateful to this hard working fellow who enabled us to create the homes that sheltered our growing families for so many years to come. In fact, Kathryn occupied hers until her death in 1997. (The residence later became home to Rob and Kathy Tracy and their children, Haylee Marie, Kathryn Alyse, and Trent Jackson Tracy. Kit would have been pleased—it was her hope that her family would continue to enjoy her home.) Here, as neighbors, we lived in peace and harmony for fifty-two years.

The Tilton Tracys moved into their home in 1945 when their son Bill was three. He was followed in rapid succession by three little towheaded brothers: James Tilton was born June 16, 1945, John Richard on October 13, 1947, and David Grant on September 13, 1949.

By nature, their mother moved quickly. She would now find this characteristic to be to her advantage, for with each new baby the treadmill of her life quickened to a heightened pace. One baby was barely out of diapers before another took its place. Now there were five "cowboys" to feed and care for. Mounds of Levi's took their turn on pant stretchers while the ever-present basket of ironing seldom reached bottom. These were challenging days on Jerry Slough. No diaper service or Pampers to lighten the load of a nursing mother. No telephone to consult with the doctor about a feverish child; no drugstore; no McDonald's; no Denny's! (After they put Interstate 5 through the ranch and built the truck stop on Tracy Avenue, complete with fast-food outlets, Kit lamented, "Where were they when I had four little ones and was up on a stepladder trying to paint the bedroom?")

The Baby Fund

By the time the third little Tracy boy was due, it was evident that Kit and Tilton could hardly afford the delivery of a new baby on their monthly $100 salary. It was then that Jack Frey suggested that a "Baby Fund" be created to solve the problem. John Tracy was the first recipient of this arrangement. Years later his Uncle Jack would show him the receipt of payment to the hospital for his delivery. David Tracy, too, owes a bit of himself to Tracy Ranch. I can't help but think that this gesture of thoughtfulness on the part of my husband was one of the things that endeared him to Kit for the rest of her life.

John, in center, proudly displays his new watch as David, Jim, and Bill pose for the camera

Good Ol' Family Fun

Before long, corrals for sorting and branding cattle appeared adjacent to Kit and Tilton's home. A modest display of silver bridle bits, chaps, and lariats hung inside the doorway of their home. Tilton's roping friends soon began to arrive on weekends, and on Sundays the yard would be filled with horse trailers and pickups. A kettle of beans simmered on the stove as Kit graciously received their friends and made them welcome in her home. Friends from all walks of life brought donations of cakes and pies to add to the Sunday fare. All this made extra work, of course, but she knew that it made Tilton happy and that was all that mattered to her. Like his father, Tilton was not a drinking man and, as far as I know, no bottle passed among this high-spirited circle of weekend cowboys. None was needed!

With the corrals just a few yards from her home, Kit suffered from the attending flies. A congregation of the pesky insects enjoyed the warm area above the entry to her home. They peppered the white paint with their indelible tattoos, and awaited their chance to slip through the door. One time Tilton installed an electric fly killer in the doorway, and we listened with a kind of malicious satisfaction to the sizzle of their bodies as they hit the hot wires. No tears were shed as we stepped around their crematorium.

Four little sons (and seven little nieces and nephews) enjoyed the weekend western festival as they trudged about in their cowboy boots or climbed to the highest rail of the big corral. They would thrill to the deft toss of the cowman's rope and follow the colorful action below. The air was filled with the sound of thundering hooves, calves bellowing, the hiss of the branding iron, and the pungent odor of burning hair. This was heady stuff! The sights and sounds of the rodeo combined with the shouts of good humor and fellowship among this hearty bunch created an exciting and wholesome childhood for the four little Tracy boys and their many cousins.

As time went by, the children would learn to help with the branding, try their hand with the rope, and occasionally attempt to ride a calf. Wes would become proficient with the rope, but only John was destined to become a cattleman like his father. He became a skilled roper though, unlike his father, his roping arm was on the right-hand side of his body. But none were to become daredevils or ride the "broncos" as their father, Tilton, or their uncles, Darrel and Cecil, had done.

Tilton

Tilton and his favorite roping horse, circa 1960

From left, Jimmy Tracy, Ted Holloway, Bill Tracy, and Tom Holloway enjoy a mud bath

A weekend roping party at Uncle Tilton's

Our Mom and Dad

Bill Tracy remembers:

It's curious how life takes on more vivid dimensions and colors in the shadows of "the hard times" than while basking in the sunshine of ease. We have a greater appreciation of those who ride with us on that roller coaster careening between terror and the elation of merely surviving. Dissatisfaction seems to set in when we finally "have it made" and sail into a safe harbor. In Kern County, more farms and relationships seem to run aground on the golden sands of success than splinter on the rocks and shoals of a shared struggle. Perhaps our real enemy is complacency. Life lived on the brink allows little time to entertain boredom.

In mom and dad's young family, despite our limited income, we seemed to have everything we required. I know now we certainly had much more than the average youth of today. If deprived, I never knew it. Unlike many contemporary parents who wish to give their kids "all the things we never had," I would deem myself most successful if I have given our children half of what we received: love in full measure, respect, security, worth, sense of purpose, and a good name. Most important, mom and dad planted the seeds (though it was upon rocky soil) that ultimately gave root to a solid base upon which a better life could be built. Mom and dad were remarkable people. I assumed all children were so blessed until I, too, became a parent!

While I've never allowed bloodlines to obscure family foibles nor the facts to hinder a good story, I can say without qualification that my parents were by far the finest people I have ever known. Mom came closest to successfully practicing the teachings of Christ while dad, blessed with a love for life and all of nature's dominion, came closest to successfully practicing the teachings of the church with seldom a presence therein. It was a blessing to have but known them; to have been their son is a reward beyond measure.

The Footrace

Our mother was not by nature a disciplinarian. Due to her kind and gentle nature, she was not cut out for such a role. With four boys to raise, that was somewhat of a handicap! My father used to tease her by saying to us, "Now when your mother stomps her foot, you boys had better behave or something terrible might happen to you!"

One day I was to learn that my mother could rise to the occasion when necessary. I must have been about four when I challenged her to a footrace. Naturally, she allowed her small son to win. Perhaps to test my newfound skill, I became naughty enough to provoke her into action. Much to my surprise, her running ability had improved remarkably!

Lazy Susan

I don't know where they got the idea, but mom and dad always had a Lazy Susan dinner table. The first was a simple platform with a male and female pipe for a pivot. That quickly proved too small for a growing family that usually had a couple of supper guests, so dad built the current one using a truck wheel bearing. The table was a great way to serve a large family; all you needed was a good "lead" to intercept the food plates. But we discovered, when visiting friends with conventional tables, that we lacked the social skills of passing the serving dishes!

Another great feature was that the dinner plates didn't need to be cleared when dessert was served—one simply pushed them under the Lazy Susan! Children were mesmerized by the contraption. We even got my little brother David to sit in the center, and then set the table to spinning. We would then lift him to the floor and watch him stumble about in a dizzy stupor! (Didn't take much to be entertained!)

The Lazy Susan hosted many guests from around the world, as well as simply kids on the ranch. Dad's old FFA instructor, P.D. Spilsbury, often had supper with us. He'd survey the folks' four towheaded boys around that table and exclaim, "Tilton, they sure are an improvement over the old stock!"

I now realize that the dinner table was the only democratic center in the family. Mom and dad always had the final word; they were the parents. But at the table, everyone was equal and was therefore afforded the opportunity to speak their philosophies. We learned to think as a family there.

My One Cowboy Event

Dad was a natural cowboy, and it was difficult for him to realize that his firstborn had not inherited similar traits. When I was five he decided it was time I learned the joys of "rodeoing." It took some convincing to get me up on the calf—the clincher was my father saying I'd only have to try it that one time. He was convinced that, after one ride, I'd be hooked.

After making sure everything was okay, he swung open the chute. Out came one startled calf with an even more scared five year old who bailed off after the second halfhearted buck of the calf! Coming up with a mouth full of corral dirt and covered with alkali dust, I headed toward the house, more scared than hurt, crying at the top of my lungs. Dad caught up with me and made an offer I couldn't refuse. If I wouldn't tell mom he had put me on a calf, he wouldn't ever ask me to ride again. Dad was a brave man, but he feared mom's wrath!

Dad's Strip-Down

Dad's first car was a "strip-down." He obtained a decrepit Model T Ford and removed all fenders, doors and other body parts, leaving only the frame, transmission, wheels and exposed engine. He added an extra milk crate for the examiner's seat so he could take his drivers license test. When the examiner saw the contraption, he declined to ride and told dad, "You take 'er to the end of the parking lot and make a U-turn and come back here and make a U-turn, and I'll give you your driver's license!" When the brakes needed relining, he didn't have the money to purchase new shoes, so he used bacon rinds.

Driving to Wasco High School one day, he hit a skunk. The impact spread the skunk not only onto the car, but also all over dad. During class, the students were getting up and looking out the window to see if they could locate the skunk they smelled. Dad participated in this action so his classmates wouldn't discover that he was the source of the odor!

One night after taking mom back to her home in Semitropic in another automobile (this one had a coupe body), dad saw something pass him up. His car veered off the road, and a quick investigation revealed that the left rear wheel had come off and careened out into the desert. He searched the sagebrush, but with little moonlight and no flashlight, he couldn't locate it. He napped in the seat until it was light enough to locate the wheel and proceed to the ranch.

Snapshot Memories

Mom was always home when we returned from school. The house was always clean, and so were we. (I still remember the Vaseline Hair Tonic she used to slick down our hair.) When we were beyond control, no greater words of fear could be uttered than, "Wait 'til your father gets home!" We knew we had pushed her too far, and there was no reprieve with dad.

Mother was an eclectic and avid reader. She was contented in Buttonwillow, but traveled the world. She had many friends, some of them were real "nuts," but she was non-judgmental (definitely a trait she didn't inherit from her mother).

John Tracy remembers:

We probably lived the most idyllic children's life possible. Horses to ride, animals to hunt, and a slough to ride your bike up and down in. We had few rules: don't talk back, and no matter how late you got in or how tired you were, show up for work early.

Both of our parents were even-tempered and fair. You could get your way with a logical argument, but "no" meant "no." There were no further discussions.

We went on most vacations with mother; dad might show up for one or two days. You didn't see much of dad unless you worked or shared one of his hobbies. I think he came to one of my football games in Coalinga because he needed me for work the next day!

Practical Jokes

We had a good-natured relationship with mom. Once, I told Mike Knecht that mom was almost totally deaf. When introduced, he almost blew her over backward with his, "Hello." Another time, mom thought the windshield on my car was too dirty, and she was washing it for me when my friend Walt Fisher showed up. I took after mom with a whip, yelling, "Get back in the house! You know I don't allow you outside when we have company!"

When I Know Everything…

At one point, I yearned for the time when I would finally know everything and just settle down to work and play. I found out that you never reach that point, and the fun is in learning more and improving on everything you do.

David Tracy remembers:

Mom's Passions

Besides her family, mom had three great passions in her life. Gardening was a great release for her. It was someplace where she could be alone with her thoughts and could have a bit of solitude in a life filled with non-stop activities and responsibilities. Reading was another favorite pastime. She once mentioned that, as a young mother with a husband and four children, about the only time she had to read was the thirty or forty minutes between getting breakfast for dad and waking up her boys to go to school. It was a short period of time she relished.

Travel was mom's other great love. Whether it was a trip to Ashland, Oregon, driving 600 miles with four young boys, or a cattle-buying trip to Arizona with dad, mom always enjoyed the experience. After dad's death, she combined all of her interests with several extended trips to England. The books she had read about the great gardens of England and the personalities who created and lived among them led to her actually visiting those areas. As much as mom loved to wake up on the ranch and pursue a busy life, England added a new and rewarding dimension. After she became ill, the thought of just one final trip to England kept her going when her frail body continued to let her down.

Dad's Candor

I think that dad was the most forthright person I have ever known. He could tell someone what he honestly thought and, even if it reflected negatively on the person, he or she could not take offense because it was true. Mom had a bit of problem with that concept, as do most of us who try to be diplomatic. About the only time dad deviated from that aspect of his nature was while playing poker or trading livestock, and I am sure that worked to his great advantage in those pursuits.

Hunting

While mom had her interests, dad's one great diversion from ranch life was hunting. He worked hard and played hard. Aside from an illness or a migraine headache, dad was always active. He was either full speed ahead or asleep!

Jim and the Dead Steer

One of the stories often repeated in the family concerned the time dad had to haul a dead steer to a dumping spot on the ranch. We were used to having to push or tow vehicles to start them. When young Jimmy saw dad tie the dead animal to the back of his truck he asked, "What's the matter, dad, won't it start?"

Bill Tracy remembers:

Old Red Flatbed

I have fond memories of heading home with dad in the old red flatbed, windows down, warm alkali air swirling through the cab, and dad singing the chorus of "The Beer Barrel Polka." When you heard that, you knew he was happy and all was good in the world.

Dad and Jim (about six years old) were driving his old red flatbed truck into Bakersfield to pick up his first new truck. A patrolman stopped them around Rosedale School. He proceeded to point out a list of the old truck's deficiencies: no break light, broken windshield, slick tires, when Jim piped up and informed the officer, "The horn doesn't work, either." The patrolman laughed and told Jim that maybe he should give him the ticket!

Grandma Tracy didn't like dogs because they'd dig in her garden. When we visited his mother, dad would always have one of us tie the family dog in the truck bed. Jim felt sorry for our dog Tippy, so he tied him with enough slack to let him roam the bed. When we finished our visit, we discovered Tippy had decided to jump off the truck—his feet lacked about six inches from hitting the ground, and there he hung. Dad had one dead dog, one unsympathetic mother, and three crying boys on his hands!

Going to the Brushfield

After 1952, much of the ranch's effort went into developing the Brushfield. We'd beg dad to bring us along. He'd get busy working the 'dozer clearing brush, and we became captives. Eventually we had trapped all the lizards, horned toads and rabbits we could stand—got tired of playing cowboys and Indians and hunting arrowheads. We discovered there was no shade, and that when you drank or spilled the one jug of water, there wasn't any more. It didn't take long to decide it was no longer fun to go to the Brushfield with dad. Dad realized this, and when he didn't want to take us (even if he wasn't going there) he only had to say he was "going to the Brushfield" and there were no takers!

A Surprise for Earl

We stuffed David into our big mailbox on Tracy Avenue one time with the intent of scaring Earl, the mailman. We were disappointed because David wouldn't stay; it was probably a little too cramped, hot and dark!

David Tracy remembers:

Ranch life required that everyone pitch in, regardless of age, to do what he could to contribute. As the youngest child, I got to spend a lot of time with dad before starting school. Often, that meant steering the truck while dad fed hay off the back. My feet didn't reach the pedals, so he would adjust the throttle and point me in the right direction.

Sometimes a cow would need to be doctored. After roping it, he would leave me in charge of the horse while he tended to the animal. All of the brothers and cousins helped in the round-ups and branding of cattle. Fun times were hanging jerky in Grandma Tracy's canning shed with an eye to the treat that would be ready a few days later!

The Walk-In Box

The walk-in refrigerator at Grandma Tracy's house was where the abundance of summer produce was deposited. Anything in it was free for the taking by family members. The only problem was that when you added fifty or sixty watermelons, along with crates of grapes and peaches and sacks of potatoes, it became a bit overcrowded! When dad decided to build one in our new garage, he made sure to build it big enough to handle several dressed deer along with all the bountiful seasonal produce. It is still the family supermarket, and I never visit the ranch without checking out the contents!

Snipe Hunt

Dad loved practical jokes; mom failed to find humor in them. One joke dad played on me when I was a young boy almost got him in trouble with mom. Knowing how curious and trusting I was, one night dad set me up with a sack and a stick by the garage with instructions on how to tap the sack with the stick and call out to the snipe. I had visions of soon catching one of these wonderful birds, which helped me to temporarily overcome my fear of the dark. Over and over I called out to the snipe and tapped the bottom of the bag, to no avail.

Finally my fear of the dark overcame my lust for snipe, and I retreated to the house in tears, crying, "Dad, the snipe won't come." You can only imagine mom's reaction as she glared at dad. Always a quick thinker, dad grabbed a flashlight and an old birdcage, and we headed out to a shed where a woodpecker made his roost. We soon had the "snipe" captive, and dad not only had a delighted young son, but also a wife who would speak to him. I think the snipe "escaped" by morning.

Yellowstone

The only vacation we took as an entire family was a trip to Yellowstone Park and most of the other attractions in the western third of the country. We borrowed Grandma Tracy's new station wagon and rented a trailer to tow behind it for our adventure. One morning somewhere out in the wide-open spaces, we stopped for gas. As usual, dad informed us that if we had to use the bathroom, this was our last chance until lunch. I stayed behind in the back of the car under a blanket while everyone else got out.

A few moments later I decided to explore the store attached to the station, and I thought it best to use the restroom as well. When I returned to the car, I saw it driving away, a couple hundred yards down the road. Immediately I gave chase, waving my arms to no avail.

As the station wagon faded from view, I saw a pickup coming towards me and I stood right in the middle of the road, still waving my arms. The driver and his passengers had no choice but to stop. While I was explaining that I needed a ride to catch up to the family that had just abandoned me, I saw the familiar station wagon heading back to retrieve me. The ranchers in the pickup looked a bit dumbfounded at the situation. Once again, mom was not too pleased with dad, but he was certainly not the only guilty party. A mother hen is supposed to keep track of her chicks! The incident has provided many laughs over the years.

Dogs, Cats, and Other Creatures

Mom went along with, and sometimes encouraged, all the animal life that shared our lives, even though it always meant more work for her. A pet lizard I kept tied in the front hallway; a duck won at the county fair I kept alive by warming it in the oven. (It survived to make messes on the sidewalk.) Countless 4-H projects, pigs, calves and lambs often required bottle-feeding.

I think the true test of the marriage, though, were the many dogs dad expected to be great cattle dogs to help in his work. Mom loved her yard, and more often than not these "cattle dogs" became "diggers." Once a dog began digging up mom's yard, dad knew that he had to find another promising breed.

Ravens

When we were youngsters, dad related stories about the pet raven he had as a boy, and what a marvelous bird it was. One spring we set out on a mission to capture a young raven so we could share this experience. Setting out in the truck, dad eventually spotted a nest at the top of a tall power platform several stories high.

Though Jim was only about ten, he wanted to climb the tower and retrieve the birds. Up and up he went, clear to the top where the nest was. One slip would have been fatal, but he came down with the prize. Thank goodness mom was not there to witness the feat—and I am surprised dad allowed it. He had confidence in his young son. We had a great time with our new pets.

A few years after our first pet raven, I was eager to procure another. Dad almost always went along with such adventures. He called his friend, Carl Twissleman, and asked if he knew where a raven could be located. Carl did, and he gave us directions to a sandstone cliff in the foothills of his ranch. We spotted the nest, but the only access was from the top of a high cliff.

Dad explained that the only way to accomplish our mission was for him to tie a rope to me and lower me into the nest. My fear was overcome by my trust in dad and my desire for the raven. Over the edge I went, and into the nest. Next dad had to pull the rope up and lower a box for the birds. After that, he retrieved me and all was well.

Ravens are remarkable birds, and I think it is now against the law to keep them. They are very docile when hand-raised from an early age, and they are extremely intelligent. Ours were allowed to live a free life outside and always flew down to land on a shoulder for a handout. Many visitors were startled when the large but friendly bird landed on them as they got out of the car!

Raccoons

After reading the book *Rascal* about a pet raccoon, I had to have one. Once again dad obliged, with mom going along with the plan—a good sport, as always. We ordered it by mail from somewhere in Arkansas, I think. It did not bode well that we had a call from the post office that it had broken out if its cage and they could not deliver it until it could be recaptured. It finally arrived, hissing and snarling—nothing like the cute little "Rascal" I had expected. As I recall, dad traded him to a man looking for such an animal (to train his dogs to 'coon hunt) in exchange for two adorable baby raccoons that we had to bottle-feed. Once again we had intelligent and curious little creatures, but raccoons do not make good pets as adults, and are best left in the wild.

The Pack Trips

Mount Whitney or Bust!

The author remembers:

Nancy sat on our high kitchen stool as I carefully braided her hair; this would have to do for the next two weeks, for she and Larry were to join several of their cousins on a pack trip.

By 1950, the Tracy Ranch partnership had survived eight years of struggle, but there were many years of struggle yet to come. Consequently, it came as a surprise when Tilton announced that he would like to take our children on a pack trip into the High Sierras. He enlisted the help of Darrel and brother-in-law Skip, and soon a trip was in the planning. (During the years 1950 to 1956, there would be three pack trips. Although they were referred to as the "Whitney Trips," it was not until the third trip, in 1956, that Mt. Whitney was actually reached. Dick Jeffries accompanied Tilton who kept a log, and that trip was composed of boys only: Bill and Jim Tracy, JB and Wesley Selvidge, Larry Frey, Paul and Bill Evans, Jamie Salari, and Steve Jeffries. The Evans boys were sons of our foreman, Ed Evans.)

By 1950, there were eleven cousins in our family, and seven were considered old enough to participate in the journey: Darrelen Tracy and Nancy Frey who were both ten, JB Selvidge who was nine, and Marilyn Tracy, Larry Frey, Wesley Selvidge and Bill Tracy, all eight years of age.

It was not until years later that I fully appreciated Tilton's courage in planning such a trip, for several of the children were barely old enough to lift a saddle let alone heave one onto a horse's back. Although all had some riding experience, none had any experience riding in the mountains. Nine reliable saddle horses, as well as several pack animals, would need to be shod. The entourage would need to be outfitted with kayaks, camping equipment and bedrolls. And stocking food for nine people for a two-week period could not have been a small matter! I doubt we had enough horses on the ranch at the time but, like our mother Fannie, Tilton would find a way.

Tilton and Darrel were no strangers to the pack train, and soon all was ready. One day in June 1950, the group of boy-girl cousins took off for an unforgettable journey in a spirit of high adventure.

Pack Train Impressions

Marilyn Tracy remembers:

Eight was the magic number! One had to be eight years old to go on the Mt. Whitney pack trip. I waited anxiously to reach that magic age.

I practiced tying knots with a piece of rope around the leg of the dining room table. I remember making a loop and then, "Go into the squirrel hole and back out." I think that was the knot for tying up your horse. The rope marks from my practicing still remain on the leg of the table. Mother's advice as we prepared to leave: "If a bear gets after you, climb a tree and stay there. But don't go to sleep and fall out of the tree!"

I don't remember packing all that much for the trip, but I do remember we were good about always brushing our teeth. I don't know how "clean" we were other than that.

We each had our own horse to saddle and ride. Saddling a horse when you are only eight years old took some doing. First, the saddle blanket goes on the back of the horse. Next, throw all the "things" (stirrups, cinch, ties) on the right side of the saddle across the seat of the saddle to the left to get them "out of the way." Then, with both hands and all your might, precariously balance the saddle on its left edge and hoist it up on the horse as best you can—invariably succeeding only in pushing the blanket off the other side. So, try, try again, all the while being vigilant that the horse doesn't step (or stand) on your foot! After finally getting everything in place and cinching up with your boot against the cinch ring to pull it as tight as you can, Uncle Tilton or daddy [Darrel] would give a check and pull it really tight with a simple tug. I was always so in awe of Uncle Tilton and the way he could lift his big saddle effortlessly up and over the back of his horse. The saddle just seemed to fly into place all by itself.

On the way to our departure point at Quaking Aspen, we stopped at a spot to have lunch and let the horses out of the trailer for some exercise. Sandwiches appeared from somewhere—no doubt Aunt Kit's handiwork. I don't remember where it was, or the name of the place, but there was a very large rock outcropping—almost like a cave. We had lunch in there with admonitions to watch out for rattlesnakes.

I remember how beautiful the meadows were, and I remember seeing quaking aspens for the first time. I thought they were so pretty, the way they "quaked." They are still one of my favorite trees. Daddy liked them, too.

We were taught to lean forward in the saddle going up steep grades to help the horse, and to give him plenty of reins. Watch out for tree limbs lest you get brushed off. Be alert that your horse might shy going across creeks or streams. Some horses just don't like doing that; sometimes, the water splashing up on their underbelly spooks them.

Daddy warned us about leaving little personal things out loose around our campsites. Things might well disappear, replaced by an acorn or a rock or whatever. Eventually, we did find a pack rat's nest and it was full of odd objects and things that it had packed away.

The boys pulled some kind of a trick on Nancy one night. I think they short-sheeted her sleeping bag. [(Actually, I think they put a long stick in my bed, hoping I would think it was a snake!-NFT]

The horses got loose one night, and Uncle Tilton spent several hours the next morning rounding them up. It seemed as though, no matter how careful we were to hobble them or tie them up, they always managed to get into some kind of trouble.

One time Uncle Tilton undertook to get some firewood by roping an old snag high up on a dead tree. Unfortunately, the snag didn't cooperate and snapped right off. We all wound up on the end of the rope, pulling and struggling to get the snag to break. When it finally did snap off, it flew straight at us like a rocket and crashed just in front of Uncle Tilton, knocking us all over and spraying us with Black Forest soil. Uncle Tilton got up laughing and looking like Amos or Andy, with black face and white teeth grinning. Of course, we all thought it was hilarious—being knocked down and covered in black dirt—we didn't realize it could have been rather disastrous. But we had plenty of wood!

I remember pancakes, beans, bacon, jerky, canned peaches. I think daddy did most of the cooking. I don't remember anything not being good to eat!

The Robert William "Bill" Tracy Family

Bill and Susie Tracy with children Todd, Stephanie, Mark and Rob, circa 1974

Same family, circa 1990

The Mark Tracy Family

Bill's oldest son Mark James Tracy marries Allison Sobocinski in Bakersfield on August 11, 1996

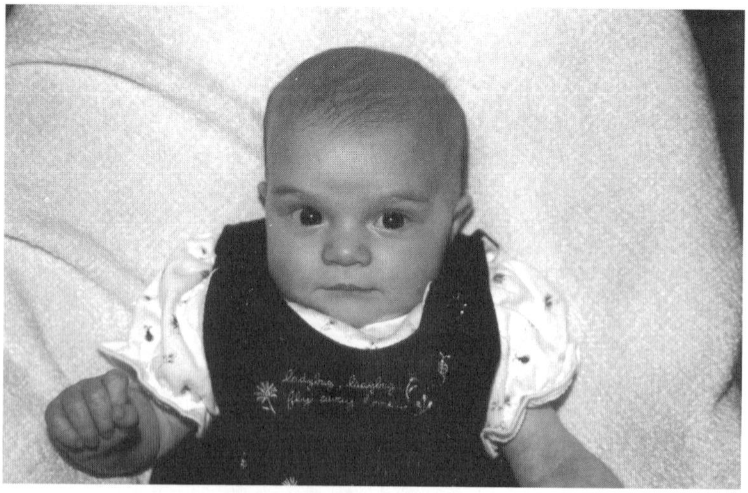

Isabel Celeste Tracy is born to Mark and Allison on February 4, 2001. She weighs seven pounds, three ounces and is 20 inches long.

The Robert Tracy Family

Bill's second son Robert (Rob) Lee Tracy marries Kathy Lee Twisselman, of the pioneer ranching family, on September 29, 1992 at the Tracy Ranch. More than 1,000 guests attend the ceremony (below) and the dinner that follows. This historical location has been the site of many weddings.

Haylee Marie is born to Kathy and Rob Tracy on September 11, 1996. Here she poses with mom and dad in front of her great-grandmother's fireplace.

Kathryn Alyse arrives on September 23, 1998. She is named for her great-grandmother Kathryn Pollard Tracy.

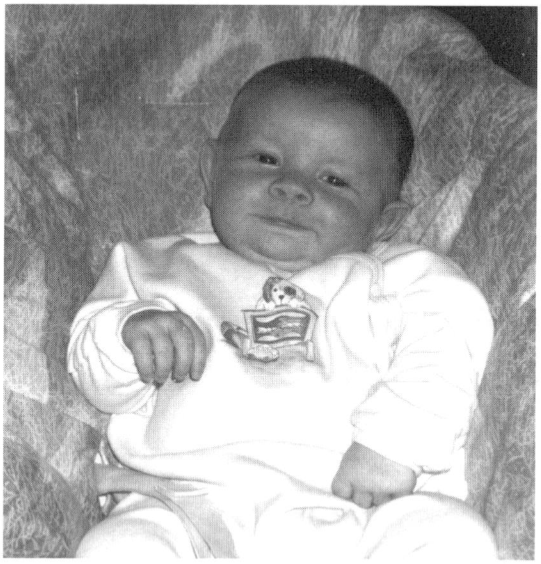

Trent Jackson arrives on February 18, 2000, the same birthday as his great-grandmother, Fannie Tracy

Grandparents Susie and Bill Tracy pose with Haylee

It was apparent that Fannie Tracy's spirit was still alive on the day a stretch limousine was parked in front of Rob Tracy's house

Little Haylee poses in front of the limo, which her dad, Uncle Todd, and some other young bloods purchased second-hand, then mounted Texas longhorns to its hood!

The Kevin O'Callaghan Family

Bill and Susie's daughter Stephanie Ann Tracy marries Kevin O'Callaghan in Bakersfield on September 26, 1995; reception at the Tracy Ranch

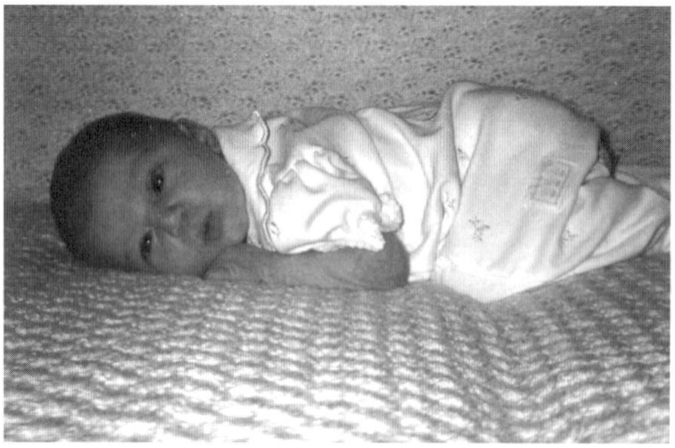

Grace Ann O'Callaghan is born to Stephanie and Kevin on March 31, 1999

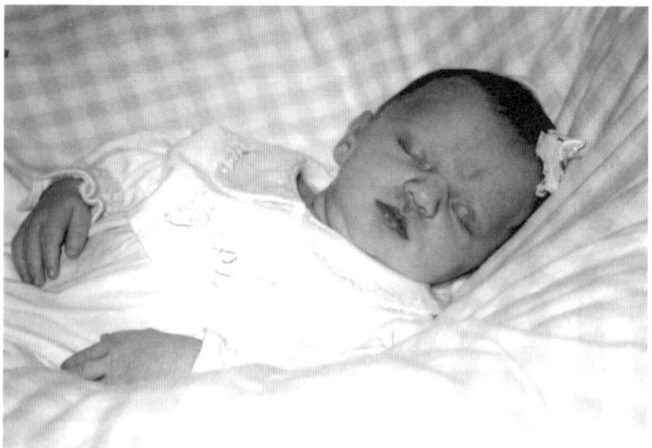

Magdelyn Kae is born to Kevin and Stephanie O'Callaghan on January 17, 2001

The Robert William "Bill" Tracy Family

Bill and Susie's son, Todd Blake Tracy, marries Annette Clarisse Easton on September 15, 2001, in Bakersfield; reception at the Tracy Ranch

Clarisse Ann Tracy is born to Todd and Annette on January 15, 2003

The John Richard Tracy Family

John Richard Tracy marries Donna Kay Allen on January 10, 1970 in Redondo Beach

Donna and John with their sons Jimmy (J.T.), left, and Allen in 1981

Jessica, age 4, and Allen, age 6

Little Jimmy (J.T.), the boy we lost

Jessica, John, Donna, and Allen, Christmas, 1999

Kit and her boys, clockwise from back left, David, Bill, and John;
Christmas 1995

Cecil and His Women

Florence 1929 - 1930

During Cecil's senior year in high school he fell in love with Florence Williams. She was blond and had a then-fashionable "boyish" haircut. When Florence was in a good mood, she was fun to be with and, when she laughed, her face displayed a pleasing array of dimples. But when she was in a bad mood, it was best to stay out of her way! Cecil and Florence were married in 1929. Unfortunately, the marriage lasted just over a year, for they separated in 1930 and divorced in 1933.

Had this marriage worked out I think that Cecil's life might have been much different, for Florence was his one true love. He was brokenhearted when they parted, and throughout the rest of their lives there was a bond between the two.

For many years after this marriage Cecil remained single, but he was by no means romantically idle. During this period, and in between three later marriages, a veritable assembly line of women passed through his life. Cecil was a creative lover; an artist in the field of romance. He knew exactly how to make his way into a lady's panties, and he didn't need a Cadillac car (or even a clean shirt) to do so. Errol Flynn had nothing on Cecil; they even looked alike. But eventually, following World War II, he was involved in a romance that was more intriguing than all the others.

Sigrid

Sigrid was an extremely beautiful and talented German actress with a charismatic personality, and she easily dazzled everyone who came within her presence. Soon after the end of WWII, this attractive divorcee had advertised in the *U.S. Farm Journal* for an American husband. She wrote that she was "fed up" with German men and wished to bring her small son with her to the United States "and make her home in California." She invited all eligible American men to write to her and describe their qualifications.

Out of the 4,000 letters she received, guess who wrote the best letter?

Cecil borrowed money from Uncle George, Aunt Dee's none-too-rich but kind-hearted husband, and he sailed for Germany where he met Sigrid and her little boy. They were not disappointed in one another and, after a three-week courtship, they agreed to marry. And this marriage might have taken place had Sigrid's former husband not kidnapped his son. When the German newspapers got hold of the story, a court battle ensued and, naturally, the German father was awarded custody of the boy.

Still, Sigrid did not give up; she planned to get the child back somehow and eventually travel to California and marry Cecil. By now the *New York Times* had the story, and when Cecil arrived in New York on the way home, he was greeted by a crowd of reporters. When they asked his occupation, he said he was a cattleman and owned a ranch in California. When asked the size of his ranch, he truthfully told them that it was approximately 2,000 acres. This must have made a fine impression, for he was now written up in the *Times* as the "wealthy cattleman from California," who was going to rescue the

Sigrid receives 4,000 letters

beautiful German actress with her little boy and bring them to California. What Cecil had neglected to mention was that his ranch in California was a totally undeveloped piece of property, and a few head of half-starved cattle provided his only income.

Imagine the embarrassment at home when we read the headlines in our local paper: "Kern Cowman Wins Fraulein!" The article continued: "Cecil Tracy, wealthy Buttonwillow farmer and ex-cowboy champ of Kern County, announced yesterday he had wooed and won beautiful German actress Sigrid Haessler away from 4,000 other admirers." It was about this time that our mother arranged to take a trip to Europe with a "Farmers' Friendship" tour. She left the tour behind in Germany where she stayed for three months, met Sigrid, soon succumbed to her charms, and gave her hearty approval to the marriage.

During the following year, as the drama unfolded and Cecil was living out his fantasy, Sigrid was frantically trying to regain custody of her child. Years later, Wesley came upon old letters which, along with many other papers belonging to Cecil, had been dumped into a barrel and stored in the old white barn. As he was clearing out the barn and about to burn the barrel's contents, he came upon the letters. Only then did we learn that Sigrid, after failing to regain custody of her son, was planning to abduct him. She had made arrangements to meet Cecil at a designated time and place, and flee with him to the United States. (The letters had survived after the home place burned to the ground.)

Cecil had purchased a second ticket to go to Germany, but when he reached New York he apparently changed his mind and returned home. No one will ever know exactly what happened to cause this change of heart, and his family was no less than astounded when he showed up on the ranch with a woman named Aileen and announced that they had just come from Las Vegas where they had been married!

The family had met Aileen only once; she had attended a small rodeo at the ranch and we were not favorably impressed with her behavior at that time. When Cecil arrived at mother's to introduce his bride, mother, in Aileen's presence, said, "Cecil, if you had taken an arrow and driven it straight through my heart, you could not have hurt me more!" (Aileen and mother never became friends.)

Cecil in his Errol Flynn pose

Glamour shot of Sigrid

In retrospect, there may have been more than one reason for Cecil's change of plans. For one thing, by abducting a child of German descent and bringing him into the United States, Cecil would be entering heavy legal territory; secondly, Cecil may have come to realize that Sigrid, a strong and domineering woman, might be a little too much for even him to handle. Consequently, because he was not actually the "wealthy cattleman" described in the *New York Times*, perhaps he had best extricate himself from a rather sticky situation by quickly marrying someone else.

Aileen 1952 - 1962

Aileen was a small brunette and, although she was not unattractive, she had a rather dour personality. She brought with her a young daughter from a previous marriage. In addition to the usual household pets and a horse for the daughter, an entire menagerie of exotic animals followed during Aileen's regime. At one time an alligator resided in the empty swimming pool, and a monkey who had the run of the house freely left his droppings wherever he pleased.

One day as I was paying Aileen a rare visit, we were sitting in the living room having a conversation when something landed on my shoulder with tremendous force. When I turned to see what had attacked me, I discovered a full-grown bobcat sitting there. I then remembered that someone had told me that Aileen and Cecil now had a pair of bobcats, and that these wild creatures had been tamed from kittens and they, too, had the run of the house. They were beautiful cats and each must have weighed some thirty pounds. But they were very gentle, and I was grateful this particular cat hadn't buried his claws in my shoulder.

The only thing that had suffered any damage was my nervous system.

Cecil did not ask much of his wives over the years, but he did insist that the annual family Christmas party be held in their home, the old home place. One year as I was depositing my Christmas pies in the back room of their house, I was surprised to hear Aileen's voice calling, "C-E-C-I-L!" and then to hear Cecil's voice answer with his typical response "YEA HO!" This seemed very strange, for there was no one in sight. Then I heard it again, "C-E-C-I-L" and another "YEA HO!" There was no mistaking the sound of their voices, and yet there was something strange. The voices, though seemingly nearby, were faint, and I looked around for speakers. I spotted a cage and when I uncovered it, I found a myna bird inside. Another surprise!

Somehow, Cecil's marriage to Aileen lasted for ten years. He seldom complained about any of his wives, but he could not resist sharing one incident that took place near the end of this marriage. Aileen had been trying, apparently without much success, to provoke a fight with him. The harder she tried, the angrier she became, until nightfall when she managed to lock him out of the house. Since it was a cold winter night, Cecil built himself a fire in order to keep warm, and he was just getting comfortable when he was drenched by a great stream of cold water. For a moment he thought it was beginning to rain, but he soon realized that Aileen had turned the fire hose on him! Since the water had put out his fire, Cecil located a pile of leaves, crawled into them, and claimed he stayed warm and slept comfortably the remainder of the night.

After this marriage drew to an inevitable close, Nanette replaced Aileen.

Nanette 1962 - 1972

The family had met Nanette twelve years earlier when Cecil was dating Nanette's divorced mother, a striking bleached blond, who went by the name of "Margo." Her daughter, Nanette, was an attractive seventeen-year-old girl when she and her mother first visited the ranch—a petite young brunette with a pleasing personality. Twelve years had passed and during that time Nanette had been married, divorced, and was herself the mother of a daughter, Leslie, who was seven or eight years old.

By the time Nanette became the object of Cecil's intentions his courting skills had been honed to perfection, and he did not fail to employ them with finesse in making this new conquest.

I will digress here to tell more of Cecil's personality. Cecil worshipped his father and favored him somewhat in appearance; in many ways he tried to emulate this fine man. Although our father was orphaned at an early age and did not attain a formal education, he had somehow developed a great love of poetry and, like our father, Cecil had committed Sir Walter Scott's poem, "The Lady of the Lake" to memory. As children, we listened with rapt attention as our father recited passages from this long and rhythmic poem. I still remember a few lines:

The stag at eve
Had drunk its fill,
Where danced the moon
On Monan's rill.

Anyway, by the time Cecil began courting Nanette he, too, had memorized "The Lady of the Lake." One of his favorite ploys, when entertaining a lady friend, was to place Scott's poem on a nearby table. At an appropriate time, he would pick up the book, read a few verses; then lay the book aside and continue to recite the entire poem by heart.

Clockwise from above: Nanette and Cecil. The Tracy home circa 1940. After the fire, only the brickwork remained.

Nanette succumbed to the charms of this older man, Leslie fell in love with the ranch, Nanette fell in love with Cecil, and they were soon engaged to be married.

A Bower for the Bride

By this time Cecil, our mother, my husband Jack, my sister's husband "Skip," and my youngest brother, Tilton, had formed a family corporation, and were beginning to develop the ranch. These were tough years. We allowed ourselves meager salaries on which to live and, this way, got by the best we could. Cecil approached the members of the partnership with a request for a few thousand dollars to spruce up the old home place for his lovely new bride. He wished to make sure the toilets were in working order and perhaps freshen up the kitchen a bit. The partners granted his request, but almost before they knew it an entire sixty-foot wall was being swathed in brick! A brick mason of no small talent had been employed, and bricks and stones of grand proportion now adorned the old home from one end to the other. A path of stone led to the entrance and, on the landing, bricks radiated in boundless circles from our father's "91" brand. From here they encased a round window before traveling on to a raised hearth in the kitchen. Best of all, there was an enclosed patio outside the master bedroom. Here the new bride could enjoy her sunbaths! This was beautiful stuff! No ordinary craftsman could have accomplished it! Cecil, however, soon found himself out of funds. The toilets were still not flushing, the kitchen hadn't been touched, and Cecil needed more money! It was with some chagrin that he was granted additional funds.

Nanette's daughter enjoyed having a horse to ride plus the usual array of pets, and several happy years were shared before the joys of country life wore thin. After ten years, this marriage went the way of Cecil's previous two.

A large moving van pulled into the yard to pack Nanette's belongings, and Cecil told her that she might take anything that she wished except the set of sterling flatware that his mother had brought him from Germany. As he helped to load the van, he noticed that a box labeled "pillows" emitted an unusual sound and, upon investigation, Cecil discovered his set of silver, tucked between the pillows!

Shirley 1976 - 1978

As Cecil grew older, his brides grew younger and younger until, by the time he married Shirley, there was an enormous gap between their ages. When Cecil first met her, she was sitting alone inside a car in a parking lot. Always on the lookout for a pretty face, he spotted her and, walking up, he employed what was by now one of his favorite MOs. "You, without a doubt, " he told her, "are the most beautiful woman in the entire world." Shirley replied, "Why, thank you!" Then she told him she was sitting there because her car wouldn't start. Now Cecil was good at a lot of things, but a mechanic he was not. However, he released the hood to see if, by some stroke of luck, he might be able to ascertain the problem. Soon another fellow stopped by, and then another, until a small circle of men were gathered at the scene. They were eventually able to start the car. Now Cecil thanked them for their trouble and politely dismissed them by informing them that he had been there first.

After a brief chat, Cecil addressed Shirley with, "By the way, would you, by any chance, care to have dinner with me sometime?"

"Well," Shirley replied, "that would be very nice, but I have a young daughter."

Cecil replied with, "I would be very happy to include your young daughter." (This was a technique that Cecil had discovered worked very well.)

When he turned to leave, he realized he had not obtained Shirley's phone number. She gave it to him, he departed and, because he could not believe his good luck, he thought Shirley might have given him a false number. However, when he called the same pleasant voice answered the phone, and Shirley agreed to a dinner engagement.

Within a year Shirley's divorce became final, and she became the fourth Mrs. Cecil Tracy. But this was a marriage in name only, for Cecil later confided that the marriage had never been consummated—Shirley never made her home on the ranch, and when he stayed at her apartment he slept on the couch. This foolish old man had simply bought himself a showpiece! After a year or two Shirley began to tire of this arrangement, and she filed for divorce.

INCEPTION OF THE TRACY RANCH

The First Partnership

After Jack and I arrived on the ranch in 1942, I was surprised to learn that Tilton and Frances's families would soon be following. Tilton left the security of his job in San Diego, and he and Kit arrived with their new baby plus a dachshund with a basketful of pups. They settled northeast of the ranch while the oil field bunkhouse Fannie had designated for them was being transformed.

Frances's husband, Skip Selvidge, resigned from his job as a projectionist for Fox Studios, they left a lovely home in the foothills of West Los Angeles to cast their lot with the rest of us, and they arrived with their young sons, JB, and James Wesley. Because we Freys were already occupying the family home with mother and Cecil, the Selvidges settled in makeshift quarters—three nearby buildings which included our father's homestead cabin, the "mess hall" where they slept, and an adjacent building used as a bathroom facility. On October 24, 1945, while Frances and Skip were enduring these crude circumstances, their daughter Tracy Ann was born.

During 1942 a rather loose partnership was formed. My husband became general manager, Skip was shop foreman, and Tilton was in charge of the cattle. Mother and Cecil were "ambassadors-at-large." (Originally, the responsibility of managing the ranch was to rotate among the men. However, this idea never took hold, and the others were happy to leave that role to Jack.) Several years later, the relationship would be formalized into "Tracy Ranch, Incorporated," precursor to the present-day "Buttonwillow Land and Cattle Company."

As I mentioned earlier, in addition to running a grocery store in Buttonwillow, Darrel was also postmaster there, and he and his wife, Marie, had two daughters.

Now, in addition to mother and Cecil, the partnership consisted of a football player, a movie projectionist, and a cowboy. Darrel decided not to cast his lot with this motley crew in such a risky venture, and he requested that his share of the ranch be set aside and not be used as collateral with the rest. Mother opposed him because she felt that, if the partnership were to succeed, all of the property would need to be used as security.

Bessie Baker

Through one of my school chums, Margaret Bridges, Jack and I became acquainted with Bessie Baker. After losing her husband, Bessie ran their business, Baker's Transfer and Storage. Perceiving Jack's youthful ambition and desire to farm, she generously advanced us a $10,000 loan. This was the nucleus of our farming operation, and for the next three years each partner drew a subsistence wage of $100 a month. Somehow we all managed to survive on that meager salary.

The Sills

During those early days while we were trying to get the ranch established, the Sill family also offered assistance. My husband and Jack Sill had become friends when they attended UC Davis Agriculture School together, and it was Jack and his eldest brother Ben who had enticed my husband to take a leave from his job in the oil fields to run their potato shed in 1941. Jack Sill was one of the sparks that triggered our move to Jerry Slough and, as a member of the Board of Directors of the Shafter-Wasco Ginning Company, Jack used his influence to help us obtain financial backing.

Jack and Betty Sill on their wedding day. We counted this handsome couple amongst our closest friends.

Roe Brady

Roe Brady was a faithful employee who worked as an irrigator on Tracy Ranch for many years. He and his wife raised their four children here. In fact, their young son Ray was born after they made their home on the ranch. During cotton-picking season his stalwart mother wrapped her infant son in a blanket and placed him at the end of a row of cotton while she picked her way down that row and up the next.

To relieve the monotony of his life, Roe sometimes indulged in the bottle. During one of these occasions, his boss approached him. "Roe," Jack said, "you know that you are not supposed to drink while you are on the job." Roe thought about this for a moment, then replied in his deliberate Southern drawl, "Well, Mistah Jack, the thing of it is, you works me seven days a week, twelve hours a day, an' twelve months o' the year. If I'ze gonna drink a'tall, I'ze 'most gotta do it while I'ze on the job!"

After looking out over the perfectly groomed field with the water under control and properly regulated, Jack could only shake his head and silently acknowledge the truth of Roe's observation.

Oil painting of Roe Brady

Ray Brady

Ray Brady, Roe's son, attended Wildwood School with our son Larry, and occasionally his influence was felt at home. Although Ray was a little younger than Larry, he was large for his age, and one day after school when Larry seemed to be out of sorts, I asked him what was wrong. "What is the matter with you? What makes you so cranky?' "Well," Larry replied without hesitation, "How would you like it if someone was calling you 'Sho'tie' all the time?"

Another time when we were driving to Grandma Tracy's, a rabbit bounded across the road. My small son popped up from the back seat and exclaimed, "Yonder goes a rabbit!" Larry was no doubt imitating his friend's vocabulary!

Ray Brady later played football for Wasco High School and was awarded a football scholarship to attend Washington State University. He graduated as an aeronautical engineer and was hired by Northrop Aviation. Ray has been with Northrop from its inception more than forty years ago, and he was once engaged to work on the B-2 Stealth Bomber when it was a secret mission. Ray still maintains close ties with our extended family, and he brings his lovely wife Avis to most of the important ranch functions.

Stephen Clifford

Stephen Clifford is the son of our good friends, Mary and Gerald. When he was in the seventh or eighth grade, we invited him to spend a summer on the ranch as Larry's companion, while Nancy spent the summer with the Clifford family, attending summer school in Bakersfield. Jack soon put Steve to work as an irrigator, and we became exceedingly fond of this engaging young fellow who spent a few summers enlightening our lives with his droll wit and dry sense of humor.

Jack, Ray Brady, and Stephen Clifford: the "Watermaster" and his two apprentices forty years later. The "apprentices" are now an aeronautical engineer and an attorney-at-law.

One day Stephen, who was very small for his age, told Jack that Ray Brady had been challenging him to "Indian wrestle" with him, but that he wouldn't do it because he knew that Ray could "put him down." "Tell you what you do," Jack advised, "Tell him okay, you will do it. Then, just as you are in starting position and before he even has time to think about it, put him down REAL QUICK. Then don't ever agree to wrestle him again."

Stephen followed Jack's instructions and, to this day, Ray has been trying to get Stephen to Indian wrestle again.

When Stephen entered Bakersfield High School, he was embarrassed because he was so small. His parents discovered he was deficient in specific hormones, and this was impeding his growth. After undergoing a series of injections, Stephen started growing by leaps and bounds. Each time we saw him he had grown another two or three inches. We began teasing him, telling him that if he didn't stop growing, perhaps he would become the tall man at the circus. With typical Stephen Clifford humor, he referred to his inoculations as his "Grow Pup" shots.

Not only did our young friend grow to be over six feet tall; but he grew in stature as well. He now heads a large firm of attorneys, and he enjoys the respect and admiration of all who know him.

Land Regained

Tilton and water troughs, 1969

In her book, Fannie describes regaining Section 5. This property had been sold during the depressed years in order to save the ranch. Uncle George Taylor, who was now in real estate, had succeeded in interesting a group of wealthy businessmen from Los Angeles to invest in the property for use as a gun club. With the organization of Tracy Ranch, Inc., there was now an opportunity for the family to retain the 640 acres that were previously sacrificed. Fannie's description from *The Tracy Saga* follows:

> Today, July 29, 1961, I watched the disappearing dust of Cecil's car. A feeling of elation encompassed me, for the papers going this day were into the hands of attorney Mack for their legal correctness. Section Five, a part of the Tracy Ranch from 1862 to 1924, would soon be restored…

My husband had never been able to expand in those early days to the point where he could put a plow into this rich soil, so it continued as wild pasture under fence, helping to maintain a sizeable herd of cattle. Changing trends, coupled with human frailties, made some losses obvious during the era of depression. The country was still in the equine epoch. The enormous undertakings of today were not only impossible, they were not thought of. The ranch had kept Section Five rented for many years, so the equilibrium of our present lands will not be disturbed. Its continued use by us had made its return inherent.

...the far-flung ravages of depression took its toll. Its ingress into our lives at the time our children were small, and we were dreaming of their future made the intrinsic value of any part of the ranch desirable. The maturity of a loan at that time if not met would have led to complete destruction of our assets. It seemed prudent therefore that Section Five, in its unimproved condition, should be sacrificed even though it was the favored field for cattle.

Those who purchased Section Five organized a gun club, coming each week to shoot during duck hunting season. To accommodate the hunters I opened a boarding house over weekends, teaching during the day at nearby Bowerbank School. I wanted the children to complete their high school, but it was possible for them to come home weekends and help. They cleaned house, made beds, and assisted me. To prepare for the big meals I cooked all day Saturday, baking up to twelve pies when the number of hunters called for it. Saturday and Sunday were both days for feasts of turkey or fried chickens. Saturday nights I worked long past midnight, for many club members did not arrive until morning hours. Some often brought guests without letting me know in advance.

My husband's illness coming at the time of financial troubles was a greater shock than all else. The ranch was in no shape to finance the heavy investments he had made earlier at the hands of unscrupulous men. The fact that I was completely serene regarding this calamity had enabled me to join in the battle of a lengthy trial that ensued. This suit added six thousand dollars to our already heavy load of liabilities.

The ability of the improved part of the [current] ranch to negotiate and finance the required loan to close a transaction of this magnitude makes me more and more aware as the days go by of the resourcefulness and harmony that exist within the families of the Tracy Ranch. It is the wise and profitable use they have made of the parcels entrusted to them that has made possible this expansive enterprise.

Riding a Brahma Backward

In 1950, dining out was a rare treat for Jack and me, and we looked forward with much anticipation to dinner at the "Saddle and Sirloin," a restaurant that had recently opened in Bakersfield. As we entered we were impressed with the unusual rounded shape of the building and its tasteful decor. Rough-sawn redwood had been used for paneling, western photographs decorated the walls, and heavy wooden planks formed a circular stairway leading to a second floor. As we were shown to our booth, a very large photograph displayed behind the table immediately caught my attention.

Tilton Tracy riding a Brahma backwards circa 1936

There, mounted on a wildly bucking Brahma, rode a cowboy. Not only was he riding the fierce-looking animal—he was managing to ride him backwards! But the thing that fascinated me most was that the cowboy was the very image of my brother, Tilton. Either this *was* Tilton, or he had an identical double! Still, I wondered how it could be, for I knew of no association between our family and the strangers who owned the restaurant. Puzzled, I could hardly wait to see my brother and ask if this could possibly be a picture of him.

Returning to the ranch I learned that Tilton was not aware of the photograph, but sometime later he stopped at the restaurant to examine it. He was surprised to see that it truly *was* his picture, and he remembered it had been taken while he was doing a demonstration ride for the Wasco High School FFA (Future Farmers of America) Club. The photographer was a woman from a studio in Los Angeles, and she had taken the photograph while attending a rodeo at Wasco High School, where she had the good fortune to capture this spectacular image. We discovered that it was an award-winning photo, and that it had been used extensively on a western brochure. Now the mystery was solved: the owner of the restaurant had purchased this and other photos to add to the western theme.

This powerful photograph is of perfect composition; so perfect that no artist could have created a better one. The symmetry and balance of animal and rider holds the viewer in suspense as the wildly bucking beast, with nose to the ground and tail high in the air, lunges forward in an attempt to rid itself of the rider clamped to its back. We can detect a slight grin on the face of the rider as he, determined as the beast, resolves to ride it to the finish. The image of a teenage boy standing in the background with clinched fist adds to the scene's drama

In 1970 Tilton Tracy was lost to us in a tragic accident while tending herd on a cattle range near Coalinga. As the years went by, the "Saddle and Sirloin" closed, and the property changed hands more than once. The photograph of a young man riding a Brahma backwards had been almost forgotten; then one day Jack stopped by the old restaurant building and inquired about the photograph. The owner gave him permission to search for it and there, stored in a rain-soaked attic but only slightly yellowed by time, Jack found and recovered the picture.

It now hangs in a place of honor in another cattleman's residence, the home of Tilton's son, John, and John's wife, Donna.

There, frozen in time by the lens of a camera, rests the image of a young cowboy riding a Brahma backwards.

Pests

Flies

"It stands out like a fly in a pan of milk!" I still remember hearing mother's sister, our jolly Aunt Hazel, make this remark. In earlier days, pans of milk were set out in order to allow the cream to rise for making butter. Invariably, a few flies would become trapped on its surface. No record of our lives on Jerry Slough would be complete without including the troublesome insects that plagued the life of the housewife.

Over the years, inventors have tried various means of dealing with the common housefly. One method standing out in my childhood memories was "fly-paper." A piece of paper about 12" x 18" and coated with a sticky yellow substance was set out to trap the flies. As the flies conspicuously collected, the flypaper was not a thing of beauty! However, the flypaper did provide my teenage brothers, Darrel and Cecil, with a diabolical idea. They cut two small pieces of flypaper and stuck them to the hind feet of a cat. Well, it seems kind of mean to tell about it, but I must confess that it provided the most hilarious form of entertainment imaginable. As the cat did his best to free himself from the sticky paper, he would alternately shake each hind leg in a frenzy of irritation while we, convulsed in laughter, nearly rolled on the floor watching the cat's gyrations.

Stinkbugs

As sure as the ducks would fly in the fall, so would the brown stinkbugs! They arrived in great numbers and, wherever congregated, permeated the air with an obnoxious stench. A stinkbug could be hiding beneath any inanimate object in any annoying spot: between the sheets, under the bed, or in the coffeepot. I remember one Thanksgiving when two people were required to stir the turkey gravy; one stirred while the other shielded the pan with a lid, thus preventing the stinkbug from seasoning the Thanksgiving gravy.

Mosquitoes

Spiders were a year-round nuisance to the housewife (they still are), and crickets paid their annual visit. But king of all insects, the mighty mosquito reigned supreme! There are two species, night flyers and day flyers. This enables them to stand guard around the clock. [I have read that it was the mosquitoes that prevented the native Indians from establishing permanent settlements in the area.-NFT]

When we arrived back on the ranch in 1942, the mosquito population was in full swing and our babies wore angry welts, victims of their poisonous needles. Out in the fields, we would observe a gray cloud over the head of Waynie, our one-armed irrigator. Even his gauze mask could not deter them from their malicious intent.

Not until 1949 when Tony Wegis and Jack Frey managed to enlist the help of the Kern County Mosquito Abatement program did we get relief. At the present time, forty-five men in pickup trucks patrol the county seeking places where mosquitoes are likely to breed. In order for us to enjoy a mosquito-free environment, these men plant mosquito fish and spray oil on standing ponds where needed. Today, when I hear someone complain about one or two pesky mosquitoes, I can barely conceal my annoyance and keep from delivering a lecture on what it was like in *"the good old days."*

THE ARMY WORM

Army Worm 1968-1998

By nineteen hundred and sixty-eight,
In our new home we celebrate.
But we suffer, on this occasion,
An Army of Worms: a worm invasion!

He considers himself a handsome fellow,
This army worm of green and yellow.
Designs composed in bands of black
Decorate his pudgy back.

But underneath his colors bright,
He hides a stomach purely white.
His brilliant coloring tends to be
Of psychedelic intensity.

In legions of ten thousand troops
Upon our step he leaves his poops!
He comes right in, beneath the door
And then he craps upon the floor.

He soon upon my kitchen walls,
In undulating rhythm crawls.
When he is there for all to see
He gaily thumbs his nose at me.

At times we slow his sure advance
And step on him at every chance.
Of course, this makes an awful mess
But truth to tell, I could not care less!

We fill a trench with pesticide,
But nothing seems to stem the tide.
When thirty days have finally past
The nasty worm is gone at last.

Thirty years have come and gone
Since the army worm's withdrawn.
An we're as happy as can be
To celebrate his anniversary.

In nineteen hundred and ninety-eight,
President Clinton almost met his fate.
Now at this time we're glad to see
All worms like him are history!

GOOD TIMES

Buttonwillow Chamber Presents...

The First Banquet

In 1952 the Buttonwillow Chamber of Commerce held its first annual banquet. George Parker had just opened the first bank in Buttonwillow, and he and three other movers and shakers—Dick Adams, gin manager, Ford Williams, implement dealer, and Darrel Tracy, Postmaster—instigated the banquet, setting a pattern that has been followed for over fifty years.

The purpose of the occasion was to elect officers for the chamber, promote community spirit, and raise money for The March of Dimes polio fund. The dinners cost $2.50 a person. Voting could not have been more aboveboard: votes cost 10¢ each, and people could buy as many votes as they wished!

Joe Taziola

Darrel asked me to create table decorations for the dinner, and because 1952 was an election year (Adlai Stevenson was running against Eisenhower for president) we were provided with a theme for the evening—donkeys and elephants adorned the tables. The ladies from the school cafeteria prepared the dinner, and Joe Taziola was voted first Honorary Mayor of Buttonwillow. A feeling of good fellowship prevailed throughout the evening as spontaneous speeches erupted from the floor. The banquet was a success, and the idea caught on. Soon the social life around the little town of Buttonwillow improved considerably.

The Mayor's Race

In February of each year, five candidates were selected (or induced) to run for Honorary Mayor of Buttonwillow. Each candidate promoted money-raising events, and each dime counted as one vote.

A campaign manager was selected and, on the evening of the banquet, he was allowed to give a three-minute speech on behalf of his candidate. Standing on a small raised platform, the candidate simply looked on as his manager entertained the audience with his humorous jibes. Over time, little skits that often proved to be quite spectacular accompanied the speeches.

The Play's the Thing

The following year, 1953, my sister Frances and her husband Skip put on their first production in conjunction with the banquet. As a projectionist for Fox Studios, Skip was familiar with show business and he and Frances became partners in producing the show. For the next ten years, a large portion of their time would be devoted to this activity.

After writing a script, Frances and Skip would first design, make, or borrow costumes; acquire props; plan stage scenery; rehearse the cast; promote the event; and attend to the other myriad details.

Show produced by Frances and Skip

At first, the shows were quite amateurish, but a seed had been planted and people quickly responded to this community effort. Frances called upon members of the Italian community with whom she had attended the Swamp School as a girl. Some, like Jennie Riccomini, were shy and reluctant at first but eventually became permanent members of the cast.

Pantomime

None of the Tracy offspring had inherited musical voices but Frances discovered that by depending upon a record player, she could perform without actually singing. Recalling the pantomime she had first encountered at UCLA, she decided to try her hand at lip-syncing. After selecting a suitable song and memorizing it, Frances practiced for hours until she had perfected every gesture. She usually performed her numbers during "curtains" between acts, and they were excellent. These acts added a touch of professionalism to our amateurish productions.

The 1953 Banquet, "Come Old Fashioned"

Frey Family, 1953

Dick and Leota Adams

Ford and Doris Williams

The Old Grange Hall

From 1952 through 1957, the chamber affairs were held in the old grange hall east of Buttonwillow near the cotton gin. In the beginning I worked on posters and stage decorations by myself, and I remember painting roses on stage wallpaper at two in the morning. I knew my husband thought I was crazy, but he was too busy with his farming operation to argue about it.

Soon people began volunteering their help; Doris Williams was one of the first. Doris was married to Ford Williams who was the distributor for the Case Tractor Company in Buttonwillow, and Ford had just received the "Eagle Award" for selling the most tractors in the company.

The grange had been deserted for several years and was filthy. Together, Doris and I scrubbed the dirty toilets, reactivated the ancient refrigerator, and swept the mouse droppings from the spacious floors. For the next few weeks, the hall would become our second home.

One particular incident stands out in my mind. Ford was running for honorary mayor and I had been working on a large poster for him; I worked hard to finish it during one of my nocturnal sessions. Because Ford had a distinctive profile and an irresistible mustache, I managed to produce a rather close likeness.

The poster depicted Ford, in all of his glory, mounted atop one of his Case tractors. He appeared to be whizzing along at top speed, and above his head soared an Eagle carrying a sign: "Eagle Award, Vote for Ford!" For a slogan, I had printed: "T'will Take a Case to Win This Race!" When Doris arrived the next morning and stood gazing at the enormous poster of her husband, her eyes welled up with tears. Apparently I had captured the spirit of one of Buttonwillow's most admired citizens. I felt rewarded for my effort.

First Woman Mayor

In 1957 John Jelletich was appointed as a judge in Buttonwillow. He and his attractive wife, Sammye, quickly joined in community activities. Blonde and vivacious, Sammye was chosen to run for honorary mayor. This was the last mayor's race to be held in the old grange hall and, posing as a sultry Marilyn Monroe, Sammye became our first woman mayor.

New Auditorium

In 1958 a new gymnasium/auditorium was built at Buttonwillow School. From then on, the banquet and mayor's race would be held there. Although we were excited about moving into these new, spacious accommodations with all of the latest sound and lighting equipment, we wondered if we would lose some of the fun and spontaneity we had enjoyed at the grange.

Our Buttonwillow banker, George Parker, warned us that it would be difficult, if not impossible, to sell tickets for the 500 seats now available. After all, the price per ticket had been raised to $3.50! To avoid an empty feeling at the event, Frances invited the Horseless Carriage Club to attend, gratis, and to park their antique autos around the interior of the hall. To our surprise, the event was a sell-out and, in the end, we had to park the antique cars outside. We ran out of dinners for the club members!

The theme for this show was old time movies, and the title was *The Uncovered*

Wagon. Although Sammye Jelletich was now several months pregnant with her first child, she was a good sport and agreed to run for mayor a second time. She made a wonderful subject, and we had a lot of fun creating her posters. We painted her dressed in a fringed doeskin maternity outfit with "2-4-1" printed on the costume. Playing up her outgoing personality, dimpled smile, and fluent speech, we pictured her on a platform that said "Freedom of Speech." Her slogan was: "Vote for Sammye, She's a Barrel of Fun; Shows Her Dimples to Everyone." In spite of our artistic efforts, Dante Giannini became honorary mayor for that year and the year to follow.

A Roman Holiday

In 1959 Frances and Skip produced what turned out to be their best show. The theme was *A Roman Holiday* and the title was *Ben Hur Rides Again.* Who can forget Gladys Delfino and others with their legs painted purple up to their knees stomping grapes for wine! Or the handsome Judge John Jelletich in his robes, weighing the scales of justice while Cecil Tracy's Nero played the harp as Rome burned!

Can you believe that Ford Williams and Tony Andreotti actually built Roman chariots for this occasion? Pulled by Kenny Ringsdorf's beautiful little ponies, these two "Roman gladiators" circled the town and staged a chariot race on Main Street in Buttonwillow. Television cameras from Bakersfield rolled as "Roman" citizens in full costume posed for the event! What fantasy! What fun! What a great time was had by all!

Mayoral Skits

After moving into the spacious new school auditorium we had to increase the size of our posters. A twenty-foot ladder was used to attach them to the walls. In addition to the posters, slogans were carried around the hall on ten-foot poles. For Dante Giannini's campaign, we played upon his fondness for gambling by painting a square box white, adding black dots to simulate a gigantic die, and placing it on top of a pole. One side read, "Handsome, Young and Debonair," while the other side read, "Vote for Dante, He's On the Square." Dante wore the mayor's hat two years in a row, 1958 and 1959.

We had a Cuban theme in 1961 that we entitled "El Toro Grande." "Fidel" Osborne was one of the candidates, and we had just been seated when an ear-splitting blast nearly knocked us off our chairs. A group of Fidel's revolutionary soldiers blasted their way into the auditorium. In their full beards, authentic Cuban Army uniforms, and blazing pistols, they nearly scared us out of our wits! Feathers later flew above their heads as they staged a chicken-plucking episode for our entertainment. That was the year I was elected mayor.

Another skit that stands out in my mind is the one that took place in 1976; the year Dante Giannini's son Dennis threw his hat in the ring. This time it was not guns that startled our senses; instead, it was a group of Hell's Angels roaring into the auditorium on motorcycles. Dressed in black leather jackets, full beards, and earrings, they created quite a sensation. With their engines revved up to an earsplitting pitch, they circled the auditorium's interior. Dennis won the crown that year.

New Talent, New Friends

After a decade, a new infusion of talent arrived. In 1962 I once again took up my role as art instructor at Buttonwillow Elementary School. As I was looking out the window in the art room, I spotted a small car with the head of a huge dog protruding through its roof. Our new music teacher, Tom Anthony, his wife Pat, and their Great Dane occupied the car. They had just completed a long trek from New York to Buttonwillow in their little Volkswagen with a sunroof to accommodate their animal.

Pat and Tom were a handsome young couple. Slight of build and immaculate in his Ivy League attire, Tom always wore a dark suit, crisp white shirt, and tie to school. It seemed unlikely that these sophisticated New Yorkers would become close friends to a country school art teacher and her farmer husband, but we were to enjoy many happy times together.

When preparations for the banquet rolled around, Pat and Tom joined the crew at the grange. Tom was talented in art as well as music, and he created many of our best posters. Together we laughed and painted away as we thought up jingles and complimented ourselves on our cleverness.

The Millennium

We are now in the twenty-first century. Fifty years have passed since that first chamber of commerce dinner in the old grange hall. Finally outlasting its usefulness, the building was set aflame and burned down. An enormous edifice has replaced the old Farmer's Co-op Gin, just west of the grange. In 1973, Carlton Hooper became the manager of the Farmer's Co-op Gin and, under his careful supervision, the new, state-of-the-art gin was born. In time it would house a second gin and a cottonseed storage unit. Because it occupied such a prominent space, it was dubbed the "Hooperdome" in honor of Carlton Hooper.

In the early seventies, cotton modules replaced the cotton trailers that were used to transport cotton to the gin. Each module is equal to approximately fourteen bales of tightly compressed cotton. It is then covered with tarp to protect it from the elements and stored in the field before it is transported. The cotton grade is consequently improved and the tiresome bottleneck at the gin yard is avoided.

At an appropriate time, trucks with mighty engines and specialized equipment arrive to slide beneath the giant modules, scoop them up, and deliver them to the gin. These "box cars" of white lined up and silhouetted against the sky create a most spectacular sight!

Over time, hundreds of obsolete cotton trailers have been left to rot beneath the sun and rain along Highway 58, just east of Buttonwillow; mute testimony to the wheels of progress and a reminder of a bygone day.

For twenty-seven years, Carlton Hooper, with careful expertise, guided the business of the Farmer's Co-op. In 1990 his son, Michael Hooper, took his place. Michael continues to guide the destiny of the Hooperdome today; like the cotton trailers, Carlton has been replaced by a newer model.

The Author's Photos and Posters Supporting the "Mayor's Race," 1954 – 1967 (Proceeds to the March of Dimes)

1954 – Old Time Circus

Shooter Frances JB

1955 – Old Time Buttonwillow

Goodspeed with wife Hazel

Postmaster Darrel Tracy in a postage stamp hat

1957 – Bridey Murphy

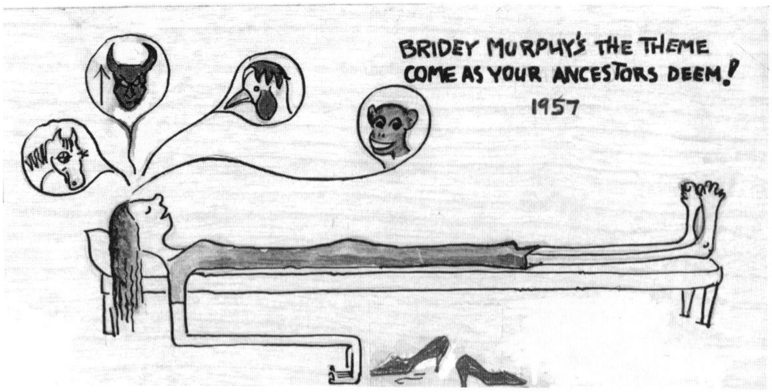

Poster for restaurant owner
Louie Ricci

1958 – The Uncovered Wagon

Banker Dick Mundy

Sammye Jelletich

Dante Giannini

Authentic wagon

Moving the wagon onstage

1959 – It's in the Bag

Banker George Parker

Dante with his sign

Martha and
Olympia Torrigiani

Ford Williams

1960 – A Roman Holiday

Dick Mundy poster

Roman chariots…

and real horses

Jackie Becker, left, Jenny Riccomini, Lilly Toretta, and Gladys Delfino

Martha Frey

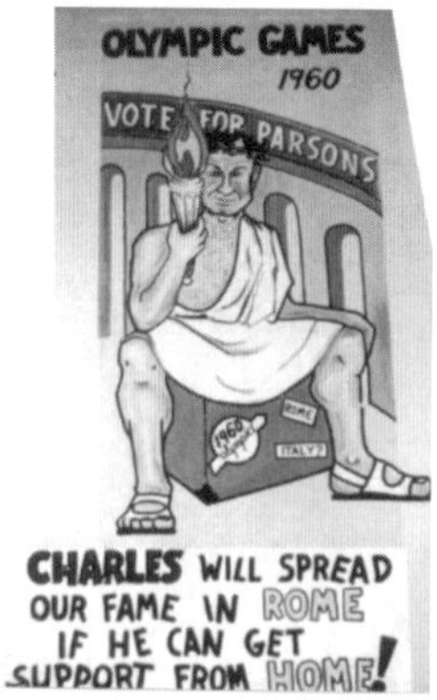

Charles Parsons

Cecil Tracy

Frances and Skip look on

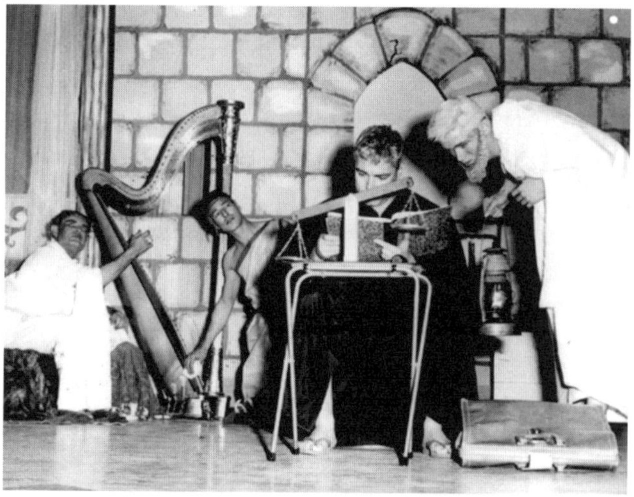

John Jelletich as judge and Cecil Tracy as Nero

Poster on roller skates

1961 – El Toro Grande

Othar Young

Wally Cannon

Roy Romanini

Vivian Wegis and Martha Frey

Poster for Martha Frey

Poster for Othar Young

Frances Selvidge and
Mayor Martha Frey

1962 – The Spell of the Yukon

Table decorations by eighth graders

Alda Kemp

Dick Mundy

Roy Belluomini

Jim Freiburger

Lee Isola

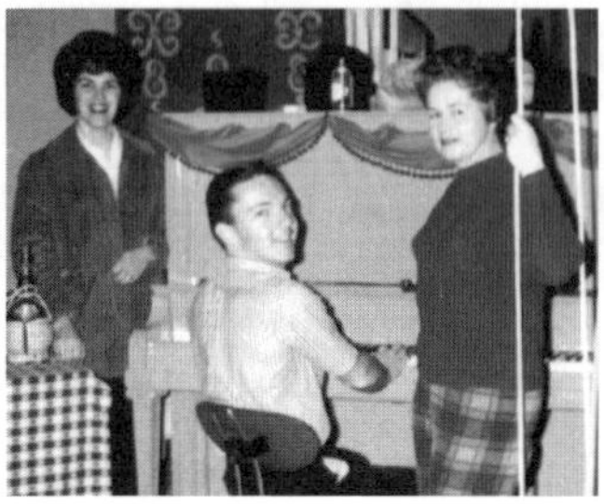

Tom Anthony in center

Alene Parsons

Tom Anthony

Dolly Hoggard and Julia and Bruna Belluomini

1964 – The Man From Buttonwillow

Wally Schroeter

Olivia Antongiovanni

Vince Bittleston

1965 – How Buttonwillow Was Won

Ernie Ghilarducci

Patty Page

Dutch Houchin

Art Torigiani

Patty Page, Vivian Wegis, and Jean Mundy

Teacher Mary Young

Skip and Frances as Mammy and Pa Yoakum

1966 – The Moon by June

Dennis Giannini

Table decorations

Alene Parsons

Dutch Houchin

Dick Mundy

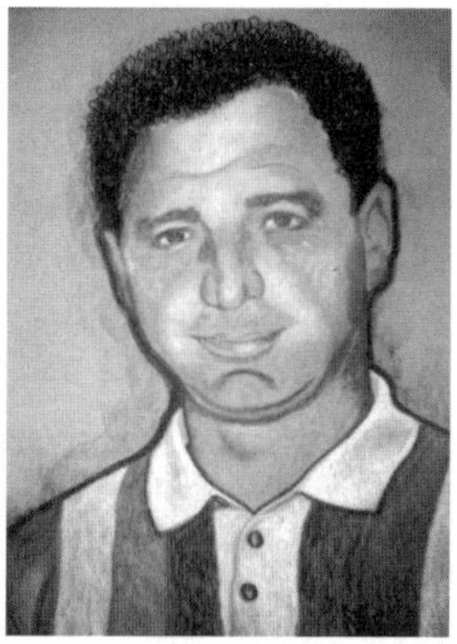

Roy Romanini

1967 – Mod Cotton

Dino Antongiovanni and
Dutch Houchin

Larry Frey as Hitler on an
authentic tank

With Fond Memories...

Kenny Wegis (at mike) impersonates Bing Crosby while Dick and Jean Mundy (left) and Nanette and Cecil Tracy sing, "I love those dear hearts and gentle people who live in my hometown."

Buttonwillow Chamber of Commerce and Agriculture
Presidents and Mayors to Present

YEAR	CHAMBER PRESIDENT	HONORARY MAYOR	LIONS CLUB PRESIDENT
1941			Rod Show
			Van McCutcheon
1942			John Overton
1943	L. E. Myers		Ken Calkins
1944	R. A. Benson		C. T. Kemp
1945	R. A. Benson		Francis Kimball
1946	Bill Orr		Bruce Crawford
1947	E. G. Boyd		C. A. Johns
1948	E. G. Boyd		Ray Blair
1949	Allen Bottorff*		Dutch Boetcher
1950	Allen Bottorff		Don Hughett
1951	Charles Parsons		Ford Williams
1952	Charles Parsons	Joe Tazioli*	Howard Jeffries
1953	Don Hughett*	Gino Torigiani	Allen Bottorff
1954	W. Darrel Tracy*	Oscar Baldwin	Charles Parsons
1955	L. Ford Williams	Jack Frey	Wally Schroeter
1956	T. C. Baldwin*	Darrel Tracy*	R. S. Hughett
1957	K. R. Ringsdorf	Sammye Jelletich	M. C. Becker
1958	George Parker	Dante Giannini*	R. G. Mundy
1959	E. J. Selvidge*	Dante Giannini*	W. C Peltier
1960	Howard Jeffries	Tony Andreotti	Jack Frey
1961	Kenneth Wegis*	Martha Frey	Bill Napier
1962	Otis Page*	Lee Isola*	Karl Muehlberg
1963	L. Ford Williams	John Cauzza	Tony Andreotti
1964	W. F. Schroeter	Wally Schroeter	Rowland Kemp
1965	E. E. Hoggard	Dutch Houchin*	Wally Cannan
			Woody Hoggard
1966	Jerry Buerkle	Dutch Houchin*	Rex McLeod
1967	Jerry Buerkle	Dutch Houchin*	Ernie Ghilarducci
1968	Rex McLeod*	Wes Selvidge	Charles Vogel
1969	Arthur Ghilarducci	Dutch Houchin*	Jerry Buerkle
1970	Arthur Ghilarducci	Rex McLeod*	Woody Hoggard
1971	Lyndon Payne	Bonnie Giovannetti	Rex McLeod
1972	Vince Bittleston	Robert Williams	Bill Williams
1973	Wesley Selvidge	Elmer Andreotti	Lyndon Payne
1974	Earl P. Williams	Patsy Quary	Joe McColloch
1975	Henry Alvidres	Michael Torigiani	Frank Riccomini
1976	Joe McColloch	Dennis Giannini	Lou Wolfe
1977	Gary Riccomini	Ernie Ghilarducci	Dennis Giannini
1978	Louis Ornelas	Blaine Willard	Ron Torigiani
1979	Anita J. Merlo*	Bill Clark	Chris Young
1980	Anita J. Merlo*	Shirley Hooks	Roy Pierucci
1981	Albert G. Pacini	Gary Franceschi	Blaine Willard
1982	Albert G. Pacini	John Tracy	Ken Torrigiani
1983	Ted R. Page	Barbara Blair	Cary Fowler
1984	Ron Torigiani	Jeff Roberts	Bill Clark

YEAR	CHAMBER PRESIDENT	HONORARY MAYOR	LIONS CLUB PRESIDENT
1985	Richard Young	John Cauzza III	Ray Pasquini
1986	Barbara Blair	Peggy Tazioli	Don Toretta
1987	Barbara Blair	Del L. Garcia	John Cauzza III
1988	Barbara Blair	David Parsons	Sam Ackerman
1989	Del L. Garcia	Phillip Williams	Jerald Page
1990	Albert J. Ghilarducci	Joe McColloch*	Allen Pierucci
1991	Sharon Page	David Torigiani	Michael Cauzza
1992	Leland Scott	Jim York	Cary Fowler
1993	Gloria Ramos Byrd	Mike Cauzza	David Parsons
1994	Gloria Ramos Byrd	Tim Thomson	Steve Cauzza
1995	Louis Andreotti	Albert J. Ghilarducci	Walter Parsons
1996	Louis Andreotti	Don Jeffries	Ron Cauzza
1997	Cynthia Torigiani	Regina Houchin	Julien Parsons
1998	Regina Houchin	Tim Dobbs	Paul McErt
1999	Regina Houchin	Brad Bittleston	Russell Martin
2000	Tina Neri	Junior Montemayor	Michael Tilton Frey
2001	Nina Neri	Albert Ghilarducci	David Lynch
2002	Mary Young Williams	Victor Rodriguez	James Torigiani
2003	Mike Torigiani	Russel Paterson	Mitch Milwee

MORE GOOD TIMES

German exchange student Erika Tronika is welcomed back by Martha and Fannie

Erika Tronika and Martha each pose for a photo

Martha in 1952 as the Queen of Mardi Gras

Las Vegas – Martha and Jack with Vivian and Ken Wegis

Las Vegas – The Jack Freys with the Sills and the McNamaras

At the Frey home

HAWAIIAN PARTY - BAKERSFIELD

Bill and Pat McNamara — Martha and Jack Frey — Margaret and Jim Bridges

Afro – Mardi Gras Ball – Pages of History
Bakersfield Country Club

Charming Frank Ghezzi designed our new home, which was completed in 1970. The builder was Al Pagni.

Martha and Jack at home in 1968

The home's interior in 1970

Jack in 1972 aboard *Pyrenees Queen,* which Jack fondly called, "a hole in the water to pour money into"

Jack Frey in La Paz

Jack's first marlin

Larry Frey in front of his Cessna 210

The mountain hunting "lodge"

Duck hunting Freys in 1990 are, from left, Larry, Jack R., Jack, Stephen, and Mike

Larry, Bobo, and Jack duck hunting in 1983

Martha and Jack's 35th wedding anniversary, 1973

JB Selvidge steers the "Red Okie." The "New Barn" in background was built by Will Tracy in 1915.

Bill Tracy and Larry Frey

Jack: "It was a very good year."

The Jerry Slough Bridge Club in 1979 is comprised of, clockwise from back left, Mary Lou Thomson, Mary Clifford, Clara May Lundy, Astrea Rogers, Martha Frey, Bessie Cassidy, Connie Schroeter, and Bee Blair

Bridge buddies visiting Shell Beach are, clockwise from front, Jo Reeder, Evelyn Koenig, Marie Yost, Patty Tkac, Diana Mosier, Martha Frey, and Patty Bass

Bridge buddies Martha Frey and Patty Tkac

Attendees of a tennis party hosted by Jim and Bebe Burke are, from left, Wally Schroeter, Bebe Burke, Ken Wegis, Martha Frey, Jack Thomson, Mary Lou Thomson, and Vivian Wegis

Guests at Jack Frey's birthday held in 1983 at Patty and Vick Tkac's Sunriver, Oregon resort home are, clockwise from back left, Jack and Mary Ann Snodgrass, Mary Lou and Jack Thomson, Patty and Vick Tkac, Ken Wegis, Wally and Connie Schroeter, Jean Mundy, and Jack and Martha Frey

Fishermen Ken Wegis, Jack Thomson, Jack Frey, and Vick Tkac display the daily catch

Vick Tkac and Jack Frey

Feasting at the Wegis' home in Sunriver, Oregon in 1989 are, from left, Ken Wegis, Mary Ann Snodgrass, Patty Tkac, Martha Frey, Mary Lou Thomson, Jack Frey, Vivian Wegis, and Vick Tkac

Jerry Slough potluck group members celebrating New Year's Eve at the Thomson's beach home are, clockwise from front left, Jack Thomson, Jack Frey, Pat Webster, Martha Frey, Ray Blair, Mary Lou Thomson, and Ken and Vivian Wegis

Jack's sister Effie and Lawrence Parker

Barbara Fritchie visits Jo Reeder at La Costa

Martha Frey becomes Barbara Fritchie in this patriotic Civil War poem. Through the years it provided entertainment for several Fourth of July parties.

CHRISTMAS AT THE RANCH

Will and his brother Theodore had built the original ranch house around 1900. By the time the Tilton Tracys, the Selvidges and the Freys returned to the ranch around 1942, the old home had undergone many changes. After her marriage, Fannie had installed closets and placed a large sliding window in the kitchen for ventilation. To avoid the pesky flies, she had screened in a porch at the east end of the house. The porch contained a hand pump and a long wooden trough.

With the advent of electricity around 1920, the electric pump, and inside plumbing, the dreaded nighttime journeys to the outhouse were eliminated, as was the loathsome chamber pot! Rooms had been tacked onto the original structure, and the kitchen was remodeled more than once.

By 1942 mother and Cecil had at last been able to free the ranch from debt, and Doug Miller was engaged to undertake a major remodeling of the family home. A long porch was enclosed on the north side, and then it was combined with the dining room and breezeway to form an enormous living room. The Ampico reproducing grand piano was installed in one corner, and an upright piano stood against a far wall. In the stairwell along the hallway was Cecil's pride and joy, a full-size, fully functioning jukebox—lights and all. A huge fireplace, big enough to stand in, was built into one wall; it was a double-sided structure, and there was a fireplace on the exterior side with a spit large enough to roast a whole pig or a side of beef. Mother covered the high living room walls with tasteful wallpaper, and the woodwork was freshly painted. With Cecil's beautifully designed staircase, sturdy ranch furniture, and new carpeting, the old place acquired a certain charm. It was in this room by the warmth of the huge fireplace that our family gathered to celebrate Christmas for more than thirty years.

Nancy remembers:

Although the immediate family was large and growing, it was further augmented by great aunts and their families, in-laws, extended families of spouses, plus assorted friends and "strays" that had been invited for the holiday. The turkey was roasted at Cecil's, and the women divided up the menu with each bringing a contribution for the feast. The families then loaded up their cars with gifts for the short trek to the home place. The gifts would then be heaped under the large Christmas tree in the cavernous living room, and the children would be pressed to perform their latest musical or reportorial accomplishments.

Once Fannie's grandchildren began to appear, ingenuity lent high drama to Santa's yearly visits. After dinner, and when the dishes had been cleared away, there would be a thump on the roof followed by the tantalizing sound of sleigh bells, first from one side of the house and then from the other. Finally there would be a glimpse of Santa in through a window and, at last, he would appear, to the wonderment of the older children and the terror of the younger. As the children grew older, they would make a studied effort not to notice that Uncle Darrel somehow always narrowly missed the climactic arrival of Santa Claus!

The toys that adorned our Christmas tree were not expensive, but the children never noticed the difference. There was a mountain of packages heaped below it or tucked between the branches, and Santa would enlist several "elves" to distribute the gifts to over fifty family members, extended family, and guests.

After the excitement of the gifts wore down, the eleven grandchildren would gather expectantly on a big upholstered mattress that lay in front of the oversized fireplace. The adults perched around them on the sofas and chairs, or simply sat on the floor, and the bear stories would begin. Cecil, Darrel and Tilton outdid each other in telling stories that were mostly humorous but always contained a thrilling twist of danger or mystery. When the atmosphere became saturated with anticipation, the bear would appear! It was Cecil, bent over with an animal hide thrown over his head and shoulders. He would engage the children in wild entertainment as the ferocious bear wildly chased screaming children throughout the big house, trailed by taunting children who were pursuing the bear. This continued for some time—until the first child cried. That was the signal for the women (who, for the most part, did not look on this game with much favor) to gather their dishes from the kitchen and for the fathers to round up their over-tired broods and try to sort out the gifts for the short drive home.

From the children's point of view, no family ever had more satisfying Christmases!

The author continues:

After Cecil's home burned to the ground in 1974, only the beautiful brick masonry remained. The fire, which was caused by ancient wiring, changed forever the nature of our family celebrations. By now the children who had run in terror from the bear were grown and had families of their own. Larry, Wesley, Bill, and John had joined the family business and had established homes on the ranch. For a number of years we gathered at one or another of their homes during the holidays for the "Children's Christmas Party."

By the time of Cecil's death in 1978, JB Selvidge had returned to the ranch after nine years in the San Francisco Bay area as a mechanical engineer for Peterbilt Trucks. He had gained valuable experience there, observing the operations of a large company, and now was in charge of the ranch shop, machinery maintenance and repair. In time, he would undertake the enormous task of sorting through his Uncle Cecil's lifetime collection of antique farm machinery and memorabilia.

In 1983, like Phoenix from the ashes, a new house arose on the site of the original family home. Desiring to preserve the handsome brick masonry of the former house, JB incorporated it into the design of his new home. This was quite a challenge, but one he surmounted with patience and ingenuity. By 1984, JB's new home was complete.

Christmas at JB's

Perhaps it was the memory of Christmas past as he was growing up; perhaps it was simply his inclination to honor the family tradition. Whatever the motive, JB Selvidge invited the family to hold their annual Christmas in his new house. For many years we enjoyed the hospitality of this unique and innovative home. This was no small undertaking—the family and friends had now grown to 40 or 50 in number.

The sights and sounds of Christmas surround us as we enter JB's new home. After we deposit our contribution to the bountiful harvest on the oval dining table, we can already feel the warmth of the great fireplace: the one that "Uncle Cecil" designed to warm the kitchen of one of his young brides. Some new arrangement or Christmas lights shimmer and shine as they cast their reflections on JB's collection of antique polished brass.

Tracy Ann Gill has been here earlier to add her creative talent to the décor and see that the house is in order. She, along with Susie and Donna, is largely responsible for helping with the seemingly effortless preparation as the party begins to unfold.

Rosemary Ryan is arriving; soon the rafters of this two-story house will ring out with her glorious music. Alternately, or with one hand on the piano and one on the organ, she will sacrifice her own time to share her talent with us. Soon carolers will gather around the grand piano to accompany her in song.

Like fish in a pond, a new generation of little cousins weaves its way in and about the house. Up and down the stairway they climb to explore the wonders of their uncle's attic and look down upon the folks below. Now the sound of jingle bells pervades the air and, just as their parents did a generation ago, they dash from window to window to catch a glimpse of Santa Claus. Now he is in the doorway! Soon each will take a turn upon his lap to receive a special gift. Devoted parents and grandparents look on in admiration as cameras flash to record the magic of this moment. Sometimes the littlest among them rebels and cries out at the sight of this strange creature with the white beard, but within a year or two, they too will come under his spell.

Christmas at Susie and Bill Tracy's

In the year 2000, as a new century takes its place in history, a new setting is set for the family Christmas celebration. Susie and Bill, with traditional hospitality, volunteer to hold the event in their home. With the expansion of a new generation and six grandchildren of their own, they set out to plan the event. For additional space they tent the long patio outside their dining room windows and place temporary heating inside for comfort.

Like all of the homes on the ranch, this one reflects its own unique personality. Upon entry, one comes face to face with a large and spectacular watercolor painting. I suspect that Bill might have obtained it on one of his trips to the south, where he often attended the meetings of the Cotton Council. Obviously the painting reflects the attitude of a bygone day when, long before the advent of the cotton picker, African-Americans harvested the nation's cotton by hand. The scene is set in a cotton field, probably on a southern plantation. A group of field hands looks on with quiet resignation as the white master places their burdens of toil upon the scales. With skill and great sensitivity, the artist has captured the pathos of this powerful painting.

An adjoining room houses Bill's collection of Civil War ammunition. Bombs, bullets, and shells of all shapes and sizes line the shelves, all neatly arranged and catalogued for inspection. It is Bill's hobby, and he is an authority in this field.

In the hall there is a gallery of family pictures; the record of three athletic boys and one pretty girl—a story of two families united into one. A story of trial and tribulation overcome, resulting in happiness beyond compare.

In the living room a giant Christmas tree nearly touches the sixcteen-foot ceiling. A chorus of Christmas carolers greets us from the Tracy mantle and here, too, we can feel the warmth of a brick fireplace. This time there is a Negro "mammy" resting on the hearth. In her arms she cradles a small baby as she sits in her rocking chair. This is only one of Susie's large collections of Negro Genera from the old south. Her home is alive with all sorts of her own needlework creations, paper mache pigs, singing nuns, etc. In fact, this home represents a beehive of creative activity, and we marvel at the energy and warmth it sheds upon the entire family.

Christmas 2002

At Christmas 2002 Santa visited the Tracy home a second time. The family gathering is much the same as it was so many years ago. Only the generation marks the change. Some have moved far away, but they will be back to share Christmas here on the ranch.

Todd and Annette Tracy married and, in 2003, added another Tracy to the family tree—a girl. That means of Susie and Bill's seven grandchildren, six are girls!

Afterglow

"Uncle Cecil" is no longer around to entertain the older ones, terrorize the younger ones, and antagonize the mothers. But Cecil Tracy is not forgotten; there are those who remember him well—most of these are grandparents now.

Now and then as we gather around the family tree, we are saddened as we recall the loss of a loved one, or gladdened as we welcome a new member into the family. But whatever the past has been, and whatever the future holds, each returns to his home with a feeling of family warmth and integrity.

Many "flowers" bloom in Susie and Bill Tracy's landscaped garden

Christmas Photos

Let the drums roll out, let Santa Claus call, just loosen your belts, there's a feast for all!

From above left: Tracy Gill decorates. Gary Gill prepares hors d'oeuvres. Host JB supervises.

From left: Madeline meets Santa. Nicholas Frey, Haylee Tracy, and Justin Frey greet Santa. Evan De Fraine is so anxious he forgot his shoes!

From left: Uncle Todd holds Haylee. Jack R. Frey and daughter Madeline. Uncle Ed De Fraine and Ian Koll. "Santa" with son Allen Tracy.

Party planner Susie Tracy poses with picture-taker Bill Tracy

Kevin and Stephanie O'Callaghan encourage a dubious daughter to sit with Santa

Tracy brothers Bill the politician, John the cattleman, and David the travel agent

Larry Frey, left, congratulates son Stephen on his engagement to Melissa Midzor

Martha smiles during a surprise visit by Tracy Koll (holding Madeline)

Evan, Ed, and Carla De Fraine

Back row left to right: Wes, Janet, Jason, and JB Selvidge. Middle: Casey, Gary, and Tracy Gill.
Seated in front: Frances Tracy Selvidge.

Standing left to right: Rebecca Lake, Larry Frey with granddaughter Madeline, and
Nancy and Randy Trahan. Seated: Jack and Martha Frey.

Isabel Tracy

JB Selvidge graciously hosts the family Christmas party. Talented Rosemary Ryan alternates between piano and organ for musical background.

BACK TO THE CLASSROOM

In 1962, with both of our children away at college, I was faced with the empty nest syndrome. There was an opening for an art instructor at the Buttonwillow Elementary School, and I taught here for the next five years.

The art room was a happy place. We experimented with many types of media but particularly enjoyed working with mosaics, a media that was so popular at the time. We gathered materials of all kinds for our mosaics. We collected colored bottles and smashed them; we used shattered windshield glass, coffee beans, white, pink and red beans, beads, seeds, and buttons; in short, we used anything that could be glued to a piece of plywood. Marbles became eyes for animals; broken coconut shells became texture for a monkey's coat.

Eighth grade students were allowed to purchase Venetian tile for creating their own coffee tables, and Buttonwillow's Ford Williams created wrought iron frames for them. We decorated old disk blades from abandoned plows and created murals that still hang in the hallways of the Buttonwillow School. We used plywood of various sizes for the murals, accompanied by many gallons of good old Elmer's white glue.

One of our first murals was "The Buttonwillow Tree"; a tree growing near the site of the old Miller and Lux headquarters that has been designated an historical monument.

It is fitting that another mural portrays Jelsey Romanini, Buttonwillow's remarkable kindergarten teacher who, for more than forty years, has helped get the community's little ones off to a good start. The mural portrays Jelsey and her little flock crossing an intersection with their STOP and GO signs. Fortunate was the boy or girl who had the privilege to carry one of those signs!

For the most part, our murals were created from the drawings of children from kindergarten to the sixth grade. Seventh and eighth grade students converted the drawings into murals. Cotton and the insects that invade it, watermelons, circus freaks, animals of all kinds, and anything else that struck our fancy became our subjects as we happily "glued away" in the art room.

Enter the Parakeet, Gaucho

Our daughter Nancy spent her first three years of college at the University of California at Santa Barbara. She spent her fourth year away in Boston, and she bought a little blue parakeet to keep her company. She named the little bird "Gaucho" in honor of the UCSB mascot. When she came home that summer, she brought Gaucho with her. He did not lose much time in charming his way into our hearts.

Nancy had taught Gaucho to say "Pretty boy," "Hello," and "Boston bird." Soon he learned to say "Buttonwillow boy," "Gaucho," and "I love you." He was a quick study, but sometimes he would mix his phrases together, so he might say, "I love you, pretty boy" or "Hello Buttonwillow bird."

His cage hung adjacent to the fireplace, but its gate was never closed. Gaucho was allowed to come and go as he chose which, of course, meant there was extra cleaning to be done in all his favorite spots. One of these spots was by the toaster, where he paced back and forth making friends with the handsome bird reflected in its shiny surface.

Another favorite spot was the top of the buffet, where he took possession of some smooth-surfaced gray rocks. Perhaps he thought they were eggs. At any rate, he claimed them as his own, and if I came too close to them he ruffled his feathers and scolded. One day when I dared to pick up a stone, he attacked my wrist like an eagle! He clung so tenaciously with his sharp beak and little claws that I was barely able to shake him loose!

A favorite pastime was watching Jack shave. Gaucho would station himself on the top of the medicine cabinet and wait for the performance to begin. Jack was using a safety razor at the time, and Gaucho was fascinated by its shiny surface. When Jack began to shave, the bird would descend from his perch, hop onto Jack's hand, and attack the razor. Angry shouts would be heard from the bathroom: *"Down Gaucho!" "Damn bird!" "Get outta here*!" There would be a brief silence, and then the entire routine was repeated until the shaving procedure was finished.

Nancy and Covell were married at the end of the summer, and when they returned to Boston in the fall I inherited "The Bird." Due to my fastidious nature, I was not too sure I wanted him but, as anyone who has ever owned a parakeet will tell you, these little birds can soften the hardest heart.

Little Mascot

And so it was that when I returned to school that year, I put Gaucho in his cage and we went off together. The art room was large with a high and spacious ceiling. When I released him from his cage, Gaucho sailed around the room and finally perched on the tall cabinet behind my desk. As he surveyed the area below, I could tell he approved of his new domain.

One day our principal appeared at the door of the art room to introduce a visiting official. As we conversed, Gaucho flew down and landed on my shoulder. Paying no attention whatsoever and acting as if it were perfectly normal for an art teacher to have a little bird prancing back and forth across her shoulder, I continued our conversation!

A Close Call

One day after selecting some colored paper from the tall cabinet in back of my desk, I had begun to close the cupboard door when I heard a faint sound at my back. It was barely audible, like a quick intake of breath. The sound stopped my hand from entirely closing the door—thank goodness!

Gaucho, noticing the open cabinet, had seen a chance to explore new territory. He was flying to the top shelf of the cupboard at the same instant that I was closing it. Several students, noticing that an accident was about to happen, reacted instinctively. There had not been time enough for them to shout, but that faint gasp combined with my sixth sense was enough to prevent Gaucho from being one crushed parakeet!

Neighborhood Kids

The Rodriguez family lived adjacent to the school. When I first arrived Mr. and Mrs. Rodriguez had three sons, but during the five years that I taught, it seemed as if there was a new little boy each year. (I say "boy" because the Rodriguezes appeared incapable of producing a girl.) Because the family lived so near, the school became a second home to the children, and we soon became well-acquainted. After school when the boys dropped by, I allowed them to explore the many wonders of the art room. Well-behaved, they never abused my hospitality.

Richard was the oldest of the trio, and I grew very fond of him. Richard was not a handsome boy, but when he smiled his radiant smile, one was sure to admire that set of prominent white teeth. Richard was apparently fond of me as well, for a few months after each new baby arrived, he would show up in the art room with the new little Rodriguez in his arms. Barely large enough to hold the baby and beaming his magnificent smile, Richard presented the family prize while his little brothers looked on with equal pride.

The McCormicks lived but a stone's throw from the Rodriguez family. Tom McCormick was vice principal of the school and he and his wife, Barbara, had three small boys. These little boys were just as Irish as their name, and the Rodriguez youngsters traveled with the McCormick kids in various combinations of three to five in number. I thought of them as the Little Rat Pack.

Barbara McCormick with her three boys and husband Tom. Tom was vice principal of the school, and Barbara was a teacher.

The Little Rat Pack

The Inevitable

Gaucho had been entertaining his audience of newfound friends for several months when the inevitable happened. One Friday afternoon as the last student left the art room, I saw Gaucho fly out the door! My heart sank, for I knew that our little friend could not survive long in the wild. Some larger bird or animal would make short work of him.

Because it was Friday, I was in a hurry to get to my appointment with the hairdresser. Gaucho had no sooner flown out the door than three members of the Little Rat Pack showed up. I handed Gaucho's cage to Richard and told him of our plight. "Please take the cage," I said, "and see if you can find Gaucho!" This was but a desperate gesture, for I had not the slightest notion we would ever see the little bird again.

As I settled in under the hair dryer to write my lesson plans for the following week, my mind kept returning to Gaucho. I thought how bravely he had fought with Jack over the razor. I remembered how this tiny little fellow had become David to my Goliath as he defended his precious stones. I felt sorry for the many children who would surely miss their little mascot.

Just then the door swung open and there stood Richard, beaming his wonderful smile and holding the cage with Gaucho inside! The rest of the happy pack was smiling, too, as they trailed behind. With relief and much appreciation, I handed each boy a half-dollar and they went happily on their way. The Little Rat Pack stood ten feet tall that day!

I later learned that our little heroes, in search of Gaucho, had spotted him in a tree. They set the cage down, and Gaucho lost no time in returning to his home. *But of course!* His precious rocks were inside! After that, we learned that we could walk outside with Gaucho and, as long as we held one of his stones, he would never leave our hand!

Gaucho Moves On

When I first assumed my teaching job I felt entitled to some household help, and Irene Bradshaw came into our lives. Years later, after I left teaching and Gaucho was at our house full-time, I felt a little guilty asking Irene to clean up after him. I decided it really wasn't fair to her, so one day I said, "Irene, if you know of anyone who wants this pesky parakeet, they can have him!" In her calm, quiet voice she said, "I'll take him." Gaucho had worked his way into her heart as he had worked himself into mine.

So Gaucho went home with Irene. Her husband worked in a harvest crew and often came home with seeds lodged in his hair. Irene told me that Gaucho enjoyed "harvesting" the seeds from his curly hair.

Gaucho lived happily with the Bradshaws for several years. By then he was getting on in bird-years, and one morning they found him dead in the bottom of his cage. His brave little heart would beat no longer, but his memory was to live on in the hearts of a multitude of children—and not a few adults.

GAUCHO

A Poem for Gaucho

Now here's a little story
That you may not have heard,
While Mary brought a lamb to school
Our teacher brought a bird!
We liked to call him "Gaucho,"
For Gaucho was his name.
But if you called him "Pretty Bird"
He answered just the same.
He looked about the art room,
For this was his new home,
And here from floor to ceiling
The bird was free to roam.
When students came to art class,
Gaucho showed no fear,
He flew down to their shoulders
And talked into their ear.
Oh how the children loved him,
And he was happy here,
But when they reached to catch him,
The bird was nowhere near!
When we set mosaic out,
Gaucho was ecstatic!
Casting seeds and beads about,
The bird became erratic.
In this paradise of pebbles
One thing was very clear:
This Santa Barbara mascot
Was now the master here!

With children off to college, Martha takes a position as art instructor at Buttonwillow Elementary. After being elected honorary mayor, she presents Superintendent "Ken" Wait with a Buttonwillow tree, while sister Frances looks on (circa 1961)

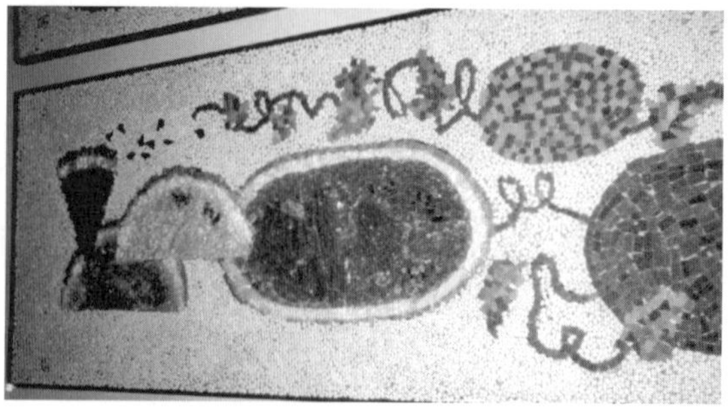

Shattered windshield glass overlaid the painted surface of this watermelon design. Tile and white beans completed it.

Eighth grade students' decorated plow discs

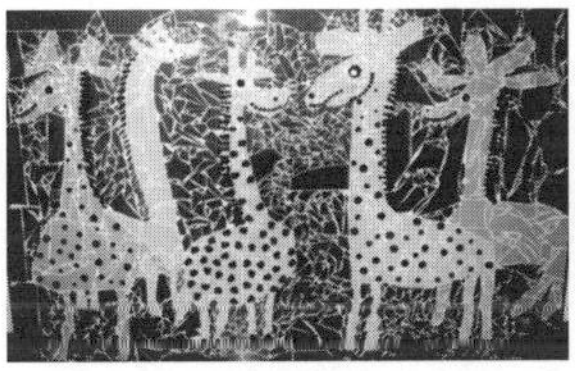
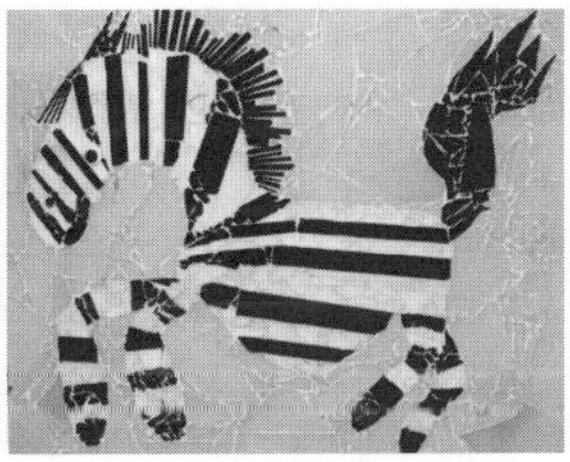

Windowpane mural animals were drawn by fifth and sixth graders; backgrounds were covered with broken glass, then completed with tile pieces

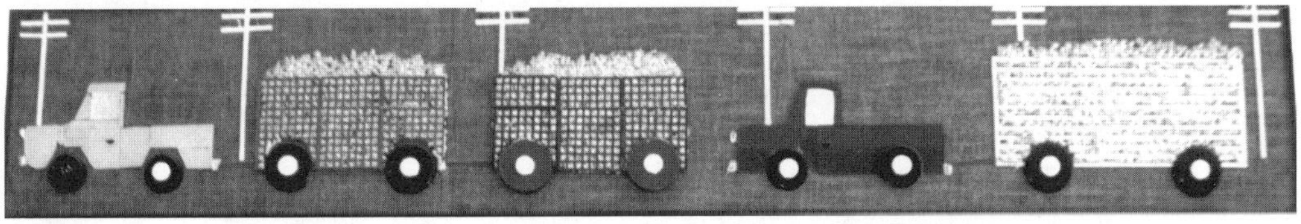

Burlap-backed cotton gin depiction

Cotton motif for coffee table

Circus

The Rodriguez and McCormick families lived right next to the school. The children traveled in packs ranging from three to five and were frequent visitors to the art room. They became my special friends.

Halloween parade

Jelsey Romanini is immortalized in this large 4' X 16' mural. It depicts her and her kindergarten class as they cross a school intersection complete with "stop" and "go" signs. Jelsey is one of those exceptional teachers that has dedicated almost 40 years to the profession.

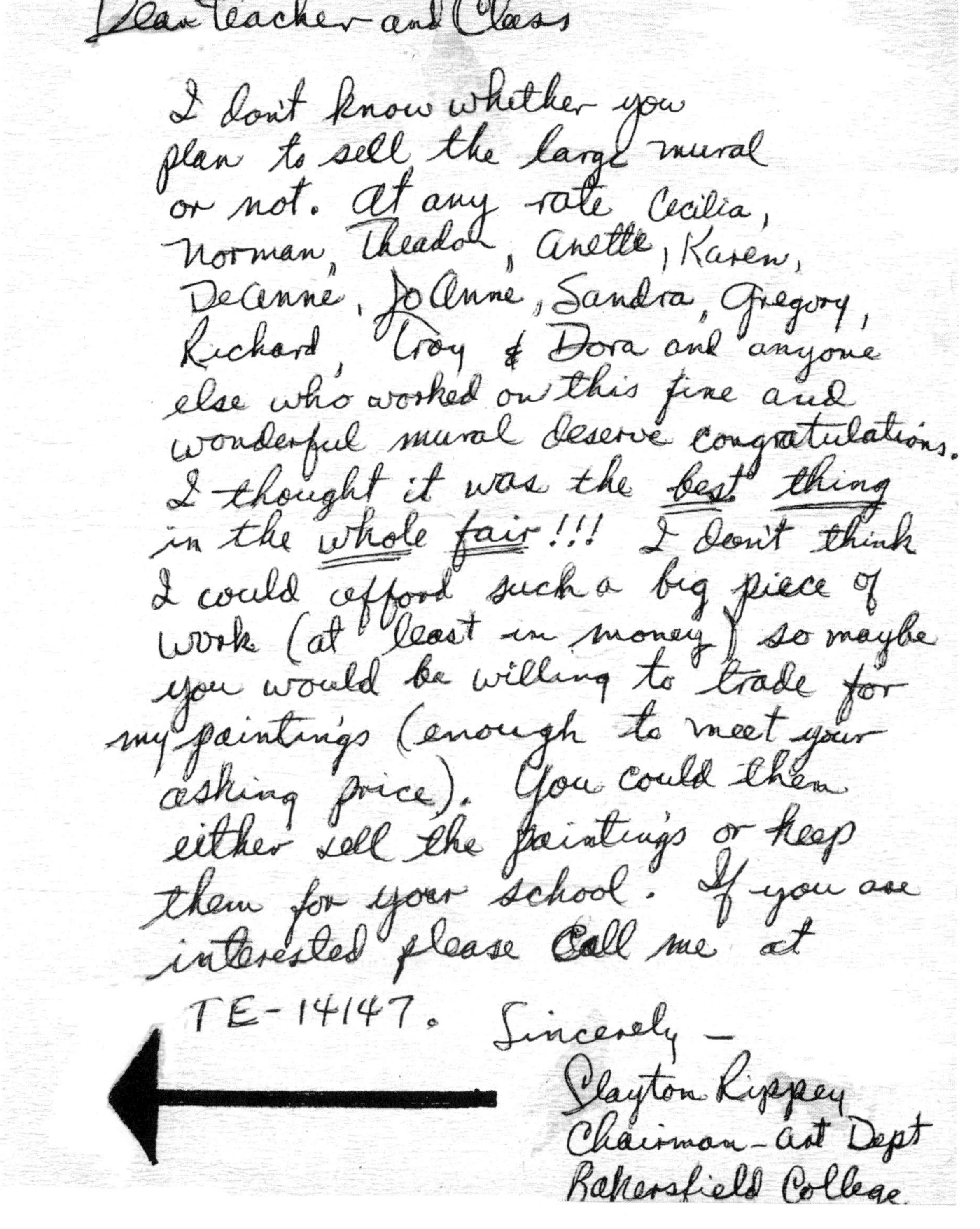

Dear Teacher and Class

I don't know whether you plan to sell the large mural or not. At any rate Cecilia, Norman, Theador, Anette, Karen, DeAnne, JoAnne, Sandra, Gregory, Richard, Troy & Dora and anyone else who worked on this fine and wonderful mural deserve congratulations. I thought it was the best thing in the whole fair!!! I don't think I could afford such a big piece of work (at least in money) so maybe you would be willing to trade for my paintings (enough to meet your asking price). You could then either sell the paintings or keep them for your school. If you are interested please call me at TE-14147.

Sincerely –
Clayton Rippey
Chairman – Art Dept
Bakersfield College

Outstanding Bakersfield artist Clayton Rippey paid tribute to this mural when he visited the fair in 1965. When the mural returned, we found this message written on a paper towel.

Transitions

Aunt Dee

Our gentle Uncle George Taylor, Aunt Dee's husband, patiently put up with the Tracy Tribe for many years. When he passed away in 1952, Aunt Dee was no longer capable of living alone in her large home in Bakersfield. Because she had so often sheltered me as a child, it was only fitting that I should now shelter her. We sold her old home at 2323 Chester Lane and, with the proceeds, we purchased a two-bedroom mobile home for her directly behind our house; not the kind on wheels, but a pleasant little pre-fabricated house with hardwood floors. Her familiar furnishings were set in place, and though she did no housekeeping, she was provided with a home of her own, and our daughter Nancy moved in to keep her company.

We shared our home with Aunt Dee for seven years—from 1952 to 1959. It then became necessary to place her in a rest home, and she passed away after a stroke in 1960. She is interred at the Union Cemetery with her husband, her parents, three sisters and a brother.

Aunt Dee with dog

Gentle Uncle George

Tony

Aunt Dee had taught school for most of her life and, after moving into the mobile home, she spent her days tending her dog, Tony. She had two dogs when she gave up her home, but since we also had a dog, Denny, I persuaded her to give up one. She kept Tony because he was the older of the two. He might have been old, but this did not prevent him from living to a ripe old age. When he arrived on the scene, he and Denny took an intense disliking for one another; this hostility continued throughout their lives. For the next seven or eight years, each time we returned from a trip these two old fellows staged a dogfight for us. Since they were too old to inflict much damage on one another, Jack and I didn't worry, but their behavior caused Aunt Dee much consternation. She believed if she turned the hose on them she could break up a fight. One day when she attempted to separate them, there was a sprinkler on the end of the hose and as it went round and round, Aunt Dee was doused with more water than the dogs!

At one time Aunt Dee also had a mother cat with a litter of kittens, and while she was away from the house Tony ate a few of the newborns. Of course Aunt Dee was very dismayed but, not wanting to believe Tony could perform such a horrendous act, she excused him by saying he had probably thought the little kittens were gophers!

In the end, old Tony journeyed down the slough in back of our house to visit a neighboring bitch. He must have been at least fifteen years old by then, and the trip was apparently too much for the old dog, for we found him dead on the banks of Jerry Slough, about halfway home. Although Aunt Dee did not outwardly accuse us of any wrongdoing, we always felt she was a little suspicious that we may have done the old boy in.

A Family Tragedy

Five Cousins at College

By 1965, five young men from Tracy Ranch were attending the campus at California Polytechnic School—"Cal Poly"—in San Luis Obispo; five cousins including JB and Wesley Selvidge, Larry Frey, and Bill and Jim Tracy. The older boys were about to graduate while young Jim Tracy was a sophomore.

The boys were living in town near the Southern Pacific Railroad tracks. One day they were enjoying a few beers when someone came up with the idea of "riding the rails." With each beer the idea became more enticing, and they soon decided to ride the rails to Santa Barbara, some 100 miles away. Unfortunately, the train they caught happened to be a work train stopping at numerous intervals along the way. They could have walked to Santa Barbara faster! A little wiser, they hitched a fast freight for the return trip.

But it was enterprising young Jim Tracy who thought of putting this means of transportation to better use. Since the Cal Poly campus lay a mile or two away, he hitched a ride to and from school each day. (Thirty-three years later, your author learned of this prank. One wonders what other mischief took place amongst this circle of cousins at Cal Poly!)

School was not all play, of course, for they were laying a foundation for a future on the ranch; JB majored in mechanical engineering, Bill in journalism, and the rest in agricultural management.

Losing Jim

As the boys were growing up on the ranch, each took a turn working as an irrigator. How well I remember the tug at my heart when my eleven-year-old son pulled on his rubber boots and armed himself with shovel and lantern to spend the night irrigating!

It was partly by this means that our son, Larry, and Kit and Tilton's son, Jimmy, were able to save enough to invest in a small airplane. During Jim's freshman year they bought a Cessna 120 that they kept in a hanger on campus. They used the plane to commute the hundred miles back and forth to the ranch.

Jimmy is twenty years old when, on July 11, 1965, one of his friends asks him for a flight to visit a girlfriend. She lives on a nearby ranch in the foothills, and there is an airstrip adjacent to her home. While the plane is being fueled at the San Luis Obispo airport, the boys phone to tell her of their plans. After she hangs up her father tells her, in no uncertain terms, that the boys are not to land on his strip. He explains that a landing is too dangerous because of a small hill at the end of the runway.

The girl tries to phone the boys, but it is too late—they have already left. The father then orders her and her brother to go to the airstrip and wave them off.

Jimmy is on approach as the kids vigorously wave him off. Smiling and waving that he understands, Jim applies power to abort the landing and pull out. Though he is able to clear the hill at the end of the runway, the plane crashes into the higher hill beyond. Jimmy is killed instantly; his friend becomes a paraplegic.

Jim Tracy Remembered

Nancy Frey Trahan:

Jimmy was an excellent pilot. Had they not signaled him, Jim would no doubt be alive today, taking his place alongside his brothers and cousins on the ranch. He was on a perfect approach and would have made a perfect landing. To lose this promising young man at the threshold of manhood was as cruel a blow as fate could offer his grieving family. How vividly I remember a comment made by his younger brother, John: "He was the best one of us." I agree with John. Jimmy was the one who would have assumed the leadership role in the next generation. He was intelligent, kind, humorous, and charismatic. I can't think about him without shedding tears for our loss.

Bill Tracy:

James Tilton Tracy was buried July 14, 1965, at Wasco Cemetery. When we lost Jim, mother's concern wasn't for her crushing loss, but for what effect the funeral would have upon his friends, as it would be the first funeral for many of them.

John Tracy:

To this day, after being hardened to life's emotions, it is still hard to talk about my brother. A life so full of promise cut short.

Jim was the best of the lot of us boys. He had better grades, faster cars, excelled in sports, and was a pilot. Although we got along fairly well when we were younger, we became much closer a couple years before his death.

I thought he was a little too cocky so, to teach him a lesson, I was going to pick a fight and "cold cock" him. (I have to be excused for the language here, but this is what happened.) I got the fight going and reared back to fire my haymaker. He hit me three times before I could unload! It hurt so bad, I said, "*F... it,* I'm never fighting you again!" I didn't, and we became good friends.

He accompanied me to Cal Poly where he showed me around. He was playing football, and he was well-recognized and liked. We were to room together when I entered as a freshman; I also majored in farm management. The painful loss of Jim, coupled with the unfamiliarity of a new school, was somewhat eased by his many friends who took me under their wings and showed me the ropes.

David Tracy:

Our brother Jim was the catalyst among the four Tracy brothers. He was always active and could devise a football game even if he had only his two younger brothers to play with. Of course, he played quarterback on both sides!

Jim was younger and smaller than he wanted to be and, always striving to keep up with the "big" kids, he became a great competitor in high school and college sports. He worked hard on his weight bench to pump himself up in order to compete with bigger and stronger athletes. He didn't want to miss out on any activities.

Cars and airplanes both held special attraction for Jim. He spent many hours working on his '59 Chevy Impala, and he installed what was probably the biggest engine of the time—a 409. By the time he got to college, he had traded up to a Corvette. Not to be confined to the ground, he and Cousin Larry bought a used airplane. He enjoyed flying, and it was a special treat to go flying with my big brother!

The last time I saw my brother was while driving out of the ranch headquarters as he and a friend were heading to our airstrip; they were planning to fly over to the coast and visit another friend. There was the usual smile and wave; this is an image frozen in my memory, his expressive smile and youthful vitality. Jim did not survive the plane crash. He remains forever frozen in time, a strong and energetic youth, never to grow old.

Still Another Dark Day

During the night, May 4th, 1970, Jack and I were awakened by the telephone. It was around midnight, and we wondered who could be calling at such a late hour. When Jack answered the phone, I could tell immediately by the tone of his voice that something was radically wrong. "Slim" was on the phone. Slim, a crippled cowhand who worked for Tilton on a ranch that Tilton was leasing in the foothills above Coalinga, told Jack that he had been preparing dinner around five. After telling Slim that he would return shortly, Tilton left to check on some cattle. Now, in a state of near exhaustion, Slim was calling to tell us that Tilton had not returned, and that he had been desperately searching for him. We were greatly alarmed. Only the previous week, Tilton had been gored in his leg by a boar.

Several years earlier, Tilton had undergone back surgery and was unable to ride a horse. During that time he had acquired a Honda motorcycle. (I was surprised, one day, to find Tilton at my front door on his Honda. He was still wearing his brace and, clinging to his back like a Koala bear, was our little two-year-old grandson, Michael Tilton Frey! This combination created such a good picture that I ran to fetch my camera; hence we posses this treasured photograph of Tilton on his Honda with his little namesake firmly attached to his back—*see Color Section page R*)

After his back healed, Tilton found the Honda to be still useful; it was quicker to jump on the Honda than it was to saddle a horse. Consequently, whenever it suited his purpose, he continued to check his cattle from the back of a Honda. This time it turned out to be a fatal decision.

After Slim's phone call, Jack phoned all of the partners and notified the sheriff. There was little anyone could do in the middle of the night, so helicopters were alerted to be ready to search at daybreak, and the men set off for the mountain ranch. After some consideration, I decided not to call Kit until after she had arisen; then Frances and I went over to explain the alarming circumstances.

In spite of the tragedies in her family, our sister-in-law always maintained a positive attitude. We tried to share that optimism now, but in truth we were extremely apprehensive as the three of us drove to the little cabin on the mountain ranch. When we arrived we were greeted by an eerie silence; broken only by the ticking of a clock. We had been there only a short time when we saw four men slowly approaching the cabin. We could tell by their demeanor that the news was bad.

Tilton's body had been located at the base of a thirty-foot precipice. Entangled in the wreckage, he was still clinging to the handlebars of the motorbike. He had fallen from the sloping ledge above. The tracks in the loose shale indicated that as he wheeled the bike around to leave, it had flipped him over the edge backwards. No doubt he had paused to enjoy the sunset from this vantage point many times before; this would be his last.

Kit, Frances, and I drove home in silence. No tears were shed. Frances and I searched in vain for words to comfort our sister-in-law who had already experienced such tragedy, but we could think of none. With heavy hearts we returned her to her empty house to grieve in her own way. Her beloved cowboy was gone, her world shattered. This was the darkest day of her life.

Bill remembers:

Dad had terrific stamina. Whether it was work or play (hunting), he pushed himself to the limit. A few months before his death, we were clearing brush at the Avenal Ranch. By the time we stopped for lunch, the first work break of the day, though he had my "tongue hanging out," he lamented that he couldn't "push" himself like he once could. He was fifty and I was only twenty-seven, fresh out of the army and supposedly in fighting shape!

After we located dad's body, we had to break the news to mom. With tears flowing, she said, "I possessed him as much as anyone ever could." They had one of those rare marriages; she knew that her rival was not another woman, but his freedom. (To her cousin Cliff with whom she corresponded freely, Kathryn would later write: "Perfect marriages are not measured in years, and ours was the perfect marriage.")

John remembers:

Our father's death in the summer of 1970 was a shock to us all. Dad and his brothers, Cecil and Darrel were, by nature, fearless. This was no doubt due to their upbringing in a harsh environment. The weak did not survive. But each of them had a gentle, understanding side, plus a genuine love for people. Dad was fit of mind and body, but his fearless nature may have been his undoing.

I had been drafted into the army and was in Los Angeles with my wife, enjoying a vacation before reporting to boot camp. When I got the call that dad was missing, I headed to Coalinga to look for him. By the time I arrived, he had been found. Now I was in charge of the cattle. With cattle in three counties and summer gathering and shipping in full swing, I started back to work. We were so busy that I could not attend my own hearing for a hardship deferment from the draft. Uncle Jack Frey made a few calls, and I was immediately granted the deferment.

Dad and I had often discussed the cattle business as we worked together on a daily basis. With the help of many of dad's friends, the void in my experience was filled in. Most of dad's friends accepted the changes I made, but a few didn't care for the transition. This often happens when a young man becomes boss. Over the years, the bulk of these friendships became more solid, and dad's good friends became mine.

David remembers:

One of the toys dad had towards the end of his life was a Honda "90" trail bike. He could scout cattle on the mountain ranch faster with it than on horseback. I once went with him into the Sierra by trail bike, covering in a few hours what would have taken days by horse. Unfortunately, he had a small accident, cracked a rib, and the rest of the trip was rather painful for him. He was on this same bike a couple of years later when he had the fatal accident while scouting cattle on the mountain ranch. Like our brother Jim, dad left us with a memory of a vigorous man who would never grow old.

The Fire

One afternoon in 1974 as I was driving home from Bakersfield, I noticed a column of smoke arising upon the horizon. It seemed to be coming from the vicinity of our family home, now Cecil's residence. As I drew closer I became more alarmed, until I suddenly realized that it truly was the old home place. By the time I arrived, the house was completely enveloped in flames. As I stood and listened to Cecil's supply of ammunition blasting off, I watched the entire roof cave in, and I felt a little like Scarlet O'Hara watching Tara burn. I thought of the many changes that the old home had undergone through the years, from the old hand pump on the back porch where my mother had once doused my head for being sassy, to the enormous fireplace that had been added in later years. I thought of the beautiful staircase that Cecil had installed; never mind that it led only to an attic filled with junk. I thought of the many Christmases that two generations of children had celebrated there, and Uncle Cecil, playing "bear" and sending the children screaming through the house. To them, this was even better than Santa Claus!

Cecil happened to be away when his home burned—probably on a three-day tryst with a lady friend—and by the time he arrived home, his house lay in charred ruins with only a little patch of smoke here and there. Gone were all of his treasured antiques: the vintage branding irons, the fancy bridle bits, the spurs, his collection of guns. Gone was the enormous mahogany table that he had retrieved from the old Bakersfield courthouse, Aunt

Dee's treasured china cabinet, and one could still see the charred and twisted frame of a round, six-foot-mosaic coffee table that had been mine.

There was one small section of the house, however, that did not burn: the room in which I had discovered the myna bird many years earlier. It now contained a large charred freezer. A few days later a fireman was picking through the ruins, investigating the cause of the fire. Imagine his surprise when he opened the freezer and came face-to-face with a rattlesnake, coiled and frozen into striking position!

Cecil faced this tragic time in his life with great courage. He did not place the burden of his grief on others but accepted it philosophically and went forward with the rest of his life.

But wait! Not quite all was lost! There, still standing in its entire original splendor, stood the brick wall! A bit tarnished now by soot, it was still structurally sound and undamaged. The irony of it was, although the family had not been happy with the original expenditure for bricks, only the bricks remained! *Almost* only, that is, because resting on a partially burned leather chair sat the box containing Cecil's silverware. Though the box was charred, the silver was rescued!

Farewell to Cecil

Cooksie

Cecil was by now no means the handsome swain of his youth. Instead, he had become a portly eccentric old man who carried a briefcase filled with pills and substituted an old fashioned stereoscope for a pair of reading glasses. Still, he managed to find a couple of women to care for him during the years that were left. As he waited for his divorce from Shirley to become final, he lived with a nurse whom he planned to eventually marry. Because the nurse (whose name I can't recall) was a working lady, he often visited "Cooksie," another of his lady friends.

Cooksie may not have been the most attractive of Cecil's female friends, but she had a heart of gold. Words flowed from her mouth like an artesian well. In her pronounced Oklahoma accent, she told us that she knew Cecil had other girlfriends, but that she loved him anyway. A dedicated Christian, she had once or twice persuaded Cecil to attend her church.

By now Cecil was in very poor health. Still plagued by the family headaches that he had inherited from both our parents, he had experienced a series of minor strokes. During this time he was always welcome at Cooksie's home, where he might enjoy a steaming hot bath and perhaps a massage for his aching head.

At this time, because of a tennis injury, I was encumbered with a walking cast. This was unfortunate timing on my part, for it had been decided that Cecil would enter the hospital to undergo an angiogram. (According to Larry, Cecil's hospitalization was precipitated by Cecil's involvement in a traffic accident. This happened while he was commuting between Cooksie and the nurse, and he had hit a power pole. It is possible he had a blackout before hitting the pole.) Frances, who would normally have been in charge, was on a trip to Alaska; consequently it became my job to look after our brother.

After entering the hospital, Cecil was under strict orders not to consume any liquids for several hours before surgery. However, when Shirley's mother came to visit him, she brought him a bouquet of roses, and after she left—while no nurses were in sight—Cecil simply removed the roses from the vase and drank the water. This same day, when I had visited Cecil, a note was tacked to his door stating that anyone visiting Mr. Tracy had to first check in at the desk. I was puzzled by this but later discovered the reason. A strange-looking woman who had occasionally shown up at the ranch had heard that Cecil was in the hospital and, apparently high on drugs, she had paid him a visit. Unfortunately, Cooksie was already visiting him. An altercation ensued, and both women were ejected from the hospital!

The angiogram was performed the following day, but because Cecil had consumed water from the bouquet of roses, the operation was very difficult and extremely painful; so painful that Cecil had chewed his lips until they were raw.

After the surgery, Cecil was removed from the third floor of the hospital to the basement where, I strongly suspect, his reputation may have followed him. The next evening when I paid him a visit I could see he was in a great deal of pain. I felt that he needed a private nurse to help him through the night, but when I suggested this to the head nurse, she did not seem sympathetic. She asked, "Why?" Then she informed me that, for this, his doctor's permission would be required. When I couldn't locate his doctor or a private nurse, I asked if I could stay with him through the night; begrudgingly, I was allowed to do so.

Well, I guess I must have broken just about every rule in the hospital that night. First, a large, intimidating nurse accosted me as I hobbled in my cast down the hallway to fetch ice water for Cecil's burning face. She told me that I was making too much noise, so I went back and removed my one remaining shoe. By Cecil's bed I spotted a nice steel pitcher that I thought would be just the thing for transporting ice water. As I lumbered back to Cecil's room with my pitcher, nurse number two gave me a disdainful look and said, "I hope you know what you are carrying in your hands!" You guessed it; I was carrying a man's urinal! Cecil gratefully moaned his appreciation as I packed his entire face in ice-cold towels throughout the night.

At about two o'clock in the morning, when he had settled down a bit, I decided that this would be my opportunity for a much-needed rest. I lay down upon the empty bed in his room. No sooner had I stretched out than nurse number three sailed into the room. She said nothing, but gave me a disparaging look. I quickly got up and smoothed the bed, but she must have reported me, for the head nurse soon followed and asked if I had been lying on the empty bed. I felt like a naughty child as I confessed to the error of my ways. I continued my vigil until daylight when Cecil requested that I rub his legs. I was doing this when nurse number one reappeared, and she instructed me that this was the worst thing that I could do for the patient because he had a blood clot and I might dislodge it! Somehow, Cecil and I both survived, but it was a night I would not soon forget.

I later discovered that Cecil had devised an unorthodox way to secure a nurse's attention. When he was in urgent need of his headache medication and no one answered his call, he simply removed his gown and stood stark naked in the hallway! This seemed to get prompt attention; it might also have accounted for some of the hostility that I encountered that memorable night.

Within a week or two, Cecil suffered a massive stroke and never regained consciousness. Now there was a funeral to be arranged and, since Frances was still away, MARTHA was in charge (or so she thought), and she was going to get it over with as quickly as possible.

Bertha

It now becomes necessary for me to introduce Bertha, for Bertha plays a part in Cecil's funeral. Cecil first became acquainted with Bertha when he was one of the judges who helped proclaim her as "Rodeo Queen" in the mountain area of Kernville where she resided. More than any other woman, Bertha remained a recurring figure throughout Cecil's life. In her youth she was an attractive young lady and an accomplished horsewoman; in later years she became a crop duster and flew her own plane. None of Cecil's former wives benefited much, financially, from their marriages to him, and so it was a mystery why Bertha was the only one of the women remembered in his will. Apparently Cecil was an admirer of Bertha's mother, for we were to learn that he had written a poem in honor of this pioneer lady.

Cecil's Funeral

Cecil had no church or lodge affiliations; consequently, I was in somewhat of a quandary as to how to proceed. Then Cooksie came to my rescue. As Cecil was lying in a coma at the hospital, Cooksie had brought the Pentecostal minister of her church to visit him. The minister generously offered to perform the services, and I gratefully accepted.

Cecil passed away on August 3, 1978. I signed the death certificate and engaged the Green Lawn Mortuary to arrange the memorial services for him. My next function was to locate the nurse with whom he had been residing. This took a little detective work, but when I located her she was able to produce a few articles including the trousers to his only suit, but not the jacket!

After hobbling up the hill to Green Lawn where I met the proprietor, I told him that I was a sister of the deceased and that I would like to hold the services as quickly as possible. This was on a Wednesday, and the memorial services were scheduled for Friday—no time wasted here. The funeral director spent about an hour going over a long list of questions with me. Everything seemed to be going quite well until I mentioned Cecil's wife. When he learned that Cecil and Shirley's divorce would not be final for a week or two, he looked at me thoughtfully and said, "Mrs. Frey, it is my duty to inform you that it is Cecil's wife who has the legal authority to be making all of these decisions, and that she could sue you for doing so in her place." Well, that rocked me back on my walking cast for a moment or two. Then I remembered that Shirley was on her way home from Arizona and, by a stroke of luck, I also remembered her mother's name. When I called, I was gratefully relieved to hear that Shirley had, indeed, arrived home and that she would be happy to come to the funeral parlor to assist me.

Shirley agreed to all of my decisions, and we selected an appropriate casket. I showed her the trousers and a clean shirt that I had retrieved from the nurse, but when I showed her a white silk tie with the abstract figure of a woman on it, she said she thought she could come up with something a bit more suitable. I told her I would now try to locate the jacket to Cecil's only suit.

After his house burned, Cecil had set up temporary living quarters in a nearby mobile home. As I hobbled up the ramp of this now-deserted place, I whispered a prayer. "Dear God, please let Cecil's jacket be here." And when I opened the door my prayer was answered, for there, hanging in an open closet, was the other piece of Cecil's suit!

I drove back to town where I took the jacket to the one hour cleaners and later, with it now in hand, I returned to the mortuary. All was going quite well until I received a call from Bertha who told me that Cecil had once written a beautiful poem as a tribute to her mother. She wished to know if we would care to include it in the memorial service. Since Bertha lived some distance away, time was of the essence—things were getting a bit tight. It was eventually decided that Bertha would fly the poem to Green Lawn, where Cooksie would pick it up and deliver it to the minister.

Memorial services for Cecil Foster Tracy took place at Union Cemetery on August 19, 1978. It was an extremely hot day and, by ten o'clock, I could feel the perspiration trickling down the small of my back. As I limped onto the family plot, I found Cooksie with her camera, already taking pictures of the casket and flowers. She was upset that her bud vase with a single rose had toppled over and was broken.

A handful of friends and a few relatives attended the funeral including our cousins, the "Cooper Kids," offspring of mother's youngest sister, Hazel. Shirley's mother, who had brought the roses to Cecil at the hospital, was also there, and when Shirley arrived she was wearing a black afternoon tea gown with a large black picture hat, black gloves, and heels. She looked *great!* Cecil would have been proud!

Considering this rather strange set of circumstances, the Pentecostal minister performing the services did very well, and I felt grateful to him and to Cooksie. He read the poem Cecil had penned for Bertha's mother; Cecil had done a fair job writing it, and the poem contained a few lines from one of Cecil's favorite quotes:

"Full many a flower is born to blush unseen,
and waste its sweetness on the desert air."

Well, things had come off quite well and, with the exception of Shirley's mother who thought there should have been some kind of family feast after the service, everyone seemed pleased.

Examining the floral pieces after the funeral, I found one that had no signature attached; it simply included the lines of an old popular song: "Mares eat oats, and does eat oats..." This puzzled me, until I remembered that Nanette's daughter Leslie had ridden a horse named "Marezidoats." Somewhere in her confused childhood, Cecil must have struck a chord.

A Pile of Junk

Cecil had a keen sense of history, and he never tired of learning all that he could about the area in which he lived. He was well informed, and he often wrote articles pertaining to the history of Kern County. Also, as I mentioned earlier, Cecil collected antique farm implements and other items that he felt were of historical significance. Because these items were of no use to their owners, they gladly allowed Cecil to haul them away. This practice had gone on for so many years that by the time I moved back with my family in 1942, the home place looked like an enormous pile of junk. I certainly had no idea that Cecil's "junkyard" was of any significance whatsoever; it was simply one great, disorderly mess. And it remained so until after Cecil's death, for he would not allow anyone to disturb his things. I am sure Cecil could never have dreamed of its later restoration and importance.

With Cecil's passing, JB Selvidge had the foresight (and the patience) to undertake the task of sifting through the chaos. As the years went by, he slowly and carefully restored each piece of equipment and mounted it for display; today there is a veritable museum of farm machinery and other memorabilia on the ranch. One can almost follow the progress of farming during the past century while exploring this considerable collection. Ancient plows, harrows and rakes reflect the history of the pioneer farmer. A large harvester and feed wagon have been completely restored, as have a Miller and Lux chuck wagon and a large brick Dutch oven. These are but a few of the items on display. Since this collection lies adjacent to William Tracy's homestead cabin, his blacksmith shop, the "old" barn, and the tank house, the area creates a fitting tribute to our father. That would have pleased Cecil greatly, and he would have been happy to see what happened to his "pile of junk!"

Afterthoughts

John Tracy on Cecil:

If I hear that Cecil "saved the ranch" one more time, I think I'll puke. He used this line over and over, seldom recognizing the brother and brother-in-law who kept him in business far beyond the period in which he made a contribution. I do understand from Martha that Cecil put his shoulder to the wheel when the chips were down and really came through. This had to be before I was born, as Cecil filled his life with pet projects and women during the forty-some years I knew him. He may have saved the ranch, but if it had been left to Cecil alone, he would have surely lost it again.

Many promises he made were left unfulfilled. One time, I came in to see if he could take me on a promised drive to see some Indian burial sites. He was sitting on his beanbag chair, and he told me he was "stepping on his heels" with projects and way too busy. He promptly fell asleep, right in front of me! We never went to the sites.

On the other hand, Cecil was a great uncle in many ways. His house was always open, and he was darn sure entertaining! I'd often bicycle over just to see what mischief he was up to.

Saint or Sinner? *(the author ponders)*

From time to time, as I write about Cecil and his personal life, I feel a twinge of guilt. What right have I to do this? What would Cecil have thought? And then it became quite clear to me. Why, of course! Cecil would have loved it! Whether judged as notorious lover, gilded cowboy, or loyal son, he would have enjoyed this last little glimmer of recognition before sinking into oblivion with all those who have gone before.

Kathryn—The Later Years

By the time of Tilton's accident, Tracy Ranch, Inc. had survived a twenty-eight-year struggle. Although the ranch was making progress, it seemed that each year found it a little deeper in debt. At our annual stockholder's meetings, our spirits would sink as we listened to each dismal report. Then Jack would remark, "Well, sometimes it goes along like this, *and then it gets worse!"* This always provided a good laugh and seemed to restore our spirits.

During the difficult years up to that time, my husband was able to put his problems to rest and sleep soundly. Now I knew that the ranch must be in deep trouble, for he was no longer sleeping well.

Tilton had wisely included a double indemnity clause in his life insurance policy, and because Kit received a substantial payment at his death, she was freed from financial worries. Jack now assumed one of the most difficult tasks of his life, for he approached her to ask if she would be willing to give the ranch a loan. The loan was granted without hesitation.

The loan was paid in full interest over the years, but I hope no one in this family will ever forget that Kathryn Tracy had faith enough in the integrity of the ranch to risk her financial security for its benefit. Like Fannie Tracy and her children who stayed out of high school to "save the ranch," Kit had certainly made her contribution. And with two sons and two grandsons now engaged in its management, it turned out to be a pretty good investment!

John and Kit (Note firewood—the hired help was instructed to keep a supply of wood on the porch for Kit's fireplace)

John Tracy remembers:

Mother proved her character after dad was killed. She had received a sizeable insurance settlement due to a double indemnity accidental death clause. The ranch was near bankruptcy, so her ranch stock was nearly worthless. Nevertheless, without worrying about her own future, she loaned money to the company, confident she would be repaid.

And she was repaid in more ways than one, for she was able to live in the ranch house and benefit from the ranch perks, even though she was not physically involved in the new businesses. Her brothers- and sisters-in-law, as well as her nephews and nieces, saw that she never lacked for anything, and she lived a happy life. She died eternally grateful for her treatment.

Kit's Grandchildren

During the years following Tilton's fatal accident, from 1970 until her illness in 1996, Kit filled the great void in her life as best she could. In spite of the cruel blows that fate had dealt her, she remained firm in her Christian faith. As was her way, she worked quietly and unobtrusively within the framework of her church. For the last ten years of her life, she dedicated herself with loving care to an elderly friend who had lost her sight and eventually become bedridden. Even after her own illness, Kit seemed more concerned for her friend than for herself.

In August of 1965, a month after Jim's death, his brother Bill married Toni Torigiani, and they had two sons: Mark James, born on December 17, 1966; and Robert Lee, born on November 19, 1969. In 1973 Toni found the constraints of motherhood too much, and she left Bill with two small boys to raise. Mark was six, Rob was four, and Bill was distraught. Kit squared her slender shoulders, determined to help her son through this difficult period. Together they formed a partnership to provide for the care of the children.

In 1974 Bill had the good fortune to fall in love with Susie Jackson. Now another partnership was formed; this time for life. Susie brought two children from a former marriage into the family: Todd Blake, age six, and Stephanie Ann, age three. In Susie's capable hands and with Bill's help, the four children were melded into a family that we sometimes referred to as "The Tracy Bunch," recalling the popular TV series.

When Kit's son John started dating Donna Allen at Cal Poly, he remarked, "This girl is too good; I can't afford to let her get away." They were married near her parents' home in Torrance, California on January 10, 1970. Tilton was heard to quip, "This is my one good suit. I just need one, since I only wear it for weddings and funerals." Donna and John's first child was not born until May of 1977, and they named him James Tilton, after his grandfather and his Uncle Jim who was killed in the plane accident. In order to avoid confusion, he was nicknamed "JT." When little Jimmy was only three, he contracted leukemia, and a most stressful year followed as the family battled to save his life. Though John and Donna did all they could to spare Jimmy's grandmother, it was not really possible. Once again Kit was forced to summon all of her courage to face tragic circumstances. Devastated, she traveled the unfamiliar freeways to UCLA Medical Clinic and did all she could to help comfort the stricken child. He passed away on May 14, 1981, and was laid to rest at Wasco Cemetery alongside the grandfather and uncle for whom he had been named.

In the meantime, on February 16, 1980, his brother Allen Richard was born, and on October 25, 1982, a girl, Jessica Christine, was welcomed into the family. Kit now had six grandchildren, and they were fortunate—for who could not have loved this grandmother with her kind and gentle manner? Over the years, she gave her love unselfishly. She was always available when needed—always nearby to lend a helping hand. She never missed a graduation or forgot a birthday; and she seldom missed a football game if one of her grandchildren was involved.

Spry at 75

Kit was prematurely gray at thirty; at seventy-five, her well-groomed silver hair was a shining asset. Still active and spry as a sparrow, she held herself erect and stepped as lively as ever. She chose clothes that were becoming to her slender figure and wore very little make-up. Because she did not consider herself photogenic, she avoided the camera whenever possible. Her boys liked to tease her about this—in fact, there existed a fine bond of humor between Kit and her sons.

Although generous with her gifts, Kit was careful with her resources. Her only luxury was an occasional visit to the English countryside that she loved. These trips seemed to restore her spirit and bring comfort to her soul. She loved to read and cherished her library. She took pride in caring for her home, did all her own housework and, best of all, she enjoyed working in her garden where she spent long hours.

While looking through a box of souvenirs recently, I came across the following paragraph that I once received from Kit. She was so amused at what she felt to be an apt description of herself that she had taken the time to type it and send it to me:

"Her typical Leo ebullience keeps getting in the way of her yearnings for serenity and quiet. All that energy also makes it a little hard to follow her train of thought. She has the kind of brain that seems to skip around, jumping over a few synapses here and there, as her stream of consciousness conversation weaves from Saturday Night Live to microwave ovens to the 1960's.

"Synapse: the point at which the nervous impulse passes from one neuron to another.

"Neuron: a nerve cell with all its processes."

Indeed, it was sometimes a little difficult to follow her train of thought. Before one could finish answering a question, another question might well have taken its place!

Blessing of the Rancheros Visitadores

John Tracy remembers:

One year I invited mother to the annual Blessing of the Rancheros Visitadores at Mission Santa Ines in Solvang. Cousin Wes was supposed to fly her over and have lunch with David and me, but the weather was bad so mom drove herself over. David was taking us around, and we decided to start with a winery tour. My brother and I both have a propensity towards the grape, so the tour became somewhat extended. We bought a bottle or two at each winery and, as the day went on and the wine got better, we bought more bottles. After a particularly nice bottle of Chardonnay at lunch, I realized the mission ceremony had already begun.

We raced to Solvang in time to catch the 500 horsemen as they were leaving. We stepped to the side and introduced mom to our numerous friends, toasting each one as they stopped. When she was ready to leave, we realized she had bought lunch; something David and I were going to do but which had, somehow, slipped by us.

We thought how funny it would be if a highway patrolman stopped her with the eight cases of wine we had purchased. It would have been hard for her to explain to her church congregation! All in all, it was a grand and glorious day, one that would not be repeated because of mother's failing health.

Fading Health

In 1980, Kit had a severe heart attack. By-pass surgery was performed and she recovered, but her doctors advised her to follow a strict diet and to limit her gardening activities. Her philosophy leaned towards the quality of life rather than its length; consequently, never one to pamper herself, she largely ignored this advice and continued with her normal activities.

By now Kathryn had suffered the loss in tragic death of a sister, a brother, a son, a husband and a grandchild. But fate was not yet through with her, for during the next two years she was to endure severe physical pain as an acute form of arthritis assailed her body. Visits to the doctor became more frequent, and the medication prescribed became more powerful, until finally her life became a cycle of hospitalization and rehabilitation punctuated by brief intervals of home care. With Kit so frail that she weighed less than eighty pounds, her sons became the daughters she never had. Laying their work aside, they paid daily visits to hospital or convalescent home to provide the loving attention so badly needed at that time.

It was during one of Bill's last visits that he questioned a God who would allow her to suffer in such a way. From her bed she looked up at her tall son and, through a well of pain, she said, "Well, I guess God will not give me more than I can bear."

"She was strong in her faith," Bill stated later. "She lost a sister, a brother, a son, a husband, a grandson and, in the end, was in constant pain, but she never wavered or questioned God's will. I never realized how small she was, physically, because she was so large in spirit."

From time to time her family's hopes would rise as she was allowed to return home. Home care would be provided, but the nurse would no sooner be settled in than Kit's indomitable spirit and eternal optimism would prevail, and the nurse would be dismissed.

As I was about to leave after calling on her during what would become one of our last visits, my attention rested upon the little display of western memorabilia that had hung by the door for so many years. I mentioned her mother's distaste for this reminder of western life. "Yes," Kit replied. "You know, one time I apologized to Tilton for my mother's behavior, and he said to me, 'Honey, if there is no worse problem in our marriage than that, we will be all right'." Little wonder that she loved this western man.

John remembers:

Mother's health continued to deteriorate slowly but steadily after open-heart surgery. As the trips to the hospital in the middle of the night became more frequent, Bill and I made sure that one of us was home in the evening for emergencies. Although mother wanted to remain independent and not bother her boys, we maintained a schedule of checking on her daily "for a cup of coffee." We made a pretty fair combo. Bill was a whiz at figuring out the myriad of insurance forms, and my friendship with the doctors allowed us instant access to help twenty-four hours a day. Our wives were an immense help, taking mom to town and double-checking on us boys. Mom found instant allies in Donna and Susie, and she would try to undo what Bill and I had done if it was not to her liking. In the end, we all saw eye to eye.

The author remembers:

On the morning of July 18, 1997, I carried a small bouquet of roses to the convalescent home where Kit had been confined. When I asked the receptionist if Mrs. Tracy was still in the same room, she looked at me with some hesitation, and I knew I was too late. Kit had died in the arms of her eldest son just an hour before. At last she need suffer no more. She would be on her way to join the others that she had missed for such a very long time.

On August 1, 1997, memorial services were held for Kathryn Pollard Tracy at the First United Methodist Church in Wasco. It was her birthday. As it had been for Tilton's service, the church was filled to overflowing; many stood to listen to the service from outside.

To the end of her life Kathryn showed concern for others. She was worried that people would grieve for her, and she had requested that her sons engage a Dixieland band to play at her memorial service and arrange a barbecue in her memory.

They did.

Nancy remembers:

As I proofread this section I found myself in tears for the loss of my gallant aunt whom we loved so much. Her gentle spirit and valiant acceptance of the tragedies that assailed her time and again still lingers in our hearts. Yet I know that she and Tilton enjoyed heights of bliss seldom achieved by most. As she often said, "We were most fortunate, and we understood that at the time."

Nine of the ranch partners pose in front of the old Cat 15 in 1992. Back row, left to right: Robert Tracy, Jack M. Frey, Mike Frey, and Todd Tracy. Front row, left to right: Larry Frey, Jack R. Frey, Wesley Selvidge, Bill Tracy, and JB Selvidge (John Tracy is absent).

BLACCO—BUTTONWILLOW LAND AND CATTLE COMPANY

"We raised our own replacements."

– *Jack M. Frey*
82-year-old patriarch
Buttonwillow Land and Cattle Co.

Larry and Jack

Wes Selvidge and Jack

Foreman Alfredo Montoya with Bobo and Larry

John Tracy poses with Larry at the site of a new well

The new office in 1995

Michael

Jack R.

Jack R. and JB with a new power bandsaw in 1991

The shop crew in 2000. Left to right, Andres Gerardo, Jose Cortez, Arturo Garza, Jack R. and JB.

Jack R. atop a TD-18

"Giant tractors now rip the soil where teams of horses once did toil," from the Gooselake poem

After 25 years of managing cotton harvest, Bill Tracy turns the job over to son Todd (left) and nephew Jack R. Frey.

Four-row cotton pickers

Cotton modules

The partners at Buttonwillow Land and Cattle Co. saw too many of their peers' farms go down in flames. The business failures all too often put big responsibilities on the untested shoulders of young family members.

So when their own children began reaching an age to join the business, Buttonwillow's partners developed a plan to test the next generation's mettle before allowing them to buy in as partners. It's officially called Buttonwillow's associates program. The young guys call it "boot camp."

The five-year postcollege program familiarizes aspiring partners with all aspects of the farm from the ground up. ▶

THE FIFTH GENERATION at Buttonwillow Land and Cattle Co. need determination to make partner. "We actually try to drive them away," says Bill Tracy, center back. Past and present "boot camp" survivors include, from left front, Jason Selvidge, Todd Tracy, Rob Tracy and Jack R. Frey.

Boot camp for beginners

IT TAKES TRUE GRIT TO ENDURE THIS PREPARTNERSHIP PROGRAM

By Charles Johnson

PHOTO BY THE AUTHOR

August 1997 – the top producers

Rodeo bulls, Black Angus, Holsteins, Herefords, and Brahmas inhabit John Tracy's feedlot

Broken Frito Lay potato chips comprise the "feed lot mix"

Editor and publishers Lynne and Jim Norris visit the feedlot

Cattle dine on cottonseed, tomatoes, potatoes, carrots, and chopped alfalfa with a dollop of molasses

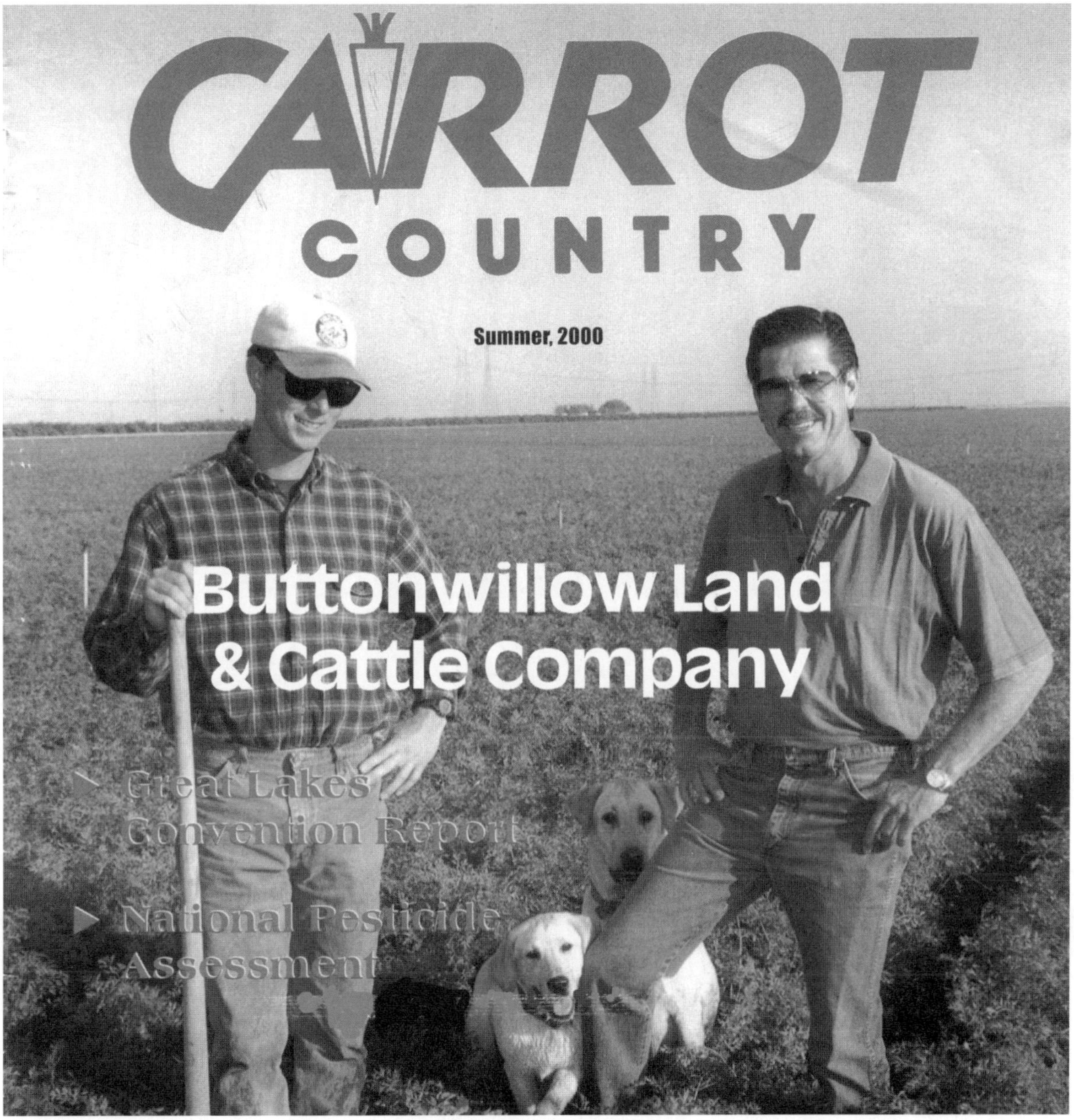

A 2000 issue of *Carrot Country* magazine features Jason and Wes Selvidge on the Brushfield property east of Buttonwillow

Todd (left), Rob and Michael surveying for the first almond trees in 1996

Rob and Jack at the squeeze chute

Planting the first almonds in 1996, Todd (left), Rob, Michael, and Jack R.

Susie and her sons

These trailers transport tomatoes

The Rio Bravo plant processes tomatoes for pizza

Family historian JB, right, with brother Wes. JB has organized and restored his Uncle Cecil's lifetime collection of antique farm equipment and memorabilia.

JB stands beside Allis Chalmers' HD-11 and IHCTD 18 crawlers from the 1950s and 1960s

JB with a 1917 Harris Harvester he is restoring. It originally belonged to the Romanini family of Buttonwillow.

A dutch oven, used by Italian families to bake their delicious bread

Buttonwillow Land and Cattle Company grows potatoes for nearby Frito Lay, on Highway 58 east of Buttonwillow

For 40 years, this little Baycraft boat rested at the bottom of Lake Tahoe

The boat after Wes Selvidge spent four years restoring it

Foreman Eduardo Montoya poses with Wesley and Jason Selvidge at the Brushfield property. Several employees, including Eduardo, make the ranch their home.

John Tracy is among the county's best ropers. Here he is on "Knothead" with Nick Siliz on Goll

Jason Selvidge Writes:

Here's a little about my racing.

I am starting into my fifth year of racing in '00. This will be my 3rd year in this type of car, which is called a "super street stock". I race mostly at Bakersfield Speedway but also travel to Hanford, Santa Maria, Tulare, and will go to Watsonville for the first time this year. All of these are dirt tracks ranging from 1/4 to 3/8 of a mile. Since I started racing I have collected 10 wins overall. Last year I was able to win races in Bakersfield and Tulare, my first wins in the Super Street Class. So far this season I have two wins in four starts at Bakersfield and hope to win many more before the end of the year. In an average year I will run about 15 to 20 races between March and November.

The car I'm currently driving started life as a 1974 Chevy Camaro. To convert it to a race car it was first stripped to the bare frame. Then is was built back with a steel rollcage and reinforced chasis. The engine, transmission, and rear end are all custom built solely for racing. The suspension is also heavily modified from it's stock form to make it perform better as well as make it much stronger. In actuality the only parts of the '74 Camaro left is most of the frame, the roof, and the front fenders. The engines we use make in excess of 400 horsepower and have to be overhauled frequently because of the high stresses placed on them. Top speeds at Hanford and Tulare reach 100 miles an hour while Bakersfield and Santa Maria reach about 90. It doesn't sound very fast but you have to remember there are as many as 22 cars on that small track all going that fast together.

Though we don't make regular pit stops at our races I normally take 2 or 3 crewman along to the races to help with the maintenance at the track. Almost always this includes Wes, who also spends nearly as much time as I preparing the car during the week. Without his help, I certainly would not race as often.

Jason Selvidge alongside his "super street stock" racecar, a 1974 Chevy Camaro

Ranch Employees

The long history of quality employees has always been an integral part of the ranch's success. The ranch's first employees were itinerate cowboys; the ranch has since advanced into the 21st century with men capable of operating monstrous, computer-assisted, and satellite-directed equipment.

The introduction of permanent crops in the 1940s also offered the opportunity for permanent employment. The security of year-round work blessed the ranch with the long-time leadership, expertise, and loyalty of families such as the Bradys, Evans', Craigs, Holloways, Burlesons, Montoyas, and Garzas. They have raised and educated their children here. Some have returned to the ranch and others, armed with the values instilled by ranch life, have ventured into the professional world.

Annual employee fishing trip, Morro Bay

Fishing at Morro Bay

Half the group suffered from sea sickness (notice "no sleeping" sign above table)

John Tracy and Bud Whelms

Quarterly meetings are an integral part of ranch operations. Each family takes turns in providing a sit-down dinner for these occasions. Here, the family is pictured at Janet and Wes Selvidge's home in March 2000.

Each year, Susie and Bill Tracy open the Christmas season with an elegant dinner and sixteen-foot tree in their home

May, 1996—The Neufeld family from Kansas arrives with their imposing truck fleet to harvest Tracy Ranch grain

Brand new trucks that transport the harvesters

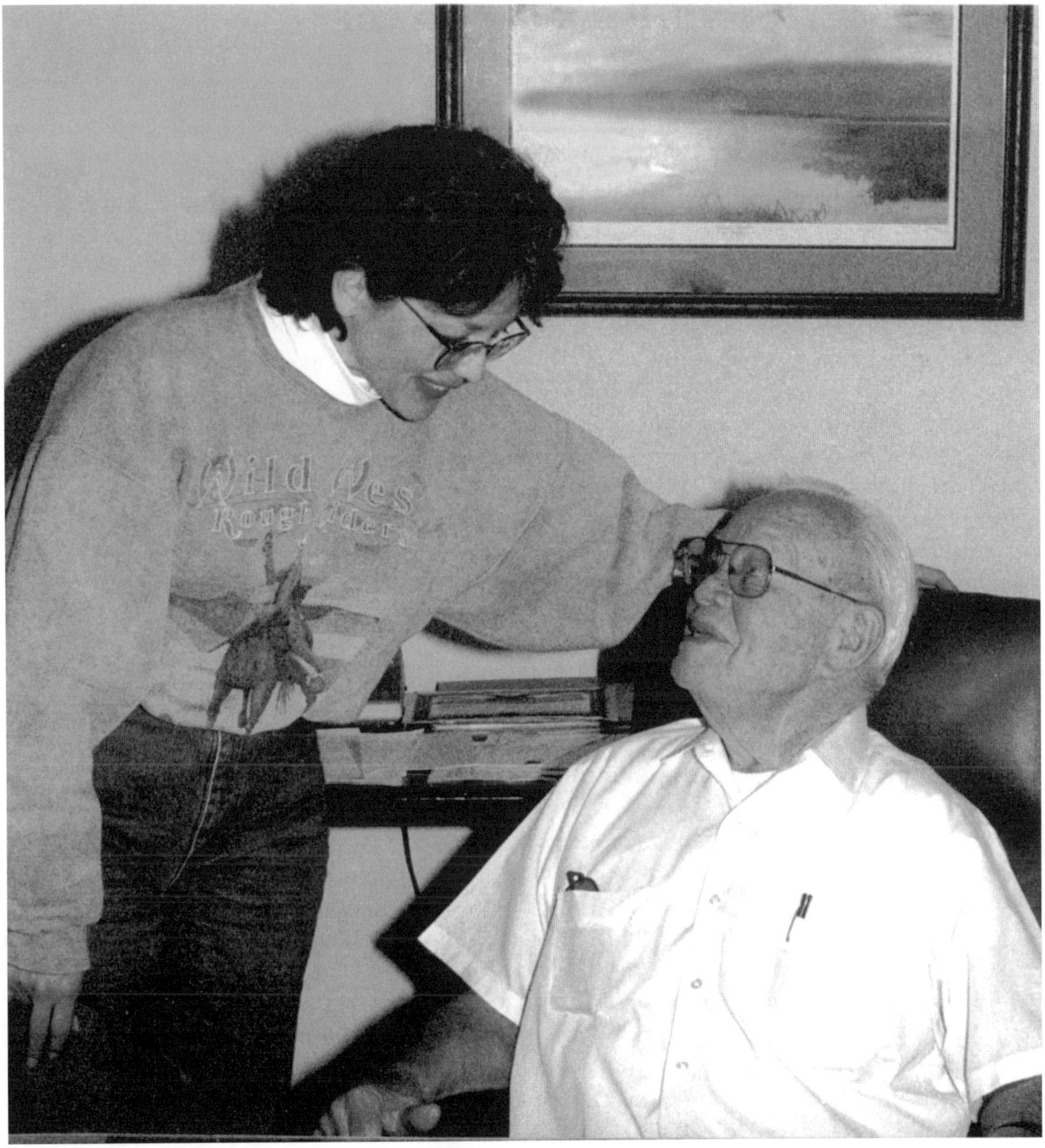

Jack Frey with Gladys Costello, BLACCO secretary for the past 10 years

"EL PATRÓN GRANDE"

Fannie Tracy's comments from the Mary Gregg notes:

We made Jack manager of the ranch at once, before he had any experience, because Cecil wanted to show that the younger ones were fully accepted as partners. The result was that they could not later say that the power was automatically reserved for the more experienced men.

At first, after the land was divided up between the families, Jack did not develop his section. His thought was that in case any of the partners ever wanted to split up and go separately he could thwart them by saying, "No, you cannot separate because my piece is still undeveloped."

In the beginning, Tilton wanted to go it alone, but he later changed completely. Jack himself was at one time tempted to pull out and join the Sill family in farming. In order to dispel any illusions of grandeur, I suggested that each of the boys go out and try to borrow $100 on their name alone without the backing of the entire ranch! In the end, all of the partners held firm, to their mutual benefit and that of their descendents.

John Tracy remembers:

I would suppose that I was closer to my Uncle Jack (and perhaps more like him) than the rest of the crew. Jack was a terrific partner; although I suspect it is easier to be a nephew than a son, daughter, or grandchild. Before I became a partner, the sight of that yellow and white Ford pick-up bearing down on me struck the deepest terror possible. You knew that, no matter how well you were doing, it wouldn't be good enough. Jack ate so many young asses that it's a wonder he didn't choke on them.

After my father's death there was a transition. Uncle Jack couldn't have been a better partner. I would go to him for advice, and he would say, "Figure it out and, even if it's wrong, you're doing something." He put a lot of stock in education and had faith that I would do right. This worked pretty well because I would analyze each enterprise over and over to justify the trust.

Many times, excellent opportunities arrived when we had exhausted our budget. Jack never failed to amaze me. In the summer of 1976 Jim Etcheverry took over the Spring Ranch in Shandon. The previous owner had 200 pairs (of cow and calf) for sale, and Jim didn't want them. I phoned Jack one evening describing the cattle, the price, and the right to graze for one year on the ranch. "If it's a good deal," he said, "get it done and I'll find the money." This would not be the last time he would come through and, if the truth were known, I think he enjoyed the challenge. We completed the deal and then set a date to brand the calves.

California had been in a drought and, although the ranch wasn't bare, the grass was short. After branding the last calf, it started to rain. Everybody headed out, as the dirt road was adobe clay and impassible in wet weather. With the exception of our one-ton stock truck, all the rigs got out. It rained twenty-four hours straight and continued to rain on a regular basis. We had to leave the truck for forty-five days! We would drive in as far as possible, then continue on horseback. Each time we showed up, the ranch was greener, and we were out of the drought.

Jack always enjoyed the livestock and, in his younger years, would accompany me to the various ranches to weigh the cattle we were selling and then take the checks to the bank. He liked action. A check in his pocket, a cigar in his mouth, and a bourbon and "branch water" in his hand always made him happy.

I always appreciated the fact that Jack said "we" instead of "I" when referring to activities on the ranch. It was one more indication of his team approach. He was fully aware that "I" couldn't get the job done as well as "we" could.

I never knew Jack to take anything out of the ranch for himself. His philosophy of sharing with the younger generation has contributed greatly to the ranch's success. He often says he wishes Grandma Tracy (Fannie) could see the progress we have made. She kept things together for future generations and would be proud of the way her sacrifices have paid off.

Bill Tracy remembers:

Uncle Jack usually had breakfast with us. He carried the responsibility of refinancing. Many times in the early fifties we'd hear him confide, with great concern, that he "just didn't know where he was going to get the money this time." After the men left for work, mom would agonize; not for the future, but for Jack. In her later years she often expressed her joy that Jack had lived long enough to see the ranch prosper, with bankers knocking at the door for our business and a group of "boys" capable of keeping the operation going.

John Tracy remembers:

Jack Frey: whose dedication to excellence transformed a desert into an agricultural empire. Jack came to the company when it had one tractor and mostly undeveloped rangeland. Under his steady hand, he guided the development of over 8,000 acres that now furnish cotton, wheat, pistachios, almonds, carrots, potatoes, tomatoes, and beef to the world. He was fortunate to see his dreams come true and happiest when touring his fields. You cannot go to a far corner of our operations without seeing his shadow. He was the last of the breed of pioneers who had no fear of failure. He saw boundless opportunities and believed that every day was a better day in America.

The author remembers:

"That Duck-legged Little Bastard"

My husband occasionally relieved the tension of his responsibilities by spending an evening on the town with his drinking buddies. Although I realized his need for relaxation, I was not thrilled by these sporadic drinking bouts. It was one of those occasions, however, that provided the family with an amusing story.

After Jack had spent an afternoon and evening enjoying the camaraderie of his friends, he was returning home at about two in the morning. Passing by a portion of the ranch where one of his employees was bailing hay, he decided to stop by and investigate the conditions. Deciding that there was not adequate humidity in the atmosphere for hay bailing, he called the driver, Brandon Evans, down from the bailer and proceeded to give him a lecture.

Now, Brandon Evans was no sissy; a giant of a man, he weighed over 200 pounds and was more than six feet tall. What's more, he had a reputation for hanging out at a Bakersfield bar called " The Blackboard," a bar where brawls and fistfights often took place. Brandon was not known to back away from a fight.

After Jack finished with his lecture he added: "Why, I ought to kick your ass!" Brandon looked down at my five-foot-six-inch husband and asked, "Do you think you can do it?" Jack's response was simple. "No," he answered before heading back to the safety of his yellow pickup.

Relating the story to a friend later, Brandon marveled. "Do you know what that duck-legged little bastard said to me?" In turn, the employee related the episode to Jack Sill, and hence we have the story of the "duck-legged little bastard."

Jack and the Mailbox

When our son Larry was about nine or ten, he and his father set out to repair the mailbox. It was attached to a pipe buried in the ground and had begun to list on one side. Using a post-hole driver, Jack drove an iron stake into the ground. That accomplished, he asked Larry to bring him some wire from the pickup—he intended to wire the pipe to the iron stake. Larry did not move. Irritated, Jack made a second demand. Once again, Larry didn't move. "Well Dad," he finally responded, "why don't you just slip the pipe over the iron stake?" The mailbox still stands, straight and true.

The Night Our House Caught Fire

It was April of 1966, and Jack and I had just returned from a fishing excursion to La Paz, Mexico. It had been an exciting trip, for Jack had landed his first Great Marlin. That night, Jack and I enjoyed snuggling down in our own bed, reliving our pleasant vacation together. I awakened at about four o'clock to go to the bathroom. Perhaps it was my sixth sense, but for some unknown reason I opened the door into the living room. To my great surprise, I discovered that the end of the couch was on fire! It was not a blazing inferno, just a gentle little fire burning steadily away with very little smoke.

I alerted Jack, and he bounded from the bed to make a dash for the fire extinguisher. Having complete confidence in my husband's ability to handle any emergency, I continued my trip to the bathroom. There I contemplated with secret pride my ability to keep my cool in such a crisis.

When I had finished, I exited the house through the back door. As I rounded the corner at the first light of dawn, I came upon a scene that I shall never forget. There was my husband, clad only in a T-shirt and crouched on all fours, peering intently beneath the house with his posterior exposed to the world. At about the same time, Tilton's pickup came roaring down our driveway. "Mother," Jack said, "for Christ's sake, get me some britches!" I scurried inside to carry out the only constructive service I was to perform during the entire frantic episode. Tilton arrived in pajamas and cowboy boots. He grabbed the nearest hose and, together, he and Jack managed to put out the fire in the floor furnace under the house.

When the smoke finally cleared, I learned of the sequence of events that had taken place during the brief interval I was in the bathroom. While I had been indulging in my secret pride, my poor husband had tried to put out the fire with the extinguisher, but he had succeeded only in filling the house with smoke. He then groped his way through the smoke to the phone in the kitchen and alerted Tilton. Tilton, in turn, had alerted the Buttonwillow Fire Department. When I found Jack on all fours, he had just returned from an eighty-foot dash to the butane tank, where he had cut off the gas to the house.

The fire was caused by a leak in the line where it had rusted through, and a sudden bout of cold weather had activated the furnace.

We later learned that the Buttonwillow Fire Department, already famous for its incompetence, had sped to our rescue, but first they had pulled up in front of Tilton's house. Realizing they were at the wrong place, they had turned around and smashed into a tree. Next, they raced half a mile past our place and dead-ended on 7th Standard Road. They turned around and retraced their path, arriving at the scene of the fire just after it had been extinguished. Nevertheless, they drenched the house with water and, using their hatchets, chopped a hole in the pretty white wainscoting behind the couch.

On the wall above the couch hung a series of chicken pictures that I had painted. I still have one of those pictures—the one nearest the fire—and its blistered surface gives mute testimony to how close we came to losing that old house and possibly our lives. The flames had begun their lethal path up that wall. In a matter of minutes they would have consumed the entire building.

As I write this, although the old house has been vacant for several years, it still stands. If you were to examine the area beneath the house, you would see a circle of black and charred wood about six feet in diameter. You would also notice the pieces of two-by-fours that had been installed to support the weakened area. In my mind I can still visualize the four-foot circle of burnt fiber in the living room where the green carpeting had burned clear through.

As so often happens, some good came from the catastrophe. Because our carpeting was the same throughout the house, our insurance company had no choice but to replace all of our worn carpeting, and a fresh new coat of white paint soon covered the walls.

Of course, Jack and Tilton were the heroes of the day, while Martha maintained her composure throughout the crisis. Come to think of it, had I not had to go the bathroom at the precise moment that I did, all might have been lost! So perhaps I did make a contribution, after all. Perhaps my bladder should share credit for saving the day!

A SURPRISE PARTY FOR

JACK FREY'S
80th BIRTHDAY

SATURDAY, JUNE 18th
AT 5:00 P.M.
TRACY RANCH HEADQUARTERS
19400 WILDWOOD ROAD
BUTTONWILLOW, CA
R.S.V.P. BY JUNE 1st
CLAUDIA and LARRY FREY
PHONE (805) 764-5375
19220 WILDWOOD ROAD
BUTTONWILLOW, CA

No Gifts Please   Casual Attire

Aunt Kit's birthday card for Jack, in reference to his hearing problem

Arriving with wife, Martha and sister-in-law, Kathryn Tracy

The birthday gift: a yellow pickup

Two hundred 200 guests attend dinner catered by Marilyn Curtis

Tracy Koll, Nancy Trahan, and Carla De Fraine

Patty Tkac, T.G. Hopper, and Jackie Cowan

Barbara Forney, Patty Bass, Frank Sill, and Dr. John Forney

Cousin David Tilton, Mary Tilton, and Ray Blair

Pete Cassidy and Barbara Forney

Allison and Mark Tracy

Gretchen Sill Zachary, Larry Zachary, Mary Sill Guerro, and Janet Funk Selvidge

Janet Funk Selvidge and Susie Jackson Tracy

Bill Tracy, John Tracy, and Larry Frey

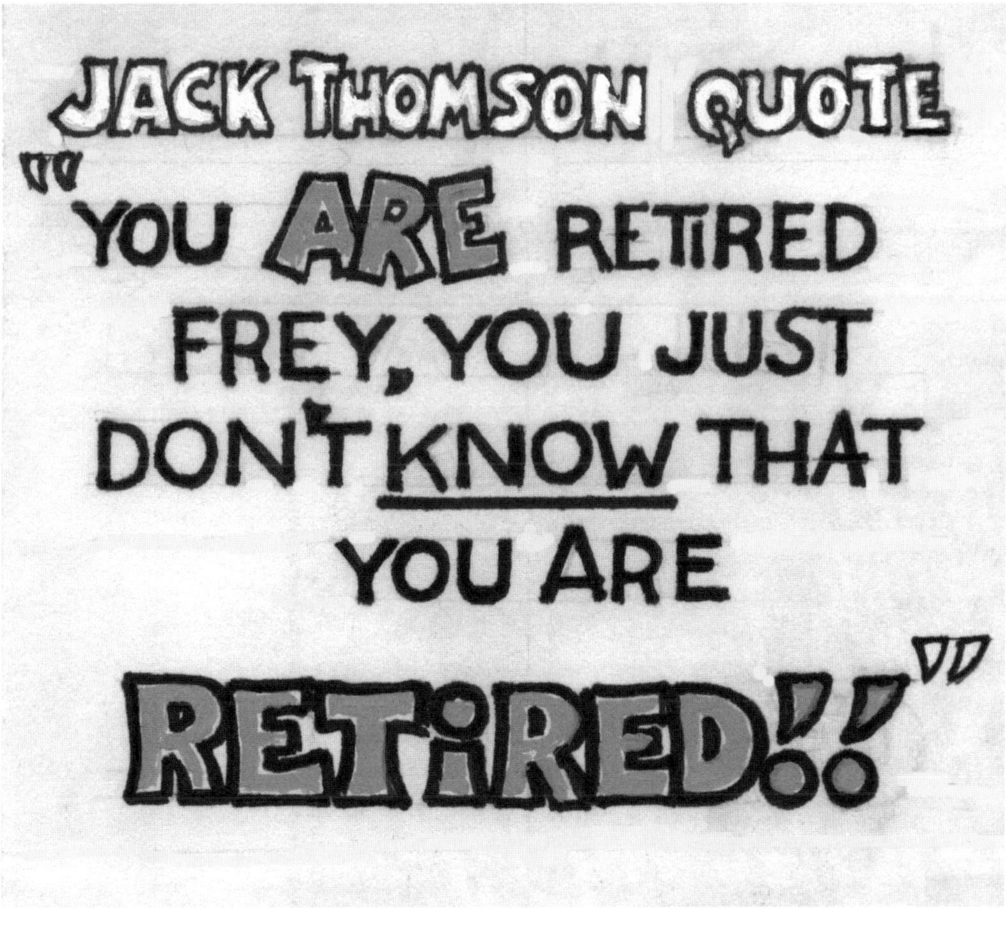

Tracy ranch celebrates 125 years

By Susan Valdez
NEWS EDITOR

A gala celebration was held at the historic Tracy Ranch in Buttonwillow Sept. 12. The sights, sounds and fun of the day will long be remembered by all those who attended. The work and preparation which went into the celebration was evident.

The grounds of the ranch were manicured, and hay bales provided seating as well as a rustic touch. Groups of guests mingled and browsed.

Wandering around the ranch, the sense of history was overwhelming. It almost seemed as if the spirits of the original settlers were there with their descendants to welcome you and show you their hard-won accomplishments.

Entertainment and music were provided throughout the day. There was a stage set up in a central location. There were various displays of memorabilia and artifacts. Several demonstrations of lost arts were also provided, including blacksmithing from a forge and the spinning of cotton by an old-fashioned spinning wheel.

Other demonstrations throughout the day included horseshoeing, calf roping and an aerobatic show. There were games provided for children, and dancing under the stars at the start of early evening.

Western style barbecue was served to all the guests. Everyone gathered 'round to watch as the pit where the meat had been cooked was opened. Liquid refreshments and soft drinks were provided for everyone.

The guests attending included people from all over California and all walks of life. It is hard to fathom the final head count; however, it was sizable. Everyone shared in the fun and remained considerate of each other, enjoying the gentle weather and cool breezes through the eucalyptus trees.

The hosts for the celebration were the Tracy, Frey and Selvidge families, and their graciousness is not often found in today's world. Somewhere up above, the original Tracy must be mighty proud of the heritage established back in 1862.

Guests at the celebration for Tracy Ranch on Sept. 12 signed in at the front, and received name tags from the Selvidge family. (Photo by Susan Valdez)

Shafter Press, September 15, 1987

BLAACO-TRACY RANCH 1862 125th ANNIVERSARY 1987 BUTTONWILLOW, CA

THE 125TH ANNIVERSARY - 1987

Fannie and Will Tracy's first picnic, which was a disaster, was followed by the more successful Red Cross Picnic in 1918, and picnics (or barbecues) became a tradition in our family. In addition to Tilton's weekend roping parties, the family began holding a barbecue at the ranch headquarters each year.

Tilton was experienced in preparing the meat for a deep-pit barbecue, and he soon taught others how to season the meat before wrapping it in burlap for the long hours of cooking underground. After the women had removed pebbles from a 100-pound burlap sack of pink beans, the beans were placed in a huge, water-filled iron caldron where they would cook slowly for several hours while being stirred with a large, wooden paddle.

To prevent the children from drowning in Cecil's pool, we recruited our close friend Earl Sargent as a lifeguard. We purchased a second-hand jukebox plus records for dancing on the ranch's outdoor cement slab, invited a few close friends, and that was it.

As the years went by, we included more and more people until the group could number as many as 900. At this point we decided things were getting out of hand, and the tradition was discontinued.

In the meantime, five little boy cousins had grown into stalwart young men with families of their own and, by 1987, they were in charge of operating the entire ranch. Happy recollections of their childhood must have prompted the desire to hold one more giant celebration. Those of us remaining from the previous generation drew in our breaths, for with the increase of family and friends we knew that this would be no small undertaking. But, just as Fannie had done before them, they proceeded with their colossal plan while their elders looked on with trepidation.

It was decided that the event would be a celebration of 125 years of farming and ranching on Tracy Ranch. John Tracy, who by now had taken his father's place as cattleman on the ranch, became chairman of the event. He organized regular meetings and assigned specific duties to each family member. In the end, it was John's proclivity for leadership that turned the day into a miracle of coordination.

Party Checklist

The following list is a partial accounting of the details in planning the celebration:

Invitations, to include time, date, local motels, map and a schedule of events.

Provide lunch for family & crew
Provide nametags for family
Provide serving tables
Engage music: Mariachi, Fiddlers, Jazz/Dixie, and Country-Western
Schedule aerobatic demonstration
Provide children's games & supervision
Set up horse racing course, judges, prizes
Set up cow plop bingo "board" in separate corral; make sure cow is fed
Organize calf roping and branding demonstrations

Ration invitations between family members
Crosscheck invitation lists
Secure insurance.
Contact Sheriff's Department for 12 security guards
Engage paramedics
Groom ranch headquarters
Prepare 7 acres of parking spaces
Provide wheelchair accommodations
Designate parking attendants
Resurface airstrip
Provide drinking fountains and hand washing facilities

Schedule crop dusters to spray for flies
Engage mime
Engage trick roper
Engage blacksmith
Arrange for horse shoeing demonstration
Arrange for spinning demonstration
Arrange quilting demonstration, display quilts
Decide on reception table, family picture displays, Family Tree, ostrich feather display
1800 pounds of beef for barbecue
? bushels of beans
? gallons of salad, salad dressing
2500 rolls & Butter
2500 paper plates, napkins, & plastic ware
Condiments
Lay out 6 serving lines
Designate 24 servers
Create a timeline
Arrange clean up

Reserve 15 portable toilets
Make appropriate signs
Bring in 400 bales of hay for seating
Install public telephone
Install extra lighting
Install loudspeaker system
Engage photographer
Provide soft drinks
Set up cash bar
Build temporary corral
Round up calves & groom horses
Invite ropers

Party Preparation

John Tracy remembers:

When the time for the party arrived, using my knowledge and observation of cattle and crowd movements, I volunteered to coordinate activities. When it comes to migration and movement patterns, people aren't much different from livestock, and I set up a flow chart similar to that used when moving cattle from livestock corrals. I intended to get the large crowd through a check-in, then allow them to flow throughout the many displays and demonstrations without having a "bottleneck" at any one particular area. For the women, one-and-a-half portable potties were ordered for every single portable potty that was ordered for the men.

JB would, of course, be in charge of getting the place ready. This job, which no one wanted, was one at which JB had become an expert. Trees were trimmed, lawns were mowed, the grounds were spruced up, and a pay phone was installed so attendees could call out—cellular phones were just coming on the scene, so most people didn't have them.

Paramedics were on hand during the day for fast response in any emergency.

The Governor of California was invited and, with his arrival a possibility, sheriff reserves provided security. (The governor did not make an appearance, but the sheriff reserves stayed for the entire party.)

Family members prepared and served the food; Wes was in charge of the meat. During my visits to amusement parks and other gatherings, I had noticed that people tend to head for the long lines and not the short ones. In order to eliminate this tendency in our food lines, we arranged three sets of tables and the food was served from both sides for six serving lines. Some of the family members were in charge of serving the food while others replaced dishes that were running low. A master-staging table held everything that was being served; a bulk service area holding salad, bread, beans, meat, condiments, and other basic supplies backed up the staging area. Each person held a specific position and knew whom to report to for more supplies.

The plan would function like this: meat would be running low on table one, the meat server at this table would yell "meat on one," and the person in charge of replenishing the meat on the table in the master-staging area would repeat, "meat on one." If table one heard no reply, he or she would repeat the order until a reply was received. A full tray of meat was then delivered to table one and the empty or near-empty tray was brought back to the meat area of the master-staging table. The person in the bulk service area would then replenish the meat tray. In order to keep everything fresh and hot, no more than three trays of meat were ever set on the master staging area—three trays were used just in case for some odd reason all three tables ran short at once (which did not happen). Most of the time only one or two trays would be ready to go.

I positioned myself near the front of the serving area and, if I noticed someone who was notoriously long-winded and liked to visit, I made sure he or she moved to line six, and I followed the same policy when someone was walking slow or was in a wheelchair. Consequently, the bulk of those who were slower moving were in one line, and the remaining five lines moved swiftly.

We served 1,800 pounds of meat to 2,000 people in one-hour-and-forty-three minutes; approximately 19.4 people per minute or one plate every 3.09 seconds. We had hoped to serve one plate every five seconds, so we were ahead of schedule.

Friends and friends of friends were enlisted for the many displays. Some of the women spun wool into yarn, others demonstrated old-time quilting, and many colorful quilts were on display. An old-time blacksmith worked iron in the blacksmith shop, and a modern-day ferrier shoed a horse.

Using family members and friends, horse-working and calf-branding demonstrations took place. And, of course, Wes performed his aerobatics. This was always a hit.

I must say that all family members contributed equally and with great enthusiasm. Kids parked cars, old-timers put together photography exhibits and gave tours, there were games for the youngsters, and everyone else pitched in to serve and entertain the crowd. No one person contributed more than another, and all were willing to take on any task that was asked of them.

I believe Mike Frey handled the unpleasant task of sending out invitations. I can't remember who designed the invitation with Cecil and Captain on the front, but Cecil would have been happy. Because there are always some whom you forget to invite, all of us handed in several "last lists" and last minute add-ons. Mike also handled additional insurance for the day.

We estimate our total crowd for the day consisted of around 2,500 people.

The country and western band started about 9 P.M. and continued until 1 A.M., and all of us returned to our homes tired, happy, and satisfied.

We hired the local 4-H to tidy up and, by noon the next day, the old home place was clean. A "harrow-bed" arrived on Monday to pick up the bales of hay used for seating, the sanitation company picked up the dumpsters, and the peaceful demeanor that our family finds so endearing, returned.

The Miracle *(the author continues)*

More than a dozen years have passed, and still the memories last—of one remarkable day in September of 1987 when our family celebrated 125 years of farming and ranching. The fears and apprehension expressed by the older generation vanished as the day unfolded.

After many months of careful planning and three-and-a-half months of intense preparation, September 12 finally arrives. Polished and groomed like a maiden on her wedding day, the old home place waits with anticipation! In addition to our two-acre parking lot, the Thompsons, our neighbors across the road, have graciously allowed us to prepare another five-acre parking lot.

Trees have been pruned, grass mowed, weeds removed, and flowers planted. Four-hundred bales of hay have been strewn about for seating. The pool is clean, fences are painted, and the airstrip is in prime condition. A public telephone has been installed, as has a loudspeaker system and additional lighting. The ancient buildings have been restored and, for the first time in seventy-five years, Will Tracy's blacksmith shop—complete with a blacksmith—is in working condition. Antique farm equipment (formerly heaped in a junk pile) is mounted for display, with signage explaining the particular function of each piece. A temporary corral for roping and branding has been created, calves have been rounded up, and the horses are groomed and ready. Several bands have been engaged to play in sequence.

Lined up like pins in a bowling alley, fifteen portable potties await the onslaught. Perhaps most unique is JB's design for drinking fountains and hand-washing facilities. Eight new pairs of faucets—sixteen faucets in all—glisten above Will Tracy's ancient cement horse trough; eight faucets are turned up for drinking and eight are turned down for the washing of hands. Suspended from a pipe running above the faucets hang eight bars of soap attached to a rope and interspersed with twelve rolls of paper towels ready for a speedy spin-off. Directional signs point to each facility.

Not One Minute Later!

According to John's schedule, the family is to meet at ranch headquarters at precisely 11 A.M., *and not one minute later!*

11:00 A.M.—Miraculously, we are all there, all thirty-eight of us! We pick up our nametags, pose for a group photo and, like obedient school children, we stand and listen intently to last-minute instructions from Wes and John. We are informed that there will be six serving lines and that each table will need four servers. We are allotted five seconds to serve each plate.

12:00 noon *and not one minute later!*—Decked out in crisp white shirts, Levi's, and denim skirts, we line up for the photographer. Pictures taken, we discover lunch has wisely been prepared for family and working crew.

2:00 P.M. *and not one minute later!*—We hear John's voice over the loudspeaker. He is welcoming our guests and friends of long standing. By now parking attendants have begun segregating the crowd, directing the elderly and the cars with young children near the staging grounds and sending others across the road. Our hearts beat with the rhythm of the mariachi band as we welcome our guests.

No one knows exactly how many people are here today, but the estimate is 2,500. If Fannie's estimate was correct, this would be fewer than half the number that attended the Red Cross Picnic. But since that was a two-day affair, perhaps it would have been 2,500 each day. So far, everything is going well! The ropester is doing rope tricks, the mime is miming, and old-time fiddlers are fiddling. A carnival atmosphere prevails as smiling people move about. They look at the displays and appear to enjoy the demonstrations of spinning, weaving, and quilting. They watch horses being shod and visit Will Tracy's old smithy where a real blacksmith demonstrates his skill at the forge. The happy sounds of children can be heard as they play their games of sack racing, water balloons, and other timeless games supervised by newly married Jennifer and Mike Frey.

4:30 P.M., *and not one minute later!*—Our Master of Ceremonies is mounted on his horse as he announces the horse races. John, "Master of the Spoken Word," proclaims the winner and moves the crowd to the corral, all the while amusing them with his clever patter.

6:00 P.M., *and not one minute later!*—John's voice again, right on schedule! He announces the dinner hour. Several young ranchers begin removing the meat from the barbecue pit, as the dixieland jazz band strikes up. The servers man their positions at the six serving stations, and the huge crowd is served in a surprisingly short time.

6:30 P.M., *and not one minute later!*—An ear-splitting roar from overhead suddenly stuns the crowd. Of course, it is Wesley Selvidge as he and his little red Pitts begin the aerobatics. Wesley is a championship performer, and it is thrilling to watch at such a close distance as he puts the little plane through its maneuvers.

7:00 P.M., *and not one minute later!*—Let the band strike up, Let the music call, While we do-si-do, 'Round this western ball.

1:00 A.M., *and not one minute later!*—Now cowhands little, And cowhands tall, It's time for bed, Goodnight y'all!

The families of Tracy Ranch
invite you to celebrate with them, their

125th Anniversary

1862-1987

Date: Saturday, September 12, 1987
Place: Tracy Ranch headquarters
Wildwood Road and 7th Standard Road
North East of Buttonwillow

Program:

2:00 Demonstration and display of traditional ranch crafts
4:30 Western style barbecue
6:00 Calf roping
6:30 Aerobatic Demonstration
7:00 until ? Dancing

Cecil Tracy on Captain - 1949

The six partners, from left, Larry Frey, Bill Tracy, Jack Frey, Wesley Selvidge, John Tracy, and JB Selvidge, September 1987

Like obedient children, the family listens to last-minute instructions by John and Wes. We are allowed five seconds to serve each plate—we did it in three.

Frances greets guests

The deputies arrive

Reception tables

Name tags: Marilyn Tracy and Kenneth Holden

JB's wash trough

Ferrier at work

Antique farm equipment

Sprucing up

JB and the "tallest eucalyptus"

After the home place burned, only Cecil's intricate brickwork remained

Cleaning up in 1982

Kathryn Tracy with Bill (left), John, and David

Susie and Bill Tracy

John and Donna Tracy with children Jessica and Allen

Janet and Wes Selvidge with children Jason and Andrea

The Frey family, clockwise from back left: Jennifer, Claudia, Larry, Martha, Jack, Nancy, Tracy, Jack R., Stephen, and Mike

Six serving lines

Hey y'all come! Jack stirs the beans while John and Wes uncover the pit barbecue

2,500 guests attend

Tracy, Casey, and Gary Gill

Casey Gill

Fiddlers

Dixieland jazz band

This tireless little couple, Mr. Twisselman and
Miss Sill, danced 'til the break of day

Entertainment

Sack race

Balloon toss

Several planes flew in to attend the party. Nancy Frey Trahan, in Cessna 172, was among them.

Wesley Selvidge thrilled spectators with an aerobatic performance in his little red Pitts

John Tracy

Roping

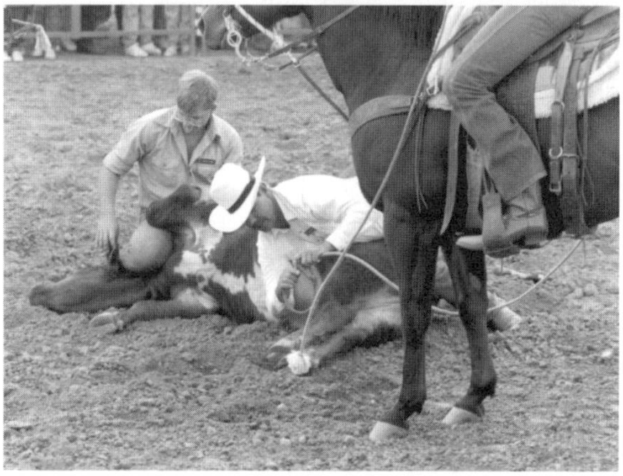

Jack R. Frey and John Tracy

Jack R. Frey branding

"Jumpin' Jack" Wes Selvidge

The old ranch shop looks much as it did when Fannie and Will Tracy were alive

Johnny Uhalt demonstrates use of the forge

The Miller and Lux wagon served as "meals on wheels" for cowboys—including Cecil—in the early days

Seating provided by 400 hay bales

THE LITTLE RED PITTS

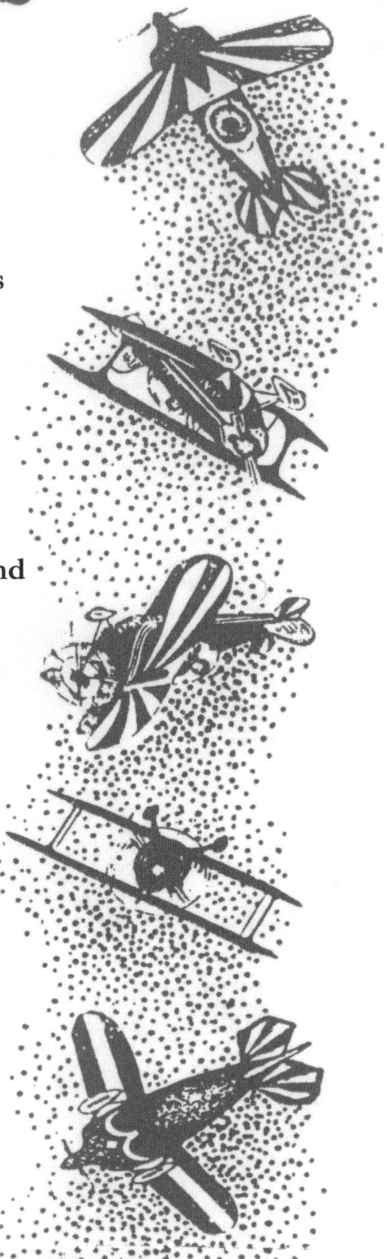

The Little Red Pitts
By Martha Frey

The little red PITTS is on her way
To celebrate this Western day.
She'll touch the clouds and thrill the crowds
And greet our western friends this way.

In a trail of smoke she'll say "HELLO"
To all the folks that stand below.
Then UP, UP, UP and away she goes—
Just how high, nobody knows.

Then down, down, down to the speed of sound
She circles low, near to the ground.
And UP, UP, UP her wings on high
She barely makes a dot in the sky.

Now all at once we hear no sound—
No sound comes down from sky to ground.
Oh my God she's in a stall!
The poor little Pitts is going to fall!

Like a helpless thing on the end of a string
A wounded bird with just one wing
That little red plane hangs up there
Oh, dear Lord, please hear our prayer.

An eerie silence pervades the air
As in disbelief the people stare.
And still no sound, no sound comes 'round
To all the people there on the ground.

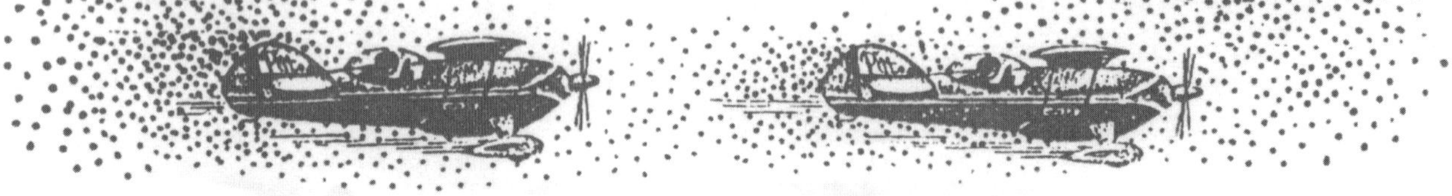

Then at last we hear that sound
From up above, the engine comes 'round.
And a sigh of relief is heard below
As that sturdy engine starts to go!

Now once again, her wings on high
She makes one last design in the sky;
Then down she comes and passes low
Over the crowds of folk down below.

She dips her wings as if to say
GOODBYE FOLKS, I'm on my way!
But hold your horses, please don't go
For there's gonna be a RO-DE-O.

Yippy-ki-yay and yippy-ki-yo!
We're going to have a rodeo show!
On a bail of hay please find a seat
For now comes beans and barbecue meat.

And later on you'll want to dance
To a western band in your Levi pants!
Dance, dance, dance, dance the night away
Dance, dance, dance 'til the break of day.

Now one last time she's on her way
And she dips her wings as if to say
Let the dance begin; let the music call;
In the western way, GOODNIGHT Y'ALL!

Martha and Nancy with a "watermelon head" centerpiece

The Morning After

Larry Frey sweeps the dance floor before breakfast the next morning

Santa Barbara cousins Mary and David Tilton are overnight guests

A Home on the Home Place—The Tracy Ranch

By JB Selvidge

William Tracy first built his permanent home on this site in 1902, four years after establishing today's Tracy Ranch headquarters here in 1898. Expanded and remodeled through the decades to meet the needs of a growing family and ranch operation, the original home tragically burned to the ground (excepting brickwork) in October 1974 due to electrical causes.

Today's structure as depicted at 19400 Wildwood Road was completed in April 1984 for the owners, JB and Diane Selvidge. Its design evolved around the beautiful and unique brickwork done as a remodel in 1968 by then-occupant Cecil Tracy. Its lines were inspired by the styles of Frank Lloyd Wright.

Its interior features massive eight-inch by seventeen-inch cross-section-exposed beams that originally served as railroad bridge timbers (date nailed 1939) and a large six foot hearth fireplace from the original home. The architect was Robert Funk of Bakersfield; contractor was Payne Construction of Shafter; Graves' Construction of Shafter completed the concrete work.

This "home on the home place" lies in the shadow of a 95-year-old majestic eucalyptus tree and mirrors the rugged past of the Tracy Ranch. Founded in 1862 as a frontier rangeland cattle operation, the ranch has so far endured nearly 140 years as a family-held, fifth generation operation.

The completed home

JB's home is located on the site of the original Tracy homestead. After the fire, "only the brickwork remained."

Ninety-one brand at the home's entrance

Brick detail

WESTERN CULTURE

A Celebration of Western Culture

Three years after the ranch's 125th Anniversary celebration in 1987, John Tracy attended a meeting pertaining to "Agriculture in the Classroom" with two members of the Kern County Cattlewomen's Association, Julie Anthes and Lou Holmes. After the meeting, they decided to stop off for a drink at Wool Growers, one of Kern County's favorite Basque restaurants.

In the course of their conversation, the declining revenues of the cattlewomen's organization came up. Memories of the Tracy Ranch anniversary celebration must have triggered their imaginations, for suddenly Julie came up with the idea of holding an annual celebration. It could be a showcase of Western culture!

"An all-day affair!" said Lou, "With horse racing, calf roping, and branding!"

"And complete with a western barbecue!" said Julie.

With a second round of drinks, the idea became even more enticing. And where could they hold it? Without hesitation John Tracy spoke up:

"Well, if Grandma Tracy were alive, she would insist that you hold it on our ranch."

Lou Holmes, left, John Tracy, middle, and Julie Anthes visit Wool Growers and plan the Celebration

Later, without the benefit of alcohol, the idea still seemed like a winner. These two "movers and shakers" of the cattlewomen's association went to their board of directors and presented the idea. It was not without debate, but in the long run the idea prevailed. Soon plans were put in motion for the first event. It was to be called "A Celebration of Western Culture." It was to be held September 12, 1990, on the Tracy Ranch.

Mona Carver remembers:

The Kern County cattlewomen had struggled for years to find a way to support their work with beef promotion, scholarships, and the fair booth. In 1990 Julie Anthes came to the cattlewomen with an idea that was new to our area: hold a "Western Culture Day," with a barbecue, cowboy poetry reading, and dancing. She had discussed it with Lou Holmes. After much discussion and some dissention, the leaders kept their enthusiasm. When someone asked, "Where would we have it?", various places were discussed and John Tracy was contacted. He never hesitated and said, "Come to the Tracy Ranch."

It was settled. We would go to the Tracy Ranch. Julie Anthes was the first chairman. She thought we needed a book for advertisers, to help with expenses. She put together the first book, and has been chairman of the book ever since. We were off and running!

We thought it would be nice to include the cattlemen, so we gave them the bar and the "Cow Plop Bingo." We did no advertising other than the book. The ticket price was $25.

After it was all over and the dust settled, we had a profit of over ten thousand dollars. We were rich! We felt generous and gave the Starlite Foundation $3,500 ($500 of it in the Tracy name). We gave Los Flojas Cancer Fund $2,000 and Teen Challenge $1,000, making a total of $6,500. The cattlewomen have continued to give to charities each year.

The author remembers:

As the years have rolled on, each chairman has tried to bring new attractions to the celebration. That first year, the barbecue was hired out to professionals. The second year, Jack and Jane Davidson did it. For the next nine years we were indebted to Rick and Martha Twisselman for this prodigious undertaking.

Nothing succeeds like success! Since the inception of that first celebration, on the second Saturday of each September cattle people from all over the valley work hard to create a happy, carefree day for the people of Kern County. Would-be "cowgirls" and "cowboys" pull on their boots and Levi's and don their western regalia to mingle with the real cattle people. Mona Carver tells us that even people not connected with the organization volunteer their assistance.

By 1998 more than 1,500 people attended the celebration. Like Helen Carver's dream of a cattlewomen's organization, the success of the event has exceeded all expectations. Thank you, cattlewomen and cattlemen for sponsoring this day of wonder, this colorful and unique Celebration of Western Culture!

For nine years Martha and Rick Twisselman supervised the western barbeque. Delicious sliced tri-tip roast, salsa, and beans are served.

Tribute to the Cattleman

When it comes to breeds, the cattleman stands apart from all the rest. By necessity, he is rugged, hard working and shrewd. In addition to his Stetson, he wears many hats. He must be a doctor, a mathematician, a nutritionist, a Jack-of-all-trades, and a gambler.

As a doctor, he must be able to deliver a calf, inoculate, castrate, or de-horn. He must be able to treat a wounded animal and to know exactly where to insert a knife to relieve a bloated cow. When the time comes, he must be able to put a terminally sick animal to death.

As a mathematician, he must be skilled in using an electronic computer as well as the one in his brain. In more ways than one, he must watch his step in his western boots. Arithmetic is part of his game. If he does not get his figures just right, he will be washed out in a hurry.

As a nutritionist, he must know precisely how to balance the menu of his herd while taking advantage of the by-products of other crops.

As a Jack-of-all-trades, it is to the cattleman's advantage if he can rope and brand a calf, shoe a horse, and mend a saddle. He must be able to walk in manure and know how to spread it!

As a gambler, he must know when to buy and when to sell; to tell a "good keeper" from a poor one. He must be able to judge when to be cautious and when to be bold, and to gauge the "fucktuating" market. No player in Las Vegas takes greater risks than the cattleman!

And even if he passes all of the above challenges, he is still at the mercy of Mother Nature. Too much rain or too little, flood or drought, varmints, vermin, blight or infestation can pose a threat at any time.

Yet with all the hardships imposed on his way of life, the cattleman survives. He is a symbol of fortitude and courage, facing adversity with a sense of humor. He is the guardian of our culture who connects us with our roots in the unique western way. It is fitting to pay him tribute with a special western day.

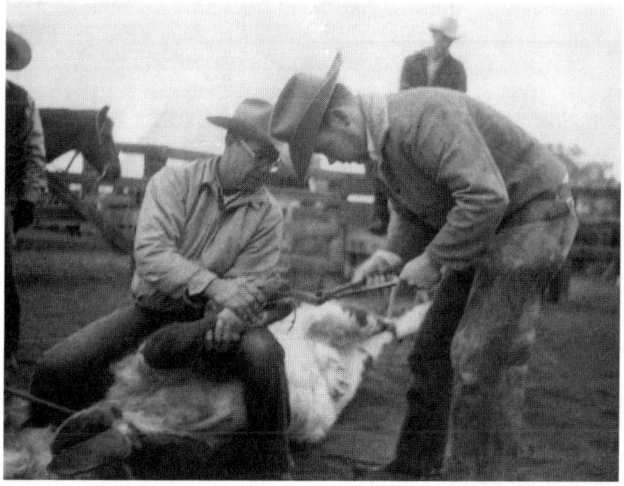

Tilton Tracy holds calf as John castrates

KERN COUNTY CATTLEMEN'S ASSOCIATION

	Kern County Cattlemen's Association Past Presidents	Kern County Cattleman of the Year
1916	H.A. Jastro	
1925	Phillip A. Klipstein	
1928	T.J. Brown	
1931	Phillip Klipstein	
1933	Bertram J. Snedden Sr.	
1936	Art Alexander	
1942	Jule Villard	
1944	LeRoy Rankin	
1945	David F. Sneddedn	
1948	Carl Carver	Marguerite Folensbee Ashe
1949		Grace Snedden
1950	Henry Bowen	Jim Robertson
1951	Henry Bowen	Phil Klipstein
1952	Jim Ben Williams	Walter Rankin
1953	Jim Ben Williams	Art Alexander
1954	Carl Twisselman	Fred Lavers
1955		Hub Russell
1956	S. Ward Woody	Carl Carver
1957		Oscar Rudnick
1958	Jack Lavers	Albert S. Goode
1959	Bertram Snedden, Jr.	Jule E. Villard
1960		Stanley Smith
1961		William Alexander
1962		David F. Snedden
1963		Henry Bowen
1964	Harold E. Thurber	Jim Ben Williams
1965		Clint McCray
1966	Grant Grisedale	S. Ward Woody
1967		W.F. Smoot
1968	Keith Alexander	Jack Lavers
1969		Mark C. Parks, Sr.
1970	John C. Hershey	Carl F. Twisselman
1971		James C. Beagle
1972	Carver Bowen	Bertram J. Snedden, Jr.
1973		Roy V. Parker
1974	Kenneth Twisselman	Harold E. Thurber
1975		Robert S. Dickson, DVM
1976	John (Jack) Davison	Kenneth Mebane
1977		H. Skinner Hardy
1978	John Beard	Grant Grisedale
1979		Glenn Record
1980	Richard Snedden	Carver Bowen
1981		John C. (Jack) Davison
1982	John App	John Hershey
1983		Nono D. Williams
1984	Jack Porch	Kenneth Twisselman
1985		Ray Holtzclaw
1986	Matt Echeverria	Mary Stone
1988	Tom Twisselman	Russ Carver
1989		Glenn Van Schaack
1990	David Lavers	Fran Simpson
1991		William (Bill) Kissack
1992	Nathan Carver	Richard Snedden
1993		Ralph Phillips
1994	Jeff Bowen	Matt Echeverria
1995		Nathan Carver
1996	Jeff Stone	John Beard
1997		Bob Beechinor
1998	Bruce Hafenfeld	John Tracy
1999		Kenneth Miller
2000	Jim Pennington	Bruce Hafenfeld
2001		Tom Twisselman

Following is the first schedule of events and the others that have been added through the years:

Chairman: Julie Anthes	September 15, 1990
Western Art Show	12 p.m. to dusk
Exhibits and Boutique	
Silent Auction	4 p.m.
Roping and Branding	5 p.m.
Western Tri-Tip Barbecue	5:30 – 7 p.m.
Aerobatics by Wes Selvidge	6 p.m.
Auction	7 p.m.
Dancing to Steve Woods	8 p.m. to midnight
Chairman: Jane Davison	September 14, 1991
Trading Post	
Horse racing and Cow Plop Bingo	
Rented a bus to run from Bakersfield to Tracy Ranch, in place of designated drivers	
Chairman: Jane Davison	September 12, 1992
Mike Moroff, actor, singer and artist	
Chairman: Nellie Garone	September 11, 1993
Indian dancers	
Horses and pack animals	
Costume contest	
Chairman: Pat Filkins	September 10, 1994
Back Country Horse Demonstration	
Cowboy Poetry	
Chairman: Pat Filkins	September 9, 1995
Monty Mills and His Lucky Horseshoe Band for dancing	
Chairman: Karin Trahan and Lynell Etcheverry	September 14, 1996
Horse riding	
Square Dancing exhibition	
Kern County Senior Fiddlers	
Tommy Hays and Western Swingers for dancing	
Chairman: Lynell Etcheverry	September 13, 1997
Gary McMahan: Cowboy singer, yodeler, poet and "general nuisance"	
Chairman: Lynell Etcheverry	September 12, 1998
Horseshoe Throwing	
Sons of the San Joaquin	
Money Mills for dancing	
Chairman: Julie Anthes	September 11, 1999
Chairman:	September 10, 2000

Wes Selvidge prepares to entertain the crowds attending Western Culture Day in his little red Pitts. Frances once said, "Wesley is the main pilot of our operation and ranked third in the nation at one time in flying aerobatics contests. He has a cute little red Pitts. I would love to go up with him, but it has only one cockpit!"

TO CATTLEMEN ALL

Now let's all drink
To the cattleman
With his weathered boots
And cheeks of tan.
He's a manly man
A right fine fellow
Who keeps his word;
His heart is mellow.
He works real hard
There's no other way
To tend to his herd
And to make it pay.
And here's to his wife,
She works hard too;
Doesn't have it easy
Like city folk do.
Here's to his horse
Both tried and true
At the end of a rope
He knows what to do.
Here's to the cowhand
Both old and new
At the end of that rope
He knows how, too.

Here's to the doggie
And his plaintive cryin'
As he feels the heat
Of the branding iron.
To the silver bit
And the old worn saddle
That rode the range
As they herded cattle.
And here's to the old
Pickup truck; may it
Last another year
(With a little bit of luck).
Here's to the beans
That simmer all day
While folks celebrate
In the western way.
Now have one more drink
Of the Mountain Dew
While "oysters" sizzle
On the barbecue.
Yes, let's all drink
To cattlemen all
May the market be good
At roundup this fall!
Now I'd better quit drinkin'
My head's in a stew
For I'm getting drunk
And my horse is too!

Tribute to the Cattlewoman

No less rigorous than the lot of the cattleman is that of the cattlewoman. Usually she must make her home far from the conveniences enjoyed by her city friends. No last minute trip to the supermarket for her! She must plan her life with clockwork precision or suffer the consequences. In winter, roads are often treacherous, making trips to town even more difficult. Getting children to school or to the doctor can be an ordeal. Meals are often irregular, but she must be able to extend her western hospitality to all that pass her way. This is her tradition, the gracious western way.

Like her husband, she wears many hats. In addition to the conventional roles of wife and mother, she is often called upon to play those of doctor, veterinarian, repairman, and bookkeeper. She may find it necessary to treat a snake bite, set a broken bone, help repair a fence or use a computer. She must be able to improvise, organize, economize, and sometimes even apologize.

To the Cattlewoman

A Tribute to the Kern County Cattlewomen's Association

By Martha Tracy Frey

**Let the drums roll out
Let the trumpets call—
To old cowhands
And ranchers all!
To cattlemen big
And cattlemen small
We're inviting you
To gather each fall.
There's a dance tonight
Gonna stay out late
Gotta look just right
Gonna celebrate!
Don't give a holler!
And don't give a hoot!
Put my hat on straight,
And pull on my boot.
Get my Levi's out
With the perfect cut
Stretch 'em good and tight
'Round my "You know what."
Gonna dance tonight
Gonna heel and toe
In my western shirt
Gonna do-si-do!
Get your turquoise out**

All you cowgirl belles,
With silver buckles
And your concho shells!
Now you old cowhands
Turn the doggies loose
They're soon gonna feel
The sting of the noose!
Let the ropes roll out
Heat the branding iron
Hear the cowboys shout
And the doggies cryin'.
Put the calves on the run,
For there's work to be done
Here today... *sorry!*
But we *name* you this way.
Let the horse-race start—
And now they're away
With their tails straight out
As we shout "HOORAY"!
When the branding's done
Have a drink or two,
And re-live the day
Like the cowpokes do.
We can smell the meat
And beans that simmer—
From the barbecue's heat
IT'S TIME FOR DINNER!
Now let us all sit
On a bale of hay
As we celebrate
In the western way.
Let the drums roll out—
Let the sunset fall—
Strike up the band!
Let the music call!
All the work has been done
In the western way—
Now it's time to have fun
With the K. C. C. A.

The Cowbelles

Alice Rankin Beard, my girlhood friend, eventually returned to her roots in the Bakersfield area. We renewed our acquaintance, and I was delighted to learn that she has been painting in watercolor. She has exhibited her paintings in Bakersfield and at the fine Historical Museum in Kernville.

The Kern County Cattlewomen's Association was formed in 1948 as the "Cowbelles." Later, the organization changed its name to the Kern County Cattlewomen's Association. In 1909, they chose Alice as "Cattlewoman of the Year." She recalled for them the founding of the Cowbelles:

"Let me tell you about how this cattlewomen's organization was started fifty years ago. Helen Carver called us together; about forty attended. And she said, 'I'm just like Mammy Yoakum of Dogpatch. I've conjured up a vision. I threw myself on the kitchen floor and kicked my feet in the air and conjured up this vision and I saw an organization of cattlewomen who would meet to plan ways to assist the cattlemen, promote beef, and from time to time schedule events to bring far-flung ranching families together.' Bylaws were agreed upon, and so it began. I wish that Helen Carver would be here tonight and realize her vision. It has doubtless gone beyond her most elaborate fantasy."

In her acceptance speech on February 20, 1999, Alice includes her reflections on growing up as a rancher's daughter. In it, her keen perception of nature and of the sights, sounds, and smells that invade the senses are all reflected. She has graciously allowed me to print it here:

"Again, thanks for this evening's honor, and to all of you who are here to share it, thank you for coming. And gratitude to Lynell Etcheverry and Lou Holmes for their special care and attention in decorating and making it all work. My thanks as well to all the cattlewomen who have had a hand in any of these details and who have pointed this tribute in my direction.
I have a question. Why would anyone go into the cattle business?

"Children early on can be introduced to the practical application of several functions of mathematics. So you arrive at the gross weight and multiply the price per pound. Wow! Looks pretty good. Might get a new pair of boots. Maybe replace the old pickup. But wait, there are some subtractions to make—the shrink, freight, death loss, inspection fees, cost of feed, taxes, interest, insurance, organization fees, hired help, living expenses, veterinary expenses, new bulls—that's not all; it goes on. So that net figure too often is a negative number, explaining the lettering on a tee shirt that reads, "Behind every successful cattleman is a wife who works in town." So you half-sole the boots and make the truck go another year.

"Also affecting that price-per-pound are out-of-reach, out-of-control variables that impact the net figure including, first of all, the weather. Next, ambitious politicians and economists who legislate price controls as in the Nixon era. Or, an act of brilliant strategy such as the dairy buy-out, dumping thousands of cattle on the market, destroying the orderly consumption and price of beef.

"A popular television personality, Oprah Winfrey, says in response to misinformation, 'I'll never eat another hamburger,' sending an already weak market into a downward spiral. The growing popularity of fad diets that exclude red meat also contributes. Now we have some real competition in something called the 'Hallelujah Diet,' which quotes the Bible. It is increasingly difficult to own property without interference from animal rights groups, environmentalists, endangered species intrusion, and water management legislation. Also, there is the frustration of trying to preserve your ability to control predators in the face of opposition from well-funded groups such as the Sierra Club. That happened in this last election with Proposition 4. Now you can eliminate such pests as coyotes only if they are threatening human life—never mind your newborn calves or sheep.

"Another impact: fickle financial institutions. Near-monopolies in feeding and marketing can influence your bottom dollar. And there are fads within the industry, such as the buyer who shows up at the ranch and says, 'Boys, I'm going to have to bid you a little mean on these calves, they are showing too much ear.' That's important when they are hanging on the rail? And now, for heaven sakes, black is the color of style!

"Why would anyone go into the cattle business?

"You know what? There will always be cattle-ranching families. I'll name a few of the unique reasons:

"Cattle people are optimists and they are gamblers. The reasoning sounds like this: 'These wet years could continue' or, 'As soon as we get the numbers of cattle on feed down, the price should improve.'

"Then many of you feel just like I once heard Lawrence Snow say, 'There has never been a day when I didn't want to go to work.' It is easy to understand. Every day is indeed a revelation, punctuated by sights, smells, sounds, and sensations peculiar to the business. What equals…

- The smell of damp earth around a watering trough on a warm summer evening.
- Seeing a cow claim her lost calf.
- Bringing the steers to the cake from the bottom of the meadow, riding bareback on a fat-backed horse.
- Cleaning a watering trough.
- Working cattle on a horse who is interested in his assignment.
- Knowing horses that you admire more than many of your acquaintances.
- Giving a horse a bath.
- Hearing the echo of a bull's statement in a high mountain meadow.
- Seeing a couple of de-horned bulls paw the dirt and square off in mortal combat, complete with snorts, bellows, and a final run escaping extinction.
- Hearing the click of cattle hooves as they walk along.
- Seeing horses stand head-to-tail, switching flies.
- A nest of newborn calves with one mama babysitter.
- Watching a horse roll.
- Riding in the back of a pickup with saddles, wet blankets, and wagging-tailed, dripping dogs.
- A child's responsibility and pride of ownership in a horse, a dog, and a saddle.

"John Beard would come home from school, get off the bus, and his dog, Joe Lewis, would be there to meet him. Then they went to the corral to catch John's horse, Shorty Freeman.

All would come to the house. John built a peanut butter sandwich and they would sit under a tree. John would take a bite, give Joe Lewis a bite, then Shorty Freeman. Now, if you have never seen a horse deal with peanut butter, it is something you need to put on your 'must see' list!

- A child's opportunity to see, learn, and participate with respected adults who are sensitive to the skills involved in handling livestock in a business-like, humane way.
- Or, how about the smell of sage after a rain?
- The feeling of utter triumph of a well-thrown loop that picks up both hind feet.
- Seeing a big string of cattle line out.
- The silhouette of a cowboy against the setting sun.
- Watching a real bridle-man work a horse whose feet are gathered, ears forward, chin tucked, mouth closed. Walker Rankin was such a horseman.
- Brisk, cool, fragrant mornings and clear, starry nights.
- Silence; pure utter silence, perhaps broken by the snort of a horse, the distant low of a cow, maybe birds and leaves rustling, and the occasional squeak of a windmill—and, while my father was living, the sound of Vin Scully calling a Dodger's baseball game.
- The greatest draw of all, the association with noble people. Which brings us to this occasion and to the wonderful folks here tonight.

"This group sponsors several very generous, renewable scholarships. Two are named in memory of Christine Beard, and subsidize teachers who go on for more background in ag in the classroom. Another scholarship is in memory of an outstanding member, Ruth Smoot French.

"It also contributes toward lobbying issues affecting the industry, promotes ag in the classroom, and donates to local charities. Just this year, it published another very fine cookbook, and it participates in the Kern County Fair and sponsors cattle sales at the fair.

"Every year, to finance their activities, they put on a top-drawer event at the Tracy Ranch called Western Culture Day. The afternoon is taken up with raffles, auctions, excellent western music, an air show, concessions, arena events, horse races, a delicious barbecue dinner, and dancing. The event becomes more popular every year.

"A few reasons have been noted here, assuring there will always be ranching families. But in preserving this way of life, the eternal optimism and the willingness to take a chance must be accompanied by ingenuity, inventiveness, and additional efficiency to meet and overcome the formidable challenges. Those challenges will be met. It will be done!"

Donna's Horses

Donna Tracy remembers:

My experience with Tracy horses began when I was a student at Cal Poly in the colt-breaking class. Before John and I were married, Tilton had sent over a big beautiful sorrel filly from Jimmy Holmes. I should have suspected a problem when it took her three times as long as the other horses to gentle down enough just to get on her! Then, just as she seemed to calm down, she would explode in a bucking frenzy. It got to the point where one of the two instructors always rode next to me to try to catch her in one of her "fits." The last straw came when she bucked me off and then threw herself on the ground! After getting her up, Mr. Hadley took her into the round corral where she acted like she had never been ridden—she didn't appear to even know how to respond to the bit! After checking into her past, it was decided she probably had gotten hold of some locoweed, and this had made her a little crazy.

I fell in love with a cute little black mustang shortly after John and I married. She actually belonged to Cecil and, when I asked if I could have her to break, Cecil said he planned to break her for his stepdaughter, Leslie. I continued to ask him about the horse, but he was always just getting ready to start her. Eventually he divorced Leslie's mother and the mustang was mine! Of course, now I had an eight-year-old wild mustang that had really been enjoying her life of ease up to this point!

Despite all the schooling I had in the colt-breaking class, this wild little gal bucked me off three times in fifteen minutes. I succeeded only in scaring Kathryn to death when she saw me covered in leaves and dirt with tears streaming down my face—I was so mad! She called John, and he chewed me out for being out there by myself. But I had wanted to surprise him with my new "broke" horse. John got on her and I have never to this day seen a horse that could buck like this mustang! Amazingly, John stayed with her. He couldn't get her pulled up to stop, but he finally bucked her into a corner where she quit. He then turned to me wide-eyed and exclaimed, "Chicken feed!"

Then there was a big long-legged gelding I was breaking. He was sweet but a little jumpy, and John and I were out at Gooselake by ourselves moving cattle. He was training a new cattle dog, and the dog kept running ahead of him. John likes his dogs to follow behind, so he kept throwing things at this pup. The commotion was making my horse very nervous, and I told John that if he got me bucked off I was going to be pretty mad. About that time the pup ran between us; John threw his rope down in front of the dog, and the dog jumped right under my colt. I don't know who was more surprised—my horse or me; all I know is that colt jumped straight up and, on the next jump, he went sideways, and I was sitting on the ground. My thigh was so swollen the following day I couldn't get my jeans on.

John has always been very careful around horses, and by the age of fifty he had yet to break a bone. This is a rarity for people exposed to horses and cattle all their lives. One day we were moving cattle through an old farm field next to the Meadowbrook, and the cattle took off running toward 7th Standard Road—and the traffic. When you are running full speed after a herd of cattle, your only concern is stopping them before they reach the highway. Next thing I knew, my horse's front feet landed in the bottom of an old irrigation ditch covered with weeds, and I was tossed over his head onto the ground. As he stumbled up out of the ditch, he stepped on my ankle. I guess we were really lucky he didn't break a leg or fall right on top of me. John came running back to see how I was. My first thought was to remove my boot before my ankle could swell up. Not having money to replace the boot, I was worried about having to cut it off. But John's first obligation was to keep the cattle off the highway, so I lay there 'til he could come back for me. He did manage to remove the boot, but not easily. Since it would take too long to get an ambulance, we decided to go to the hospital by truck.

John's idea of caring for an invalid was to buy the largest size bucket of Kentucky Fried Chicken. It lasted a week; I lost five pounds and he gained five. Even worse, we eventually ran out of clean dishes. John set a chair behind me so that I could bend my leg up backwards, set it on the chair, and stand at the sink on one leg to wash dishes. What an idiot I was! I have taught my daughter better.

Someone once asked John why his wife always ended up on the rank horses. His reply was, "If I get hurt the operation of the ranch is disrupted. If my wife gets hurt, she can stay home and recuperate." But my love of horses just kept me coming back for more. There finally came a time when I saw John lose complete control of his temper with an animal. I was about six months pregnant with Jimmy, and we were moving cattle out at Gooselake. As usual, I was riding a green colt, and it suddenly began bucking. I managed to get his head pulled around and stopped; this was probably the first time I was really scared. John had me get on his horse; then he got on the colt. He convinced the colt that bucking was not a good idea. This was one of the few times I ever saw John react totally in anger with an animal. (He has less patience with humans.)

As years went by and the kids got older I did less riding. It was pretty difficult to find a babysitter for 5 A.M. However I continued to race in charity events—at Harris Ranch and on our ranch at the cattlewomen's fundraiser. My last (I hope) near disaster happened during one of the horse races along the ranch cotton fields for a Cattlewomen's Western Days celebration at the old headquarters on Wildwood Avenue.

We knew this horse (Tommy) was hard to stop, so Tom Twisselman was at the end, ready to stop me. I won the women's race but Tommy missed the catch. I knew I was in trouble. I had two choices; I could run straight into a cotton field, or I could try to make a turn onto a farm road. I chose the road, but I didn't know that there was a pile of wood and a fifteen-foot harrow parked where I needed to turn. Bless his heart, that horse managed to jump the harrow at a dead run! It still took another three-quarters of a mile to get him stopped.

I was shaking and in no mood for jokes. A bad fall is not something one wants to face in their mid-forties. John arrived in his truck, immediately jumped out, and ran his hand down the horse's leg. I yelled at him, "Not a bit of concern for your wife, all you care about is the damned horse!" He replied, "I can see you're okay; you're still on the horse. I just wanted to make sure you hadn't injured him—we've got to run him in the finals."

I was in no mood to race a runaway horse again. But John hauled me back, got someone to care for the horse, put a couple of glasses of wine in me, and convinced me it would be foolish to let another person take the glory of winning the finals after the hard work was behind me. John was convinced I did better when I was mad so he always put quite a bit of heat on me. I still had to go back and race again! One of the harder things I've had to do. It was a good time to pass the racing on to the younger generation. [Donna easily bested the winner of the men's race to win the finals. -NFT]

Donna's House

John Tracy remembers:

I often wonder why my wife didn't leave me in the early days of our marriage. (Well, maybe in these days, too!) Work was long and hard, and the death of my father was never far from my mind. We didn't need much water for the plants in those days; her tears flowed on a regular schedule. One month I was home three days. All I wanted was a little loving, some sleep, and clean clothes for the next go. I had no time to introduce Donna around, so she was left to her own devices when it came to meeting people and creating a social life. After a while she made her own friends, and now you'd have a hard time getting her to move anywhere else.

Living in Aunt Dee's old house was great; it was in great shape and was one of the perks of living on the ranch. But in the summer it was so hot and humid that the swamp cooler couldn't keep up, and we would eat dinner outside. Donna sold her car so we could buy a washer and dryer. Probably the best investment we ever made!

One of the things that keeps a woman rooted is a house. Although Donna denies it to this day, she was on me early to get a house of our own. It was apparent that we could not build a house on our income and the idea of moving a house in and fixing it up, as my parents and relatives had done, began to have some appeal.

As luck would have it, the 58 Freeway was being extended through Bakersfield, and several houses were coming up for auction. We soon learned that extensive remodeling was out of the question, so we began to look for houses that were close to what we wanted. A total budget of $30,000 was set up to move, restore and furnish the house. We figured out that $6,000 was the top price we could pay for the building itself.

It was a Friday afternoon, and we were shipping cattle off the Spring Ranch in Shandon when I got a call that Donna had located the perfect house—and it would be for sale on Sunday! We needed a $6,000 cashier's check to bid. I explained that we would still be shipping on Sunday, and the banks would not be open to get a check. Donna, her voice catching in her throat, insisted that only this house would do, and we needed to buy it because there would never be an opportunity to get "just what we want" for the rest of our lives.

Now, any real man will tell you he'd rather be kicked in the balls than make his wife cry over a matter this important. I told her to wait and, as usual, called my Uncle Jack Frey for help. In about fifteen minutes he called me back and told me to call the banker at home. The check would be waiting for me Saturday evening. Not only was I to get the money after bank hours, but the check could be paid back after we arranged for financing. Such was the bond between banker and client in those days. I don't know what Jack told him, but it would not be the last time my uncle would work his financial magic in the years to come.

On Sunday we started the bidding, and it stalled around $4,000. Then it started rolling again at a steady pace and, when I bid $6,000, I knew the next bid would take me out. No one else bid, and the house was ours!

There are enough stories about the house for another book, but the sight of our house moving down 7th Standard Road in two pieces will always stand out in my mind!

The author remembers:

John Tracy has told you about their purchasing the "Freeway House." Each house on the ranch reflects the individuality of its owner. This one is no exception.

With paint and brush Donna Tracy has added her own artistic touches to their home. In the living room the original brick fireplace beckons the visitor into a room filled with carefully selected, antique furniture. Donna and John added a playroom of no small proportions. It encompasses a bar, shuffleboard, ping-pong table, and a pool table. Truly a room to enjoy!

The most intriguing aspect of this room is the treatment of its interior walls. I have no idea where John secured the vintage wood that was used to line them, but it had to be from some ancient structure, for its deeply textured surface gives testament to its age. Gray in color, not unlike the head of the cattleman who occupies its space—bleached that way by scorching sun and drenching thunderstorms through uncounted years—it bears witness to its own story. Positioned on one wall, large and stationary, is a barn door. Barns were once more important than dwellings; no doubt it was once the pride and joy of some hard-working farmer.

Many fine photographs and paintings of western art hang on the walls, among them a collection of watercolors by Los Osos artist and cherished friend, Chang Reynolds. Chang is no longer with us, but he has left his mark upon Jerry Slough.

On one wall hangs but a single large photograph, fragile and somewhat yellowed by the passing of time; it too has a story to tell. It shows a young man gripping the sursingle atop a wildly bucking Brahma. On closer examination one can detect a determined grin on the face of the rider. As Will Tracy would have put it, "Not only is he riding him 'straight up,' but he is also riding him backwards!"

Cattle Buying

A Buying Opportunity

John Tracy remembers:

There was a drought in Arizona, Texas, and New Mexico in 1978, but California had enjoyed a good year and we had plenty of grass. We could buy cattle in the stricken areas, bring them to California, fatten them a little, and turn a profit. Very few opportunities like this come along in a lifetime: things are right, and you know it.

The only drawback was that the cattle had to be tested for brucellosis before departure and, again, one month after arrival. If any reactor was found to the disease, your entire herd could be quarantined until you sold the infected animals under special permit; you then needed two clean tests for the rest of the herd. Many ranchers were afraid to do this, as it could prevent the sale of any cattle for a long period of time. We had "holding" fields at Gooselake that allowed us to ship cattle in, wait a month and, if the cattle were clean, ship to one of the many ranches we had leased. In that way, if we had an infected cow, we would only tie up the herd she was in. Cattle that were "suspects"—not a 100 percent clean test but testing low enough to not be considered infected—could be sold and the rest of the herd would be declared clean.

This method of purchasing cattle meant many return trips to the droughty areas as each herd was cleaned up. Under this system we bought some 1,200 cows, pairs, breds, and opens. (Cows that are not pregnant.) We got through the period without any reactors, and made a good profit on the venture.

Close Calls

Cattleman Kenny Miller was negotiating for the cattle on our behalf and taking a commission on the sales. He was using his Cessna 210 to fly into the drought-stricken area, and one day we landed on a dirt crop duster strip in New Mexico to look at some cattle. We left in the heat of the day with the temperature at 105 degrees. It soon became apparent that the runway was a tad short—warm air gives less lift. We cleared the power lines at the far end of the strip with the stall horn blaring. While sinking toward the ground Kenny shoved the stick forward to put the nose down and, just before reaching the ground, he pulled back again. We covered the country at about sagebrush height for a while until we gained speed and could climb. Neither of us spoke for about fifteen minutes. Then I said, "That was kind of close." He said, "Closer than you know!" I had a few drinks that evening (Miller didn't drink) and told him that in the years to come we would laugh about this, but I don't think either one of us ever did!

During this time, I also had a close call on horseback. I had a small, cold-blooded mare—prone to buck if not warmed up—named Holly. We would ride down the purchased cattle in the pastures and rope and doctor them on the spot. Holly got pretty good at getting to the cattle in a hurry and, with a little practice, I got to "catching" on a regular basis.

Once we took out after a calf and, as it ran down a ditch bank, Holly jumped a squirrel hole. Both front feet landed in the next hole, she went down going full speed, I fell in front of her, and she rolled over me. I got up, dazed and confused, then fell back down. I knew nothing was broken, but it damn sure hurt! This type of thing is common in our business, and it can easily result in permanent injury. I went to the doctor the next day, and learned that I didn't have a concussion; however, the casing around a muscle had been split, ribs were separated, and my gait would be changed for the rest of my life. It could have been worse.

Help from a Good Friend

I still had cattle to buy and business must continue. I knew that I couldn't stand the pace we had been setting on our previous trips: long hours in the plane, riding in pickups, late nights in the auction barns, and up before daylight to ship. I thought it over and came up with a pretty good plan. I would take the pain pills the doctor had prescribed to get through the trips, but I wouldn't take anything at night so that my mind would be clear enough to look at cattle in the mornings.

I took my friend Bud Whelms along to do my legwork with Kenny Miller. Bud had worked cattle all his life and had run several large ranches. He was well liked, fit in anyplace, and was totally trustworthy. I made a deal with him; I would pay all his expenses while on the trip. Since we had worked and traveled together quite a bit, he was more than happy to help.

We would start each day by inspecting the cattle to be purchased, make the deal, then I'd hit the painkillers and be useless for the rest of the day. On shipping days, if I felt up to it, I'd go along and then rest in the back of the truck. Towards evening, I'd quit the pills in order to be fresh for the next day; the result being that I would wake up between 2 and 4 A.M, somewhere between agony and halfway comfortable.

Bud was not afraid of Scotch. Each evening he would adjourn to the bar and catch up on the local cattle trade, as was the custom of the times. Usually he would quit about two in the morning, and that happened to coincide with closing time. One morning in Alamogordo, New Mexico, he returned to our room in a good mood. He stripped down—he always slept in the nude—and we began to visit. Then we heard a commotion next door.

Bud stuck his head out the door and told me it was a woman he recognized from the bar, and she was yelling at her boyfriend from outside the closed door. Bud continued to watch the verbal action, then said, "Why don't you get back and take a run at the door?" Evidently, the woman did. I heard a loud thump and Bud started laughing. "I think she knocked herself out!" This was followed by the police showing up.

Bud gave an interview to the cops, in the nude, and admonished them for shining a light in his face and being inhospitable to visitors like us. Why he didn't get arrested, I'll never know. He awakened the following day as if he'd slept all night. His hallmark has always been his bright attitude, so traveling with him was always enjoyable.

When the three of us made it to El Paso, we were able to get a room with three beds—cheaper than two rooms. Kenny Miller insisted that he would not sleep on the pullout bed. Since it was separate from the bedroom I told him I would sleep there. I knew something he didn't—Bud could snore with the best of them! When Miller awakened the following morning, he looked as if he hadn't slept a wink—Bud, however, was as fresh as ever. Since that time, I've never known Miller to have a roommate!

Bud's Coma

Bud was in trouble: he had a heart operation, his spleen was removed, and his lungs were filling with fluid. His wife called to tell me he was in a coma in the UCLA hospital. Even though he was unconscious, I went to see him. While I looked at him sleeping with all those tubes going in and out, I thought he was done for. But suddenly he came to. He looked at me down at the foot of his bed, rubbed his eyes and said, "If I'm dreaming, I must be having a nightmare!" I know very few people who could come out of a coma and crack a joke!

Potato Chips

Frances remembers:

Potatoes. The ranch's Brushfield potato-growing operation custom-farms a special variety of potato for Frito Lay. When the plant gives the signal, the potatoes are dug from the field and sorted by complex conveyor belts before being sorted into the specific sizes destined to fill the various sized bags. They are washed and quickly loaded into the waiting trucks for the quarter-mile journey to the big blue Frito Lay plant adjoining our property. As each truckload is brought to the unloading dock at Frito Lay, the truck sits idle for about twenty minutes while fifty pounds of potatoes are extracted. A hole is poked through each spud and sent through the cooking process. Before the truck driver is allowed to unload, these identifiable slices are subjected to five tests in the lab as to sugar content, density, etc. If the results are negative, our truck goes back to the big blacktop next to our cattle feed lot nearby. Here you will see individual piles of rejected or broken chips, corn chips, and chopped hay. Potatoes are fed to the cattle; some of the cattle are ours, some belong to other ranchers. The balance of the by-products are mixed with hay and cattle feed.

We had a problem once with a shipment of our spuds being forwarded to the plant in Vancouver; on arrival they were tested and rejected. Wesley's job? Get rid of the load—sell it, give it away, or dump it—but where? All this was accomplished over the telephone.

Eagle Brand chips (owned by Anheuser-Busch) came on the market with a big campaign about quality. This forced a race between Eagle Brand and Frito Lay, and Eagle Brand was losing money.

Today the question is, "Does the public care that much about quality?" To help safeguard our production, Frito Lay is looking into finding other markets for the rejected spuds—such as fries for restaurants, dried mashed potatoes, etc.

Spacious Skies, Amber Waves of Gold, and Green Machines

JB Selvidge remembers:

It is a Saturday morning in May 1999. A convoy of endless glistening green has just arrived. Keith Neufeld tells me that his home state of Kansas is being ravaged by tornadoes. I tell him that just a few hours ago an earthquake shook California (albeit 5.7 miles east of Palm Springs, but felt on the ranch). Keith was unaware of this, as they had been at highway speed, but he laughed at the coincidence of tornadoes back home and earthquakes here. There is no escaping mother nature!

Each March the Neufeld's and Balzer's of Inman, Kansas, load up ten or more brand new John Deere combines (traded in annually) on specially-designed, home-built trailers pulled by immaculate, big-rig diesels.

Their cross-country trek ends in California where they kick off the season in rolling, dry-land fields of wheat. Tracy Ranch is traditionally the first of many base camps. Their arrival is reminiscent of the Barnum and Bailey Circus coming to town. Equally impressive is the well-orchestrated show that follows: one by one the harvester-combines take shape as tires and wheels are re-attached, rolled off the trailers, and mated to their respective thirty-foot headers.

Larry Frey recalls his first encounter with Keith: "I remember when I first met Keith. I had been through years of incompetence and over-pricing by custom operators—grain on the ground, breakdowns, etc. I was sure leery with his low price, yet he had all new equipment. I was careful not to commit to a friendship at first. He needed a place for his house trailers, so I offered the area behind the green rooming house on Jumper. He didn't use it!" (The next season Keith moved into the "Jumper junkyard" and eventually was given the more private and secure area at the home place.)

Following the fleet each season is a John Deere factory entourage including the latest Deere prototype combine (the word "combine" derived from a combined grain reaper and thresher), at times shrink-wrapped in opaque plastic camouflage while in transit. Dedicated mechanics in a totally outfitted truck are its incognito protectors. Ominous telltale signs such as the absence of decals and padlocks dangling from access doors help guard the secrecy of this machine.

From here the green fleet, with headers in tow, begins its nomadic trek to various parts of the Central Valley, devouring fields of barley and wheat as if they were locusts in a backyard garden.

The heartbeat of the operation is Keith Neufeld, his wife Sally (as chief cook), and their sons. An obsession with excellence mixed with long hours are their trademark as they reap the golden harvest of the great American West—returning to Kansas in time for Thanksgiving.

Rachel Rankin Parker, Martha Frey, Helen Cross Rankin, and Alice Rankin Beard celebrate St. Patrick's Day at Chalet Basque

Alice Beard astride her favorite horse "Louie," on the Double U Cattle Ranch in Texas, where for 25 years Bob and Joe Mendiburu managed a large cattle operation. Son Rob Beard is now in charge.

Blue Book

John Tracy

Roping

Cow chip bingo

Racing

Western Culture Day

Lynell Etcheverry with her painting

Andrea Selvidge narrates as her father performs aerobatic maneuvers

BBQ

Rick Twisselman and Tim Hunter

Martha Twisselman

1,500 enjoy the barbecue

Sons of the San Joaquin

Zora, left, and Austin, with grandfather Ken Twisselman and Patsy Richardson

Richard Snedden and Chuck Hancock

John Tracy with the Robert Marshalls

Jeff Steve and Bruce Hafenfeld

Shiela Bowan and Bert Snedden

Heather Chapman dances with Twisselman

Bill and Susie Tracy

Rebecca Lake and Larry Frey in 1997

The Legacy

John Tracy remembers:

I knew from a child that I wanted to be in the cattle business. I grew up around horses, cattle, and cowboys and I liked them all. After graduating from Cal Poly, San Luis Obispo, on a Saturday in June, 1969, I moved home Sunday and went to work on Monday.

The following summer my father was killed in a freak motorcycle accident, leaving me in charge of the cattle operation.

Times were not good. Prices were down and the diversified farming operation, which could back the cattle operation in poor years, was having its own problems.

I vacillate between a chuckle and bad humor when it is suggested that I was "handed the operation on a silver platter." My cousins were faced with a similar fate. Though we had an expanding farming and cattle operation, we were broke. Banks were looking the other way when we approached them about financing. What we had was an opportunity, but a damn poor one at the time. But to our advantage was a legacy of working through these trying times.

The visionary Ferdinand Tracy had provided the base. Fannie Tracy came along to rescue the ranch from her husband's poor business decisions and failing health. The next generation provided Jack Frey.

Jack liked tough times and tough decisions. He, more than any other, brought the younger generation on by pushing them every day to do their best and by giving them a chance to do it. He was not the type of man to look for a pat on the back. You knew you were doing well if he didn't chew you out.

When things are the worst is the time to stay in. With the sale of some rice allotments, mother's loan of life insurance on my father, and the implementation of a new government-farming program, we began a slow climb uphill. Uncle Jack Frey, having no fear, whipped our expansion fever and, as luck would have it, we hit some pretty hot markets. We have had many good years since this time and look back to these examples for encouragement in poor times.

COWPOKES **By Ace Reid**

"... and jist think, Son, some day all this will be yores"

Area Brands

Rafter "TL"
BEECHINOR CATTLE FEEDING

Sixteen
C.W. CANFIELD

Three "C"
CARMEL CATTLE CO.

"E" plus
MATT ECHEVARRIA

Bar "77"
FILBERT ETCHEVERRY

Rising sun over bar
WALTER R. FISHER, JR.

"LF" bar
LARRY FREY

"Y" lazy "S"
GRANT GRISDALE

"JHB"
J.B. HAGGIN

"HF"
HARRIS FEEDING CO.

"H" three
OKEY W. & F. L. HOUCHIN

Diamond "A"
KERN COUNTY LAND CO.

"LS"
FRED D. LAVERS

"S" crutch
KENNETH MEBANE

"S" wrench
MILLER & LUX

Double "H"
HENRY MILLER

"74" connected over bar
KENNETH D. MILLER

Arrow
RONALD REAGAN

Half circle "V"
C.W. ROWLEE

"P" six
RICHARD RUDNICK

"JHS"
J.H. SANDERS

Bar "J" bar
WES SELVIDGE

"JN"
NICHOLAS J. SILICZ

"SIL"
SILL PROPERTIES, INC.

Slash
ERNEST STILL TRUST

1/4 circle "T"
R.L. STOCKTON

"G" lazy "T"
GEORGE TAYLOR

Cross over crescent
TEJON RANCH CO.

Tee pee
THOMSON & PELTIER

"TB"
TRACY BROTHERS

Ninety-one
F.A. TRACY

Flying "T"
JIM TRACY

Arrow
TRACY SISTERS

"T" rocker
TILTON TRACY

"T"
WILLIAM TRACY

Derby
CARL F. TWISSELMAN, III.

"F" over lazy "D"
CARL F. TWISSELMAN, JR.

Three two bar
KENNETH C. TWISSELMAN

Flying "T"
ROWLAND TWISSELMAN

Ice cream cone
WEGIS & SONS

Broken arrow
GERALD M. WELLS

BAKERSFIELD INSTITUTIONS

Frager's

As I was growing up, a trip to Bakersfield was a rare occasion. One of the most vivid recollections of my childhood was dining at Frager's. So far as I knew this was the only Mexican restaurant in Bakersfield at that time, and oh, what a treat it was to dine there!

The restaurant was owned by the Frager family, and was located near the corner of 19th and Baker streets in East Bakersfield. It was housed in an ordinary frame building and, if my memory serves me correctly, we ate from paper plates. Not the slick plastic-coated ones of today, but rough-textured cardboard plates. But ah, the food! It was magnificent! To a child used to simple, ranch-style cooking, the aroma of this spicy food was enough to excite the imagination and arouse the taste buds! When my enchilada arrived, it was swimming in a sea of rich, red sauce an inch deep. I was sure I would be able to devour every bit of it. But alas, I couldn't. There were no child's plates or "doggie bags" then, and on the following day how I longed for that uneaten portion!

Eventually the Frager family opened a new restaurant directly across the street in a brand new building with an updated kitchen and pleasing décor. They called it The Spanish Kitchen and, although the setting had changed, they still served the same marvelous Mexican food.

Eventually, the hard-working Frager couple sold the restaurant, and it no longer flourished.

The Mexicali

Almost as familiar to Bakersfield residents as the old clock tower is the giant cactus that stands at 631-18th Street. Since 1955 that sign has been beckoning us to partake of Mexicali's fine California-Mexican cuisine.

Visit Mexicali West and, in addition to the food, you will enjoy the handsome Spanish architecture complemented by a display of paintings by outstanding Bakersfield artist Clayton Rippey. The collection features twenty-one early Spanish missions that dot California's coastline from San Diego to San Francisco.

This is the story of how it happened—and why these Mexican restaurants have been so successful.

From Hot Dogs to Tortillas

There once was a man named Santos Gamez who married a girl named Trinidad. She was still in her teens when they parted from their native land in Mexico to make their home in Bakersfield. For a number of years the Southern Pacific Railroad employed Santos, and it was during this time that their four daughters were born. They named them Ruth, Rachel, Lillie, and Esther.

For many years Santos dreamed of owning his own business while operating two different grocery stores. By 1937 he had saved enough to invest in a piece of property on Baker Street in East Bakersfield. There, with his own hands, he built a filling station and, later, a tortilla factory. In time he opened a little open-air stand, where Trinidad and the girls served hot dogs and snow cones during the hot summer months.

Eventually, Trinidad began serving tamales and enchiladas at the stand. They became so popular that they enclosed the stand and made it into their first restaurant.

In her later years I had the pleasure of meeting this gentle little lady. She told me that it was her family that had supplied the masa for tamales and the tortillas for the Frager's restaurant where, as a child, I had first experienced the joy of Mexican cuisine!

The Gamez Girls

By the time they were teenagers, the Gamez girls had established a strong work ethic, even returning from school at noon to operate the gas pump in order to free their father for lunch.

They were now old enough to help in the restaurant, as well. While their mother cooked and created sauces in the kitchen, the girls did all the rest. Here they learned the restaurant business from the ground up. They waited tables, washed dishes, cleaned, operated the cash register, and helped prepare the food. No other help was hired in the beginning.

A Busy Couple

Santos and Trinidad were a busy couple. Their business was expanding in many directions. In the early fifties they opened a second restaurant at the corner of California and Chester Avenues. As the lease expired on that property, they were already purchasing the Old Town House restaurant at 18th and "R" Streets, where the downtown Mexicali remains today. By 1963 Ruth's son "Sonny" Crews was already established at the 18th Street location when his mother joined him, and they managed it together.

By 1967 the Gamez girls were taking an active part in the business decisions. It was then that the family decided to purchase a third location at 3303 Niles Street, east of Bakersfield, and Rachel took over the management there. Three years later Ruth's daughter Gloria Crews Griego became her assistant and later became manager. They operated that restaurant successfully for twenty-five years, until selling it in 1992.

Nieces, nephews, cousins, and grandchildren have all worked in the restaurants at various times throughout the years, continuing the family tradition that is so much a part of Mexicali, Inc. In the beginning, youngsters in their teens are started to work in the pantry, progressing to waiting tables and working their way into management. In that manner Lillie's daughter Irene Ramirez and Esther's son Kenny Gomez cut their teeth at Mexicali Downtown.

The Centralized Office

The Baker Street restaurant was closed in 1968 and converted into a centralized office and warehouse for the entire operation. Phones ring off the hook while salesmen await their turn. This is where the staff hires and trains the employees, keeps the books, orders supplies, and solves problems. It is the heart of the business, where computers tally how much is served at each location on a given day, making it possible to calculate what will be needed in the way of sauces and other ingredients for the following day.

Mexicali West

By 1970 Bakersfield was expanding to the southwest. Large new residential districts and shopping centers were being developed. Never resting on their laurels, the Gamez family decided to get in on the action. Gathering all of their resources, they were able to open their fourth restaurant at California Avenue and Stockdale Highway; thus, Mexicali West was born.

This beautiful and luxurious new restaurant opened its doors in 1977. What a proud moment for the hard-working family! Santos and Trinidad were still active in the business, and it pleased their daughters that they were able to share in the glory of this latest achievement. Esther and her son Kenny became managers of this flagship operation, and in 1983 Sonny Crews was named general manager of all locations and Kenny became assistant general manager.

A Loving Farewell

At the age of 87, ten years after the opening of Mexicali West, Santos went to work at eight o'clock in the morning, as usual. He suffered a heart attack and died the same day. The family agreed that it was a beautiful death; this was the way Santos would have wanted to go. And so the family bid farewell to their beloved father, the humble man devoid of false pride or ostentation who had worked so hard and achieved so much; the considerate husband and father who always removed his work shoes before entering his home. To his family he left a legacy of love built on admiration and respect.

Trinidad survived her husband by ten years, also dying at the age of 87. She had worked side-by-side with her husband for more than sixty years. Up to the end she was happiest working in the kitchen, creating the delicious sauces that helped to make the Mexicali restaurants famous.

The Gamez girls at Christmas, 2000: Lillie, left, Esther, Rachel, and Ruth

"Viva Mexicali!"

When asked for some of the highlights of their careers, the girls responded: "Serving some of the most interesting and wonderful people in Bakersfield; getting to know them and becoming special friends to some; having up to five generations celebrating a family occasion together; serving customers of forty years or more, and their growing to be seniors along with us!"

The Mexicali restaurants themselves have now served the people of Bakersfield for almost sixty years. Heads of state, famous movie stars, athletes, and even a President of the United States have all enjoyed the hospitality of this fine restaurant. Like other family-run establishments, the Mexicali restaurants are a part of our culture and a source of our pride. To the Gamez family we say, *"Ole! Viva Mexicali!"*

Left: For over 50 years the great cactus sign on 18th Street has beckoned us to partake of its fine Mexican cuisine. Right: the beautiful Mexicali West opened its doors in 1977.

Dewar's Candy Shop

In our changing world of computers, fax machines, and e-mail, it is comforting to find some things that do not change.

Who, living in Bakersfield any length of time, cannot tell you of Dewar's Candy Shop and Ice Cream Parlor! Who among us has not had the pleasure of swinging up to the counter on one of the old-fashioned bar stools and savoring a chocolate shake or ice cream sundae? Who in Bakersfield has not enjoyed the famous Dewar chews! Can anyone possibly forget the unique aromas of peanut butter and caramel chews, hand-dipped chocolate, and the multitude of flavors of ice cream that envelop one upon entering the door of this Bakersfield institution? It has been the favorite hangout for young and old alike for over ninety years.

Brothers James "J.H." and George Dewar founded the original store in 1909. It was located on Chester Avenue, just north of the hall of records. George soon moved to Taft. J.H. married Lena Yancy, who bore him four children: Joe in 1912; Jim, 1914; Mary Ellen, 1916; and Rosalind, 1923.

In 1919 the store was moved to 948 Baker Street in East Bakersfield and remained there until 1928. Two years later a new store was built at the present location on California and "I" streets. J.H. worked the store with his sons, who grew up in it. Joe and Mary Ellen have since passed away, but Rosalind learned to hand-dip the chocolates and was still doing it in 1999.

The faces of the young people behind the counter may change, but Dewar's as an institution has remained constant in the quality, consistency, and efficiency of its products and service. No doubt the recipe for the success of Dewar's ice cream and ice milk lies in the use of simple, wholesome ingredients free of the common, commercial additives and preservatives.

Thinking of Dewar's brings back to me memories of a high school escapade.

A Cold Trip

Jim Dewar and his wife-to-be, Dorothy George, were my classmates in high school. During my junior year, we decided to take a trip together and visit my family's cabin at Huntington Lake. My best friend and grammar school chum, Margaret Sawyer (later Bridges), my sister Frances, a fellow by the name of Earl Ditmore, and I made up the party. Mother came along as chaperone.

It was early May and, upon arriving at the lake, we found to our great surprise that the roads around the lake were still closed due to heavy snow. Since our cabin was at the far end of the lake, it would take a seven-mile hike to reach it. As usual, Fannie found a way. She and Jim Dewar negotiated with some people who owned a raft. We loaded mother and all of our gear on the raft and the rest of us began the journey through the snow.

By the time we approached the cabin our feet were like blocks of ice. When we needed to cross a small stream I decided to walk right through it. Surely my feet could not be any colder than they already were. I was wrong!

That night we all sat huddled around the fireplace shivering as we feasted on a five-gallon container of Dewar's ice cream and a five pound box of Dewar's candy, all provided by Jim.

We were young, carefree, and full of adventure, and we had a great time despite the cold.

Recent History

In 1947 Jim took over full management of his father's store. In 1977 J.H.'s brother George rejoined the family firm and his son Mike still works there.

Bill Lee's Bamboo Chopsticks

Twenty-five cents was all it cost to purchase a giant mound of delicious chow mein in 1930. The Chinese restaurants were favorite "hang-outs" for high school kids after football games.

If you could come up with a quarter, you had it made!

Although the price has changed, Chinese food is still a bargain. Bakersfield has many fine Chinese restaurants, but since Bill Lee's Bamboo Chopsticks is the one I know best, I shall talk about it.

Bill Lee emigrated from China to the United States with his parents as a boy of nine. For a number of years Bill Lee operated a grocery store called The Service Market on Niles Street in East Bakersfield. But in 1938 he opened the now- famous Bamboo Chopsticks at 1203 18th Street and, there, like an old friend, the restaurant has remained ever since.

In time Bill married Marion Yuen, and together they raised four children: Bill Lee Jr., Russell, Diana, and Sherman.

Well-known for his philanthropic deeds, Bill Lee endowed the establishment of an elementary school in his native land. He was a long-time member and past president of the Bakersfield Kiwanis Club, and he was an avid baseball fan.

Although the venerable Bill Lee passed away in 1990 at the age of 79, his son Sherman has upheld the high standards of this family-run restaurant. Now Sherman's sons Brandon and Brian represent a new generation to carry on the family business.

Throughout the years Bill Lee sponsored young Chinese Nationals who wanted to immigrate to the U.S., paying their passage and promising them employment so they could fulfill their dream of a new life. This practice continues to this day. These young people work hard to justify the Lee's faith in them, and should you dine at the Bamboo Chopsticks you will enjoy not only the fine cuisine, but also the fast, efficient, and courteous service of its employees.

All of us who have partaken of the hospitality of this and the other fine Chinese restaurants in Bakersfield salute the hard-working people who make it possible.

Chinese restaurants have always been a part of Bakersfield culture. The Rice Bowl and Bill Lee's Bamboo Chopsticks are two of them. At right, Sherman Lee, center, poses with sons Brian, left, and Brandon, the cornerstones of high standards for the popular Bamboo Chopsticks.

The Basque Tradition

Bakersfield's Basque restaurants draw patrons who journey from San Francisco to Los Angeles to enjoy this unique cuisine! Names such as Amestoy's, Noriega's, Pyrenees, Ville de Basque, Chalet Basque, Maitia's, Benji's, and Wool Growers are all familiar to the history of this town. Among their owners there exists a spirit of friendly competition.

Mayie Maitia, half-century veteran of the Wool Growers relates, "When we opened in the1940s, that first menu featured family-style lunches for $1 and dinners for $1.50. Saturday night, steaks were $1.75. We charged thirty-five cents for well drinks, and brand name drinks were an extra five cents. When we raised prices on well drinks to forty cents, we had to first hold a meeting of all the restaurant owners.

Chalet Basque

When J.B. Curutchague arrived on the scene, he was merely a lad of nineteen. Leaving family and friends in his native land, he came to Wyoming to try his hand at sheep herding, and he continued in that endeavor for ten years.

In 1964 J.B. settled in Kern County and met Marie, a lovely woman who was also Basque. After a year's courtship, he persuaded Marie to become his wife. They were blessed with two children, son Christian and daughter Elizabeth Marie, and they dreamed of opening a restaurant. In 1969 they opened Chalet Basque in the town of Wasco. After operating successfully there for a couple of years, they were able to secure the present location at 200 Oak Street in Bakersfield.

J.B. and Marie Curutchague

Left: Phil Etcheverry, sheep grower, poses with good friend Jack Frey at the sheep pens, with

The menu at Chalet Basque is typical of the Basque restaurants in Bakersfield. The meal is served family-style. After selecting your choice of wine, you are offered the "set-up" consisting of a delicious cabbage soup garnished as you choose with salsa and pink beans. It is served piping hot and is simply delicious, especially with the warm, crusty Pyrenees sour dough bread and cold butter. Platters of thinly sliced pickled tongue in a garlic-tinged oil and vinaigrette dressing are passed around; a real treat!

Next comes an icy cold salad, followed by French fries and platters of sliced tomatoes. The fine selection of entrees includes such sheepmen's fair as rack of lamb, lamb stew, lamb brochette, or lamb chops. Of course there are perfectly charred steaks for the cattleman, excellent oxtail stew, pork chops, and various poultry and fish selections. Then there are house specialties such as the butterflied shrimp scampi, my personal favorite. The meal is rounded out with fresh garden vegetables, a plate of spaghetti, and dessert choices of Roquefort cheese, flan pudding, or ice cream sherbet.

Served with friendly efficiency under the efficient supervision of the proprietors, the meal leaves the diner with only one question: will it be possible to walk from the table to the car?

Wool Growers

No "sissies" hang out at Wool Growers Bar and Restaurant! In the bar, Levi's and cowboy hats outnumber the conventional dress of "city folk." Step in on any day of the week and the place is jumping! The newcomer can be a little intimidated at first by the cacophony of voices and hearty laughter that greets him, but he soon becomes a participant in this atmosphere of conviviality and good cheer. No monthly fees are charged for this "country club" of cattle ranchers, sheep men, and others who enjoy fine Basque cuisine, and nobody *ever* goes away hungry!

Informality reigns 'midst this organized chaos. A staff of friendly and highly efficient waitresses deliver an assembly line of tasty dishes as they dodge little children running freely between the tables. Nobody seems to mind, least of all our hostess Mayie Maitia. But, after all, she has been at this business for a long time.

Mayie (pronounced MY-EE) was just seventeen and could not speak a word of English when she came to the U.S. in 1947. During the German occupation of France in

World War II, her father would tell his family, "Don't worry, the Americans will save us. America is the key to the world." It must have been his philosophy that prompted Mayie to leave the Basque town of Etienne De Baigorry in France and seek a future in the U.S. Just nine months after her arrival, she would marry Jean Batiste "J.B." Maitia.

J.B. had immigrated from a Basque province in 1928. He was nineteen years Mayie's senior. Like J.B. Curutchague of Chalet Basque, he, too, was once a sheepherder, but his family was already well established in the restaurant business.

Mayie Maitia, founder of Wool Growers restaurant, poses with son Danny and daughter Jennie outside the now-famous restaurant

In 1954, after the birth of their two children Daniel Frank and Jennie Angela, Mayie and J.B. were able to open their own business on Sumner Street in East Bakersfield. It would become the now-famous Wool Growers. Through Basque determination and perseverance the business succeeded, and in 1976 Wool Growers was moved to the former location of the Ville de Basque at 620 East 19th Street, where it remains today.

After her husband passed away in 1983, Mayie took over the entire management of the business. Now, a generation later, her son Danny and daughter Jennie work side-by-side with their remarkable mother. Today Wool Growers serves 250 or more meals per day and up to 500 or more on weekends. The banquet room alone seats 280 people. And, like the savory dishes that flow through its dining rooms, the people just keep coming.

John Tracy invites group to lunch at Wool Growers. Left to right: Walt Fisher, Jimmy Blach, Brother Bill Tracy, and Rodney Blach.

The Wool Growers Toast

By Martha Frey

Here's to the Wool Growers
And our hostess, Mayie
To her son and her daughter
We drink to all three!

Here's to the chef
And the bartender, too
Here's to each one
Of the Wool Growers' crew.

Good spirits resound
From ceiling to floor
The moment you enter
The Wool Growers' door.

And here's to those
Who have come to dine
On fine grade "A" beef
And the fruit of the vine.

Let's drink to the men
Who follow the sheep
May the grass be abundant
And the market be steep!

To the ranchers, all
Who provide the beef
May the market be tall,
And your troubles be brief.

To this Basque institution
We have gathered to savor
The fantastic cuisine
And magnificent flavor.

Make way for the soup
With salsa and bean,
For it is the greatest
You ever have seen.

Try pickled tongue
In garlic and oil
Surely this feast
Is well worth the toil!

Now comes the salad
Named for a Caesar,
But don't take too much—
It's only a teaser.

New dishes arrive,
Are served by the score;
They just keep coming,
And still there are more!

There's succulent beef
And spicy lamb stew,
With Pyrenees bread
And cold butter, too.

We eat and we eat
'Til we can eat no more;
Our bellies so heavy
We drop to the floor.

But here comes the flan,
So sweet and delicious—
You can't miss the flan,
So smooth and nutritious!

I relish the custard,
But my clothes are so tight,
My eyelids so heavy
I must say "goodnight."

But here's one last drink
To our hostess Mayie,
Her son and her daughter
And each employee!

They all share their tips
And work hard for Mayie;
Our Bakersfield lady,
We all drink to thee!

BLACCO

BUTTONWILLOW LAND
AND
CATTLE COMPANY

In Memory of Jack

Jack Frey, 1914-2001

Jack M. Frey as remembered by nephew John Tracy and delivered by Bill Tracy on January 22, 2002:

John was supposed to do this. It's the first time I've ever seen him shun an opportunity to be the center of attention. I was, at first, surprised because he had so eloquently delivered mom's eulogy; something I, as the eldest son, was emotionally incapable of delivering. Then it dawned on me; Jack and John had a special bond known only by those who have run the good race, fought the good fight, and kept the faith together. So these words, tonight, are from my lips but John's heart.

Uncle Jack enjoyed sharing rules of life and sayings. One of his favorites was the definition of a breed: "A breed is a group of individuals having distinctive characteristics, not common to other members of the same species." With the death of Uncle Jack, we have seen the last of the breed. These men and women, who so exactly define the breed, were the people who built our county and nation. They were people who came to tame the wild frontier they loved and harness its energy for the good of mankind and to sustain a living for their families. Jack's good friends, Phil Etcheverry, Bob Beechinor, Jack Sill, and Bill McNamara were such individuals. Jack ran in good company and was the type of individual who embodied what is right in America.

Jack worked hard to transform the Tracy Ranch from a livestock operation to one of the most successful family-run organizations in the nation. During his lifetime he increased the farming operation from 1,000 acres to nearly 10,000. He did this by leveling and cultivating grazing land to farming. He oversaw the transformation from cotton grain and rice to an entity that includes carrots, potatoes, tomatoes, almonds, and pistachios. No doubt, there would be no modern-day operation had it not been for his hard work and perseverance.

While heading the company through its biggest growth years, Jack held a steady hand through the good times and the bad. During his tenure there were more bad times than good. Jack never lost his vision or enthusiasm for his final goal of success. He held nothing for himself and welcomed oncoming generations to work and share in the prosperity with open arms. While he ruled with an "iron fist," he willingly gave up control of operations to younger members. He bragged that "we grew our own replacements."

Jack was grateful that he had good, young, competent men to take over the operation, and he often stated that he wished Cecil and Grandma Tracy were alive to see the progress that had been made.

Jack loved livestock and worked hard with our father, Tilton, to keep it going. He had no qualms about John's ability to take over, even though he was only a twenty-two-year-old kid, after dad's untimely death. He never lost faith through the bad times in the cattle business, and he lived to see the operation turn a steady profit, as he was sure it would.

Jack had no fear of failure, and I think he enjoyed the tough times more than the good. He never took credit himself, always deferring to the contributions of his partners, including Uncle Cecil who had a propensity towards women more than work. He noted Cecil's contribution during the ranch's darkest days, forgiving the thirty years thereafter, when Cecil's contribution was next to zero. I asked him about this once, and he said Cecil always had the good of the ranch at heart; he just had poor work habits. He said that Cecil had great vision, and he credited him with obtaining the Brushfield Ranch.

With the passing of Uncle Jack, we have lost the last of our mentors. These people not only taught us about business but were also a great source of advice and comfort when things were not going our way. He had been through the tough times, and he knew that the good and the bad times never lasted forever. At the bottom, he knew things would get better. At the best of times, he knew that he had better keep a little in reserve. His guiding principle was that when it was tough, you got bigger so you could ride the markets up. When real estate margins were high, he knew to hold back because the fall was certainly near.

Jack said the best thing to put on your land was your shadow. His shadow shall forever be on our land, and his imprint shall forever be our guiding force.

The Lead Horse
A Tribute to Jack Frey

By nephew John R. Tracy, Christmas 2000

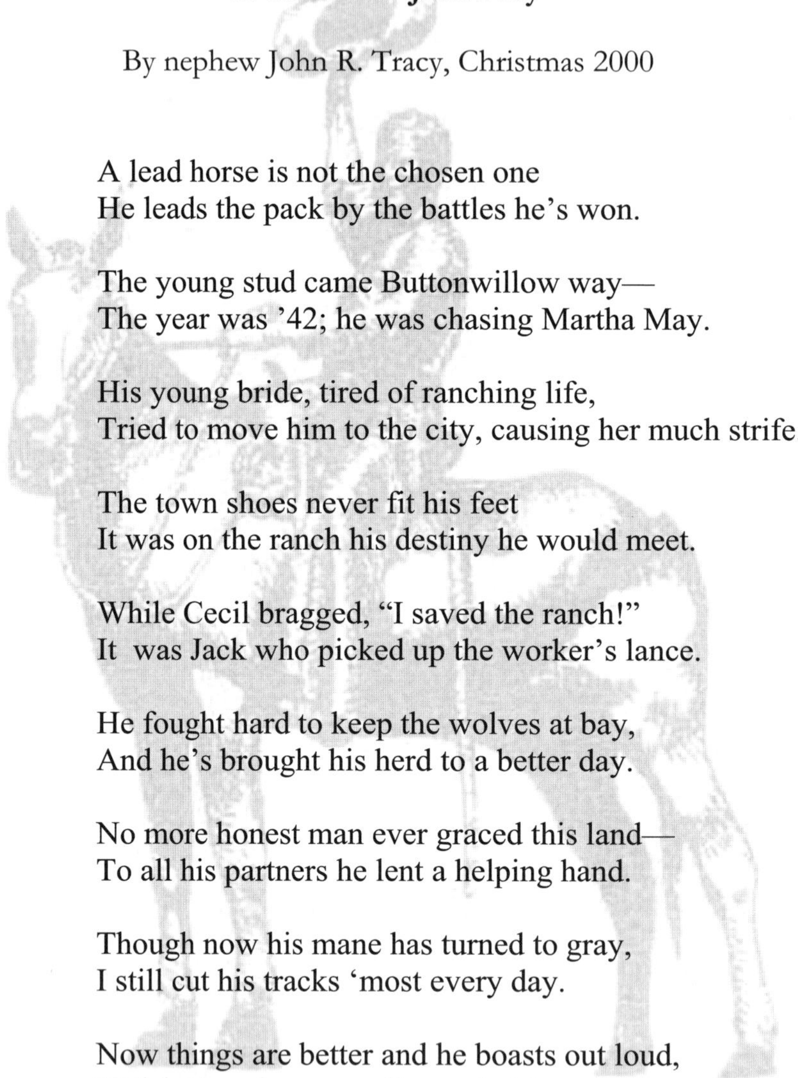

A lead horse is not the chosen one
He leads the pack by the battles he's won.

The young stud came Buttonwillow way—
The year was '42; he was chasing Martha May.

His young bride, tired of ranching life,
Tried to move him to the city, causing her much strife.

The town shoes never fit his feet
It was on the ranch his destiny he would meet.

While Cecil bragged, "I saved the ranch!"
It was Jack who picked up the worker's lance.

He fought hard to keep the wolves at bay,
And he's brought his herd to a better day.

No more honest man ever graced this land—
To all his partners he lent a helping hand.

Though now his mane has turned to gray,
I still cut his tracks 'most every day.

Now things are better and he boasts out loud,
"Wish Grandma was alive—she'd sure be proud!"

Robert Crabtree

El Patrón Grande, 1993
"The best thing a man can put on his land is his own shadow."

LAWRENCE W. (LARRY) FREY

Larry who was born on February 3, 1942, was seven months old the summer Jack and Martha moved to the ranch from the Coalinga oilfields. Like his 4th generation peers, he grew up on a ranch overrun by bullheads, mosquitoes, Johnson grass, and the rutted-out, 1940 alkali roads.

Larry oversees the northern half of the ranch today—a pristine far cry from the past, as ribbons of hard packed roads neatly bisect lush crops. A self-proclaimed "dirt farmer," his green thumb and penchant for neatness is visible in today's operations.

Larry graduated from Wasco High School in 1960, Bakersfield College in 1962, and Cal Poly (SLO) in 1965 where he majored in agricultural business management. He served on the Semitropic Water Storage District Board for eight years, and is a past director of both the Kern County Farm Bureau and the Buttonwillow Union School District. He has three sons: Michael, Jack, and Stephen. On November 17, 2001, he was married to Rebecca Lake.

Larry's motto: "Johnny at the rat hole." His favorite pastimes are hunting and fishing; his pet peeve, litigation; and his worst fear, poor health.

Michael T. "Mike" Frey
"The Money Man"

Mike was Larry and Claudia Frey's first child, and he was born in Bakersfield on September 21, 1965. Mike is the CFO—Chief Financial Officer—of the Buttonwillow Land and Cattle Company, a particularly daunting task this year as cotton yields and world market prices threaten growers nationwide. However, Mike's background, aptitude, and positive attitude are appreciated by all.

In 1987 he graduated with a bachelor of science degree from California State University, Fresno. He was a director with the Kern County Farm Bureau; a member of the Farmers Co-op Gin board; president of the Buttonwillow chapter of Lions Club International in 2000; a director with Calcot; and a chairman on the California Wheat Commission.

Mike's pet peeve is filling out surveys, and his worst fear is grain tank gravity spout failure causing huge grain spillage. He enjoys golf.

In 1987 Mike married Jennifer Goins whom he met at the university, and they have two sons, Justin and Nicholas.

JACK R. "YOUNG JACK" FREY

Jack was born on August 7, 1968, to Larry and Claudia Frey. His responsibilities on the ranch involve the increasingly arduous task of machinery maintenance and shop supervision. His job is especially tricky because he must answer the call of several field supervisors and ten partners. However, Jack's aptitude for machinery and fabrication is a positive force in the ranch's highly mechanized future.

He graduated from Fresno State University in 1991 with a degree in agricultural engineering technology. He and his wife, Kelie, have two children, Wade and Madeline.

He enjoys hunting and fishing and his motto is: "Work hard. Play hard." His pet peeve is someone being rough with the equipment, and his worst fear is running out of bullets and money, and everything breaking at once.

MID-JUNE 1990
Cotton Grower
BILL TRACY
Five Generations
On The Ranch
COTTON
90
Years
IN THE '90S

ROBERT W. "BILL" TRACY

Bill is the eldest of Tilton and Kathryn's four sons. He was born on October 16, 1942, in San Diego and was one year old when his parents returned to the ranch in the fall of 1943; Tilton had been employed by Consolidated Aircraft in San Diego, building PBY seaplanes during World War II.

Bill's education commenced in the two-room Wildwood Grammar School neighboring the ranch property, and he attended Wasco High School where he was active in track and cross-country (in 1960 he was selected to carry an Olympic Torch). While attending Cal Poly (SLO) he majored in Agricultural Journalism. He was a member of the U.S. Army Reserves and served two years active duty in the late sixties.

Bill returned to the ranch in 1965, farming the home place with cousin Larry and also overseeing the operation of the cotton harvest fleet. In addition to his farm duties, Bill is active in several organizations that serve his community. At the local level he has served as president of the Buena Vista Resource Conservation District and the Farmers' Cooperative Gin; director of the Kern County Fair; and founding director and president of the Buttonwillow Park and Recreation District. He was recently appointed to the Kern County Biosolids Ordinance Advisory Committee.

Affiliations on the national and state level include past chairman of the Cotton Foundation, and chairman of the National Cotton Council Health, Safety and Environmental Committee. He has served as director on the Acala Cotton Board and the California Cotton Pest Control Board. He is currently chairman of the California Cotton Growers Association. Bill is one of two farmers, nationwide, appointed to the Environmental Protection Agency's Pesticide Program Dialogue Committee. Past EPA participation includes the Office of Water, Management Advisory Group and the Food Quality Protection Act Implementation Committee. Between 1987 and 1991, he served with Governor Deukmejian's administration at the state capitol as Deputy Secretary of California Agriculture.

Bill and his partners were honored with the National Cotton Grower's 1995 Achievement Award; he was selected as the 1998 Kern County Fair Agriculturist of the Year; and in 1999 he was the recipient of the Kern County Farm Bureau's Ag Merit Award.

Bill's hobbies include World War I aviation, competitive shooting of antique firearms, and researching and collecting Civil War artillery projectiles.

Bill and Susie have four children, Mark, Rob, Todd, and Stephanie (Tracy) O'Callaghan, and seven grandchildren.

ROB TRACY

Robert Lee Tracy was born in Bakersfield on November 19, 1969. He attended Buttonwillow Grammar School, then Shafter High School where he excelled in football and baseball. During his formative years on the ranch, Rob irrigated crops on the home place and explored all the Gooselake back roads with brother Todd in their Volkswagen Dune Buggy. He showed pigs at the Kern County Fair as a 4-H member and participated in Little League and Babe Ruth Baseball as a catcher, establishing the reputation of never letting anyone "steal home." Rob's higher education consisted of one year at Cal Poly (San Luis Obispo) before graduating from the "University of Uncle John" at the ranch feedlot.

While working for his Uncle John, first as a hired hand and later as feedlot superintendent, Rob mastered the complexities of creating profits from small margins and learned the importance of customer and employee relationships. Through that experience he gained a tireless work ethic and a philosophy that punctuality is a sign of responsibility and respect. His ranch duties have greatly expanded, and he is now supervising the ever-increasing almond production while continuing to assist in the feedlot management and record keeping.

The focal point of his personal life is his wife Kathy, a member of the pioneer Twisselman cattle family, their daughters Haylee and Kathryn, and son Trent. They reside in the home that once belonged to Rob's grandmother, Kathryn. Because this was the home where Rob's father, Bill, and Bill's three brothers were raised, it seems quite natural to again see little ones running around the place.

TODD BLAKE TRACY

Todd, the son of Bill and Susie Tracy, was born September 7, 1968, in Tacoma, Washington. His formative years on the ranch revolved around Little League Baseball, summer irrigation jobs, and driving the red, white and blue Volkswagen dune buggy all over Gooselake with his brother and Rob. Todd always loved the country and working the ground. During his years in 4-H, he organized his own hog-breeding project and later, in college, he managed an enterprise project, growing sweet corn and selling it at the San Luis Obispo Farmer's Market Street Fair.

Todd always knew he would return to the ranch, and he prepared himself by obtaining a bachelor of science degree in agricultural business at Cal Poly (SLO). He still fondly recalls coming home from school full of new ideas with degree in hand, only to be assigned the job of night tractor driver and putting in twelve-hour shifts, seven days a week. His initial thought was, "I spent five years in college to do this?" He now realizes that this experience, combined with the many summers he spent irrigating under the tough supervision of Eddie Robbins, helped shape him into the kind of boss he is today. "Not until you've walked in the shoes of an employee can you be a good employer," Todd believes.

Todd works closely with Larry Frey on the daily management of all farming aspects on the home place, and he is currently expanding the ranch's scope with the introduction of their first pistachio orchard and the processing of tomatoes. He also manages one of the cotton picker fleets, and he realizes that much of what he knows about the operation today he learned as a youngster at his father's side. While the ranch demands full-time attention, Todd still finds time to serve his community as advisor to the Farmers' Co-op Gin. He also manages to pursue "a little golf," participate in Uncle John's local basketball team, the "Range Bulls," and manage a summer baseball team consisting of former high school buddies. He is a two-time winner of the annual Buttonwillow Iron Man event.

Todd's motto: "Live today like there is no tomorrow." His pet peeve: those who are always late. Todd and Annette Clarisse Easton were married at the home place in 2001 and, in January 2003, Annette gave birth to a beautiful daughter, Clarisse Ann.

John R. Tracy

Tilton and Kathryn Tracy's third son, John, was born in Bakersfield on October 13, 1947. His infectious smile, wit, and charm are his trademark. His mind in business dealings is like a steel trap, but he has a heart of gold once a deal is struck—though this might not be obvious until the person is down on his luck.

A year after John's graduation in 1969, Tilton was killed in a freak motorcycle accident. John's initiation into the cattle business was trial-by-fire. Today John runs the cattle operation, including range cattle, background feeding in California, and finish-feeding at Champion Feeders in Hereford, Texas.

John received a bachelor of science degree in farm management at Cal Poly (SLO) in 1969. He was awarded lifetime membership in the Cal Poly Boots and Spurs Club and in 1998 was selected as the Kern County Cattleman of the Year. This same year he was appointed to the California Cattleman Political Action Committee. He was president of the Rancheros Visitadores in 2001, and is vice president of the California Beef Council.

His pastimes include riding and training horses, hunting, and branding calves and his motto is, "Take care of family, set the pace, play only when work is done, show up early." John's worst fear is of being sent to hell and put in charge of people who are late. And his pet peeve? People who are late. John and his wife, Donna, whom he met at Cal Poly, have two children: Allen and Jessica.

JOHN BENTLEY "JB" SELVIDGE

JB was born on January 27, 1941, and was Skip and Frances Selvidge's first child. He was two years old when the family moved to the ranch from the hills above Hollywood. Before the move, Skip had been employed at 20th Century Fox Studios.

JB's interest in machinery, in addition to taking alarm clocks apart, first surfaced when he was four, found a wrench, and proceeded to remove the belly plate from "Old Rheumy," a Farm-All 'M' which was the ranch's first wheel tractor. He was later spotted sitting in the seat of a tractor loaded on a trailer, heading out the driveway bound for Los Angeles. His dad was able to catch the rig and retrieve him. JB pursued his passions and was a project engineer for Peterbilt Trucks before he became equipment manager for the ranch.

JB was educated at Cal Poly (SLO) where he received a bachelor of science degree in mechanical engineering in 1966. He has been a director at the Fox Theater Foundation, a director with Kern County Mosquito Abatement, and president of Tracy Ranch, Inc.
His motto is: "Somewhere over the rim of the world lies romance and every heart yearns to find it." His favorite pastimes: historic preservation, old roads, classic machinery, and music. His worst fear: being caught without pens, pencils, and a tape measure. His pet peeve: messes, disorder, and neglected machinery.

JAMES W. "WES" SELVIDGE

Wes was Skip and Frances Selvidge's second child. He was born in Los Angeles and was five months old when the family moved from Hollywood to the ranch in January 1943.

Wes's love of flying first surfaced when he was four, and he and JB built an airplane from two apple crates in tandem, using a 2" x 12" board for a wing, an electric motor in front, and a long extension cord. (According to Wes, they really believed it would fly.)

In 1964 he married Janet Funk of Wasco, a courtship that began in a Model A Ford hot-rod while attending Wasco High School.

His imagination and ingenuity are apparent today, as he runs the "Brushfield," the 3,500-acre "breadbasket" of ranch operations where cotton, potatoes, tomatoes, and carrots thrive. Today's Brushfield heartbeat is the potato shed which was created by Wes and which he runs with the help of his son Jason.

In 1965, Wes received a bachelor of science degree in animal husbandry at Cal Poly (SLO). He was chairman of the board for the Rosedale Rio Bravo Water District; Vice President, Valley Farm Credit; and president of the Buttonwillow Chamber of Commerce.

He and Janet have two children, Andrew and Jason. His motto: "If it can't be done with a handshake, it's not worth doing." His favorite pastimes are aerobatic flying (he's a four-time state champion), wooden boats, car racing, roping, and more. His pet peeve is with computers and his worst fear is a computerized potato shed.

JASON W. SELVIDGE

Jason was the second, and last, of Wes and Janet Selvidge's children, and he was born in Bakersfield in 1972. After graduating from Shafter High School in 1990, he attended the mile-high University of Denver where he majored in construction management. His "minor" was on the majestic snow-capped ski-slopes where he learned agility and speed that would later pay off as a front-runner on the dirt track oval, Bakersfield Speedway.

Like most of the Freys, Selvidges and Tracys, Jason was pre-destined to carry on in his father's footsteps, and in 1995 he returned to the ranch after a post-graduate stint in home construction in Colorado. His beginning course in farming was "Cotton Picking 101," where he learned the art of managing a "picker" fleet.

Today Jason is aspiring to broader management training by assisting his dad with the Brushfield operations—often a four-ring circus of potatoes, cotton, carrots, and tomatoes.

Jason's continuing passion—automobile building and racing—has provided him with the valuable skills in engineering, mechanics, and fabrication that pay off in today's ever-increasing mechanized operations.

Jason's favorite motto is: "It is much easier to do it right the first time." His pet peeve is waiting in lines, and his worst fear is the thought of ever having to live in L.A.

Kevin Fahey

Left to right: Bill Tracy, John Tracy, JB Selvidge (sitting), Larry Frey, Jack Frey, and Wes Selvidge, with William Tracy's first tractor, a 1931 CAT-15

Wes Selvidge borrowed Robert Crabtree's camera and climbed an electric tower to take this extraordinary shot of Tracy Ranch commercial cotton pickers and a crop duster biplane at Brushfield circa 1990.

Tilton Tracy
"He lived his life to the very hilt of his wondrous energy, and he reaped a bountiful harvest of the greatest of all earthly possessions—the love and respect of his fellow man. This is the way I knew my brother, the baby of our family. He was much like our father, and we loved them both dearly"—Cecil Tracy

JB Selvidge

In Memory of Cecil Foster Tracy.
One of a Kind!
Photograph circa 1972.

Kathy Merlo

The Old Barn, built circa 1900 for Will Tracy's draft horses, and the hitching post added by Cecil Tracy in 1938

Martha Frey

Tilton and Michael on Honda motorcycle

JB Selvidge

The old Tank House after a freak 1999 snowstorm. The old wheel at the top is missing.

JB Selvidge

The Old Barn after restorations and the 1999 snowstorm

Frances and Skip perform for the 1960 Buttonwillow mayor's race

Adding a little humor to "A Roman Holiday." Left to right: Claudia Hoggard, Lorenda Belluomini, Tracy Selvidge, and Sandra Belluomini

Carriage House

Martha settles into her new home at the Carriage House in Bakersfield

Martha with great grandson Wade Frey

Interior photographs by Iger Studios

Nancy Trahan with niece Madeline Frey and sister-in-law Rebecca Frey

Larry and Rebecca Frey help Martha get settled

Sammy Jo Reeder celebrates her birthday on the second of July

Celebrating Jo Reeder's birthday are, left to right, Patty Bass, Clara May Lundy, Martha Frey, Jo Reeder, Margie Ellingston, Jackie Cowan Gargano, Dixie Ford, and Diana Mosier

The Bridge Club greets Mary Lou Thomson on her birthday, 1975. Left to right: Astrea Rogers, Bessie Cassidy, Martha Frey, Mary Lou Thomson, Connie Schroeter, Vivian Wegis, Bebe Burke, and "Bee" Blair.

A group of Lee Olcott's friends pay a surprise visit to the opening of the family's new store. Left to right: Mary Ann Snodgrass, Barbara Baldwin, Jo Reeder, T.G. Hopper, Martha Frey, Lee Olcott, Jackie Gargeno, Marie Yost, Patty Tkac, and Clara May Lundy

Kathy Merlo

John at the head on "Bud" and Wes at the rear throwing on "Knothead," at the cattlewomen's barbecue

In 1989 these four Rancheros Visitadores participated in the yearly drive in Santa Barbara. Left to right: Brownie Barnard, John Tracy, Ronald Reagan, and Richard Rudnick

In memory of Elden J. "Skip" Selvidge, member of the ranch partnership from 1943 until his retirement in 1977

The yellow pickup truck Jack Frey received as his 80th birthday gift

Madeline Frey, age 4, and Wade Frey, age 5

Frances Selvidge and Martha Frey, Christmas 1997

The author, Martha Frey, and her husband Jack

JB Selvidge

Memorial garden designed by JB Selvidge. Captain is buried near here.

From ashes to ashes, dust to dust
In younger hands we leave our trust
To those that follow we pass the key
To guard the ranch posterity

JB Selvidge

Marilyn Tracy stands by the monument to her grandfather in front of his homestead cabin

JB Selvidge

Christmas at JB's, 1999. Left to right: Nicholas Frey, Evan De Fraine, Ian Koll, Grace O'Callaghan, Haylee Tracy, Kathryn Tracy, Wade Frey, Santa (John Tracy), Madeline Frey, and Justin Frey.

Left to right, Trent Tracy, Isabel Tracy, Madeline Tracy, Kathryn Tracy, Justin Frey, Nicholas Frey, Haylee Tracy, Wade Frey, Grace O'Callaghan, and Magdelyn O'Callaghan.

TRACY FAMILY GENEALOGY

1. **WILLIAM TRACY** 1866-1941
m. 1904 FANNIE CALIFERNIA ROWLEE 1877-1966

2. **CECIL FOSTER TRACY** 1905-1978
m1. Florence Williams
m2. Aileen
m3. Nanette
m4. Shirley

2. **DARREL TRACY** 1906-1964
m. 1934 Marie Fine
3. KATHRYN DARRELEN TRACY 1939-
m. 1960 Kenneth George Holden
4. KENNETH CHARLES HOLDEN
m. 1966 Maria Skytta
3. MARGARET MARILYN TRACY 1941-
m. 1963 Terry Lee McDonough

2. **FRANCES FAY TRACY** 1909-
m. 1937 Elden James "Skip" Selvidge 1906-1982
3. JAMES BENTLEY 'JB" SELVIDGE 1941-
m. 1979 Diane Marie Davis
3. JAMES WESLEY "WES" SELVIDGE 1942-
m. 1963 Janet Louise Neufeld Funk
4. ANDREA SUSAN SELVIDGE 1971-
4. JASON WESLEY SELVIDGE 1972-
3. TRACY ANN SELVIDGE 1945-
m. 1984 Gary Gill
4. CASEY GARRETT GILL 1985-

2. **CHARLES WELLINGTON TRACY** 1912-1920

2. **MARTHA MAY TRACY** 1915-
m. 1938 Jack Myron Frey 1914-2002
3. NANCY JANE FREY 1940-
m1. 1961 Covell Dorn Brown 1938-
4. TRACY ANNETTE BROWN 1962-
m. 1991 Paul Koll 1958-
5. IAN MICHAEL KOLL 1991-
4. CARLA ANN SEIDEL 1961-
m. 1993 Edward De Fraine, II 1965-

5. EVAN ANDERS DE FRAINE 1996-
5. AISLIN ISABELLA DE FRAINE 2001-
m2. 1985 Randall Earnest "Randy" Trahan 1945-

3. LAWRENCE WILLIAM "LARRY" FREY 1942-
m1. 1965 Claudia Irene Lee 1942-
4. MICHAEL TILTON FREY 1965-
m. 1987 Jennifer Ann Goins
5. JUSTIN MICHAEL FREY 1990-
5. NICHOLAS DANIEL FREY 1993-
4. JACK RANDOLPH FREY 1968-
m. 1994 Kelie Marie Wilson
5. WADE WILSON FREY 1998-
5. MADELINE KINSEY FREY 1999
4. STEPHEN LAWRENCE FREY 1972-
m. 2000 Melissa Midzor 1968-
m2. 2001 Rebecca Lake

2. **ALBERT TILTON "TILTON" TRACY** 1919-1970
m. 1941 Katherine Nelson "Kit" Pollard 1921-
3. ROBERT WILLIAM "BILL" TRACY 1942-
m1. 1965 Antoinette "Toni" Torigiani
4. MARK JAMES TRACY 1966-
m. 1996 Allison Sobocinski
5. ISABEL CELESTE TRACY 2001-
4. ROBERT LEE TRACY 1969
m. 1992 Kathy Lee Twisselman
5. HAYLEE MARIE TRACY 1996-
5. KATHRYN ALYSE TRACY 1998-
5. TRENT JACKSON TRACY 2000-
m2. 1974 Elizabeth A. "Susie" Jackson
4. TODD BLAKE TRACY 1968-
m. 2001 Annette Clarisse Easton
5. CLARISSE ANN TRACY 2003-
4. STEPHANIE ANN TRACY 1971-
m. 1995 Kevin O'Callaghan
5. GRACE ANN O'CALLAGHAN 1999-
5. MAGDELYN KAE O'CALLAGHAN 2001-
3. JAMES TILTON TRACY 1945-1965
3. JOHN RICHARD TRACY 1947-
m. 1970 Donna Kay Allen
4. JAMES TILTON "JT" TRACY 1977-1981
4. ALLEN RICHARD TRACY 1980-
4. JESSICA CHRISTINE TRACY 1982-
3. DAVID GRANT TRACY 1949-

TRACY FAMILY TIME LINE[1]

Year	Date	Event	Other
1027?		William I (the Conqueror), illegitimate son of Robert I, Duke of Normandy and a hide tanner's daughter became one of England's greatest monarchs after slaying King Harold, former Earl of Wessex at the Battle of Hastings	Rules 1066-1087
1085	-1086	W^{m} I orders survey of England, results in the Doomsday Book—a record of all property	Doomsday Book
1087		William I dies; Robert II succeeded him in Normandy, while another son, William II or William Rufus, d. 1100 (r. 1087-1100), succeeded him in England	
1068	-1135	Henry I, youngest son of W^{m} I, succeeds brother W^{m} II, as King of England & later defeats brother Robert II to become Duke of Normandy (France) His daughter Matilda & Geoffrey IV (Planagenet) are the parents of Henry II	Henry I Rules 1100-1135
1085		Gieva de Traci, concubine to Henry I, gives birth to the first William de Traci, who was acknowledged by Henry & granted title, which passed through daughter Grace to Granddaughter Eva	William de Traci born to Henry I & Gieva de Traci
1133		William Sudley de Traci, Great Grandson of William de Traci, born to Eva de Tracy and William de Traci. He is half-uncle to Henry II	Henry II born to Matilda & Geoffrey IV
1170	12-29	Four of Henry II's knights, including William Sudley de Traci, murder Archbishop Thomas Becket in Canterbury Cathedral. They are exiled.	Henry II's Queen was Eleanor of Aquitaine, widow of Louis VII
1215		King John, youngest son of Henry II, reigns 1199-1216 the throne upon death of brother Richard I 'The Lion-heart'. Richard had fought against his father, Henry II. John	King John signs *Magna Carta.* John dies next year, succeeded by his son

[1] Compiled by Nancy Frey Trahan from various sources, including genealogy research conducted by Frances Selvidge. Begun as a few notes to maintain consistency of dates, it has proven a valuable tool in arranging the story in a somewhat logical order.

Year			
		had conspired against him while Richard was away on 3rd Crusade.	Henry III.
1337	-1453	Hundred Years War, England v. France	Henry V, ("Prince Hal'), Joan of Arc, Charles VII
1455	-1485	Wars of the Roses (Lancasters & Yorks) ended as Henry VII (Lancaster)Tudor kills Richard III	End of feudalism in England
1491	-1547	Henry VIII (Tudor) born to Henry VII & Elizabeth, daughter of Edward IV	Reigns 1485-1509
1533	-1603	Elizabeth I born to Henry VIII & Anne Boleyn. She rules during time of Wm Shakespeare, Edmund Spenser, Francis Bacon, Walter Raleigh	Elizabeth I Rules 1558-1603
1566	-1625	James I, son of Lord Darnley & Mary Queen of Scots, succeeds Elizabeth I in 1603	James I reigns England 1603-25
1620	11-21	Several Tracy forebears arrive to found Plymouth Colony, sign *Mayflower Compact*, first written gov't framework in New World.	John Carver 1st Gov, Wm Bradford Gov 1622-56
1630		Henry Rowley of Shrewsbury, England (on the Wales border) arrived on the next ship after the *Mayflower*. Had a son, Moses Rowlee, who had a son Shubael Rowley. The Rowley Mansion is the oldest surviving house in Shrewsbury, and as late as 1992 was serving as the local museum.[2]	Charles I King of England 1625-49. He struggles with a Puritan parliament, signs "Petition of Right in 1628.
1641		Lieutenant Thomas Tracy marries Widow Mason	Artist Van Dyck dies
1650	or '51	Doctor Solomon Tracy is born; marries Sarah Huntington in 1676; he dies in 1732, she in 1783.	Oliver Cromwell Lord Protector of England
1670	6-17	Captain John Tracy (b. 1642 d. 7-31-1721) marries Mary Winslow, (b.11-1676 d. 8-16-1702) the niece of Gov. Winslow of Mayflower	Hudson's Bay Co chartered to operate fur trade monopoly
1679		Martha Bourne Bradford (b. 1614 d. 1679/83), widow of John Bradford (whose father was William Bradford, Governor of Plymouth) marries Thomas Tracy. See "History of Norwich Ct., Frances M. Caulkins, Hartford Ct, 1866 p. 201.	Peter I 'the Great' born 1672, rules Russia 1682-1725
1681		William Penn receives charger for Pennsyl-	Pennsylvania

[2] Source is Janice (Mrs. Wilbur) Rowlee Fay, of Fulton, NY. She referenced Plymouth Colony 1632-33 tax records.

Year			
		vania, a Quaker, religiously tolerant colony	chartered
1688	2-09	Winslow Tracy born to Mary Winslow and John Tracy. On June 21, 1714 in Norwich Ct. Marries Rachel Ripley (b. 4-17 1693 Windham Ct., d. there 4-1782, daughter of Hannah Bradford and William Ripley,). They have 7 sons, Joshua, Perez (11-13-1916), Josiah (5-10-18), Eliphalet (11-14-1720), Nehemiah (3-18-1722), Samuel and Solomon (5-23-1728).	W^{m} III, (W^{m} of Orange) son of W^{m} II defeats James II for throne of England. Accepts the Bill of Rights in 1689, rules with childless Mary II.
1697	5-10	John Tracy (b. 1-19-1672 to Mary Winslow and Captain John Tracy, d. 5-27-1726) marries Elizabeth Leffingwell (b. 11-1676, d. 10-25-1737)	
1697	8-02	Jane Tilton born in Chilmark, MA to Abiah Mayhew and William Tilton. On 6-2-1718 she married William Hunt, son of Mary Bradford and William Hunt. They had a son, Samuel and a daughter, Jane, who married a Nickerson. Jane d. 10-19-1732.	
1698		Lydia Tracy (b. 1677, d. 1757) marries Deacon Thomas Leffingwell (b. 1674, d. 1733)	Queen Anne, last Stuart ruler, 1702-07
1706		Benjamin Franklin, son of a soap maker, born in Boston. Apprenticed to brother James at 12 as a printer. Moved to Pennsylvania in 1723	Ben Franklin's birth
1730	4-09	Deacon Isaac Tracy (b. 5-25-1706, d. 1-25-1779) marries Elizabeth Bushwell, (b. 1711, d. 5-12-1764	George I, great-grandson of James I, named King 1714-27
1754	-1763	French & Indian War, England & Colonists drive the French from western Pennsylvania & Appalachia	George Washington, Surveyor
1757	-1775	Benjamin Franklin represents the colonies in England. Upon return, advocated independence.	Ben Franklin goes to England
1765	11-27	Andrew Tracy (b. 12-9-1742, d. 1801) marries Mary (Mollie) Clements, (b. 5-13-1750, d. 11-27-1794)	Grandson George III r. 1760-1820 "Mad"
1776		Ben Franklin works on Declaration of Independence with Thomas Jefferson, is a member of Continental congress. In 1776, becomes ambassador to Court of Louis XVI in France, dressed as a humble Quaker. Wins French support in 1778. Returns to Philadelphia in 1784.	Daniel Boone born to Quaker parents in Berks Co, Penn. Explored & settled Kentucky, served in the Virginia assembly.

1782	-1852	Daniel Webster born humbly in New Hampshire, educated in Dartmouth, was congressman from New Hampshire & congressman & senator from Massachusetts. Highest paid attorney of his time, argued successfully before the Supreme Court under Chief Justice John Marshall.	Daniel Webster born
1789			Geo Washington President
1790		Benjamin Franklin dies; his last public act was to urge Congress to abolish slavery	Ben Franklin dies
1797			John Adams Pres.
1801			Thomas Jefferson President 1801-09
1803	2-14	Vernet Tracy (Sr.) born to Peleg Tracy (b. 9-21, 1766, d. 3-12-1825) and Hannah Leffingwell, (b. 9-22-1773, d. 9-28, 1846)	1815 Arthur Wellington defeats Napoleon I at Waterloo
1809	-1817		James Madison Pres.
1817	-1825		James Monroe Pres.
1822		Lydia T. Moore [Martin] (Martha Martin Rowlee's mother) born in Whitewater, Ind.; father is Quaker David Moore. The Moore family followed William Penn from England to Philadelphia, Pennsylvania soon after its founding. One of the Moore or Martin family is probably the person mentioned as being tailor to Benjamin Franklin & Daniel Webster.	The hated Geo IV rules 1820-30
1825?		John T. Martin (Martha Ann's father) born in Westchester, Pa. Became an 'architect' (contractor or lumberman?) His family followed Wm Penn to Pennsylvania (?). (John Martin, Lydia Moore need more research & verification)	John Quincy Adams (1771-1832) president 1825-29
1825	11-21	Vernet Tracy (Sr.) marries Mary Ann Dilley (b. 5-31-1808, d. 6-19-1895)[3] [data uncertain]	
1827	1-13	Charles Wellington Canfield born –where?	
1829	10-21	Ferdinand Tracy born in Wilkesbarre, Pa.	Andrew Jackson Pres 1829-37; Wellington Prime Minister 1829-33

[3] Shaded entries need more research for verification.

1831	7-03	Edgar Vernet Tracy (Jr.) born in Wilksbury (Wilkes-Barre) PA to Mary Dilley and Edgar Vernet Tracy (ref military records; 10-14 given elsewhere as birthday)	1830-37 W^{m} IV succeeds brother Geo as King of England; Martin Van Buren Pres. 1837-41
1838?		Mary Elizabeth Dicks (or Dix) [Wm Tracy's mother] born (in Iowa?)[4] [data uncertain]	W^{m} IV's niece Victoria reigns 1937-1901
1845 circa	4-10	John T. Martin weds Lydia Moore. John is son of John & Ruth Martin & a cousin of Thaddeus Stevens. John & Lydia have 4 children who live to adulthood: David W, John C, Martha Ann and Benjamin L.	Wm H. Harrison pres 1841. John Tyler pres 1841-45.
1846	12-21	David Warren Martin born to Lydia Moore & John Martin	James Knox Polk pres 1845-49
1848	5-17/18	Charles Wesley Rowley (Rowlee) born in Pennsylvania, near Owego, Tioga Co. NY to Mary Ring and Lafayette Rowlee, a wagon maker. Moved west to Illinois at age 9. - Mary Ring was daughter of Levi Ring of Vermont, who participated in the War of 1812. She taught school in Vermont. She bore 8 children, Charles W., Elizabeth 'Mattie' [Deeble], Eugene, Emma [Mrs. John C. Martin], Jessie (Sacramento), Marietta ("Etta", Pasadena), Minnie (died in Oakland) and William (Oregon). Lafayette Rowlee, the son of James Rowley, who immigrated to New York State, migrated from NY to Rockford, (Winnebago Co.) IL, and from there to Rock Grove, (Floyd Co. Iowa). He died in Linn Co. Iowa at age 49. Either Charles or his father changed the spelling of his last name (common with names throughout the years). Connection to the Henry Rowley of 1630?? Rowlee an old name in England.	January 24, 1848 began 50 years of gold discoveries in Sutter's Mill, Ca-1949, Gold Hill, Col—1859 Nev—1860, Id—1861, Virginia City, Montana-1863, So Dakota-1876, Tombstone, Az-1877, Cripple Creek, CO-1892 and the Klondike River—1897 and Nome, Alaska—1899 CA Gold rush began in 1849, leading to an influx of men and a booming economy
1850		Ferdinand Tracy (Age 21) emigrates to Ca with large expedition. Commissioned lieutenant in U.S. army; quells Indian uprising in No. Ca. Mining interests in Mokelumne region.	Zachary Taylor Pres 1849-50. California becomes a state
1850	7-18	Martha Ann Martin (Rowlee) born to Lydia and John T. Martin in Westchester, Pennsylvania. One of her grandfathers	Millard Fillmore Pres 1850-53

4 Estimated date of birth; it was reported as 1808, but that seems impossible. Need more research here.

		(probably Grandfather Moore) was tailor to Benjamin Franklin and Daniel Webster.	
1851? circa		Charles Wellington Canfield arrives in CA at Firebaugh's Ferry in Fresno county on the steamer *Harriet*, commanded by a Captain Wheeler. (Information reputedly from Mrs. Canfield's scrapbook, as related by Mrs. Barnes Lewis to her daughter, Mrs. Tipton.	
1853	10-26	Vernet Tracy and Mary Elizabeth Dicks (Dix) wed in Wapolo, Louisa Co. Iowa[5] and later emigrate by ox-team & covered wagon to Ca.	Franklin Pierce Pres 1853-57
1854	2-22	Lafayette Rowley marries Sarah A. Ennis in Knox Co., Illinois. Second marriage?	
1855	1-18	Benjamin L Martin (Martha's brother born to John & Lydia.	
1857	7-01	Lafayette Raleigh (Charles' father) leaves Tioga County, NY for Rockford, (Winnebago Co.) Ill.; later to Rock Grove City, (Floyd Co.), Iowa. Dies in Linn Co. at age 49. Married to Mary Ring, who was born in Vermont but living in Tioga Co. His father James immigrated from the "old country".	James Buchanan Pres 1857-61
1857		Research note: a Joseph Moore and C. Stevens are mentioned in minutes of the Friends in Indiana as being in charge of a recently erected school	
1858	2-23	Ruth Anna Martin born to John & Lydia. Dies Oct. 7, 1858 at 7 months.	
1859	Circa	John Martin strikes out for Anderson County, Kansas to start a new, rural life for his family. When he does not return, Lydia and her young children, accompanied by her brother Benjamin pack another covered wagon and follow. Martha Rowlee remembered being hidden from bandits in a cracker barrel on her mother's covered wagon in Kansas Territory.	
1860		Ferdinand Tracy comes to Carissa Valley looking for cattle feed; settled in Bakersfield in 1862	1861—Prince Albert dies
1862		Ferdinand Tracy & Charles Wellington Canfield form Canfield & Tracy;	Abraham Lincoln Pres 1861-65

5 According to Vernet's Department of the Interior, Bureau of Pensions application for pension. Wapello is between Davenport and Burlington, near the Illinois border.

		partnership lasts 56 years	
1864		Charles Wellington Canfield weds in Boston. His wife (Miss Cooley, daughter of Judge Cooley of Boston) was a music teacher at either the Boston Conservatory of Music or Mt. Holyoke College. Her flat, rectangular piano was later shipped "around the horn" to CA and ended up in William Tracy's ranch house. After his wife's death, Canfield spends his last years with Will & Fannie.	
1864	11-15	Vernet Tracy enlists at age 33 with the Ca. 8th Regiment Volunteer Infantry, Company 'B' stationed at Fort Point, Ca. His name was sometimes spelled "Tracey." Lists hair—brown, eyes—blue, complexion—fair, 5'7" tall.	
1865	10-24	Vernet Tracy honorably discharged from 8th Regiment at Fort Point, CA. Later has a stable at Acampo. William Tracy was born 9½ months later.	Civil War ends after 4 bloody years. Andrew Johnson Pres 1865-69
1866	8-08	William (Wm.) Tracy born near Galt, Ca. to Mary Dicks (Dix, Dick) Tracy and either Edgar Vernet Tracy, Ferdinand Tracy or Charles Wellington Canfield. Fannie intimated to her children that Canfield was her husband's father, but that seems unlikely due to timing.	Russia offers to sell Alaska to the US. Wm Seward negotiated the purchase for $7.2 million in gold in 187.
1868	5-07	Frances Aurelia Tracy (b. 11-4-1846, d. 11-2-1928) daughter of Mary Ann Dilley & Vernet Tracy Sr., (Ferdinand & Vernet's sister) marries William Shelby Denham	
1869	12-24	Charles Rowlee and Martha Ann (Mattie) Martin wed in Cedar Rapids, Iowa. Her brother, John Martin marries Charles's sister Emma Rowlee in 1978, making their descendants double cousins.	Ulysses S. Grant Pres 1869-71. Susan B. Anthony founds National Woman Suffrage Association.
1870	2-16	Etta M. Bullock born to John & Lydia T. Bullock. The intrepid Lydia T. Martin has been widowed and remarried John Bullock.	Bismarck provokes the Franco-Prussian War
1870	12-29	Nellie Rowlee (Tilton) is born to Martha & Charles Rowlee	1871, France loses, forced to give most of Alsace & Lorraine to Prussia. William I Emperor of Germany & King of Prussia

1872		Charles Wellington Canfield serves as State Senator	
1872	11-24	Col. Thomas Baker dies, after asking his friend, Ferdinand Tracy, to look after his wife Ellen and take good care of her and her family	Horace Greeley, editor & political leader, dies
1873		Dolly Rowlee (Taylor) is born to Martha & Charles Rowlee	
1874	9-13	David Warren Martin weds Rebecca E. Brown	
1875	1-23	Ferdinand Tracy weds Mrs. Ellen Baker, widow of founder of Bakersfield, by Rev. A. J. Hanson, Methodist. They have two children, but lose one at birth, the other at age 2. Ferdinand leaves his holdings to W[m] Tracy, his nephew. Mrs. Baker's children contest the bequest.	
1875	12-18	Ferdinand Augustus Tracy born to Ferdinand and Ellen (Baker) Tracy. He dies of croup December 20, 1877.	
1876	12-03	Son born to Ferdinand & Ellen Baker dies at birth.	
1876		Charles W. and Martha Ann Rowlee migrate by covered wagon from Iowa to California	
1876		George Taylor (Dolly Rowlee's husband) is born	
1877	9-?	Mary Dicks Tracy dies in San Joaquin County, Ca.	Rutherford B. Hayes Pres 1877-81
1877	2-18	Fannie Califernia Rowlee (Tracy) born near Lockeford, (Stockton area) Ca.	Great railroad strike spawned sympathy strikes
1877	4-04	Newspaper cites W. Canfield as on the board of "the bank." On October 24, editor George Chester wrote, "Mr. (Wellington) Canfield has eased up on farming and looks as young as when we first saw him through the tule in 1864. He still has the same kindness of face and manner.	
1878		John C. Martin, son of John & Lydia, marries Emma G. Rowley. The John Martin's remained in Marion, Iowa.	
1879		Gilbert Rowlee born to Charles & Martha	
1881	Approx.	Emma Tracy Kilgore gives a home to her little brother, William. Her husband is Ellis Kilgore, a partner in the firm of Kilgore & Tracy of Sacramento.	James Garfield Pres 1881; Chester A. Arthur pres 1881-85

		(Ferdinand?)	
1884		The Rowlees move from near Lockeford (Stockton area) to Kern Co. and settle on Gooselake Channel, half-circle V brand.	
1886	9-01	Lawrence John Frey born to Louise Schaffer and Joseph Frey in Washoe Co, NV near Reno.	Grover Cleveland Pres 1885-89
1888		Effie Brown (b. 11-8-1869, Franktown, NV) and Ira Myron Conkey (b. 1865, d. 2-1939, Coalinga) wed in Nevada	William II, grandson of William I & also Victoria, crowned Emperor of Germany
1889	2-13	Irene Maude Conkey (Jack Frey's mother) born to Effie Brown and Ira Myron Conkey in Nevada	Benjamin Harrison Pres 1889-93
1889		Charles Wellington Canfield and his wife celebrate their 25th anniversary. A silver teapot presented by their friends memorializes the occasion. (JB Selvidge is now caretaker of the teapot.)	Washington becomes a state
1890	1-13	Hazel Rowlee (Cooper) born to Martha Martin and Charles Rowlee in the San Joaquin Valley	
1891	1-	Will Tracy moves to the Bakersfield area, homesteads with brother Theodore on Gooselake Channel on the Canfield & Tracy (Ferdinand) range (1 mile east of Wildwood Road near 7th Standard). "91" cattle brand, "T" horse brand. He joins Theodore as superintendents of the Canfield-Tracy herds.	Paul Gauguin paints *La Orana Maria*
1891	2-13	Vernet Tracy, age 58, applies for a pension under the Act of June 27, 1890. On his pension application, Vernet lists his children's births as Alice Elmar, 10-24-54; Theodore A., 10-27-55; Ida Jane, 10-21-58; Mary E., 4-5-60; Sarah, 5-10 (or 20) -62; William, 4-21-71. [We know he had other children. There were only 6 lines to list children on the form—or did he not consider himself the father of the last two girls?[6] Vernet's hair is listed as brown, eyes blue.	
1896	4	19-year-old Fannie Rowlee meets Will Tracy in the floodwaters of the Kern River, on	Grover Cleveland Pres 1983-97

[6] One of Will's sisters' names was Anna, yet we do not find Anna listed. "Alice" was the name of one of Anna's daughters; perhaps Anna was a nickname to differentiate them.

		what would become part of the Tracy Ranch. Sam McQueen courts her the same year.	
1896	9-06	Leslie Rowlee Tilton born to George Albert Tilton & Nellie May Rowlee in Tulare	1897—Victoria's Diamond Jubilee
1898		William moves one mile west on Wildwood Road to what would become the permanent headquarters of Tracy Ranch. He moves his original homestead cabin of 1891, adding another wing. Builds the blacksmith shop and horse barn. Fannie and Dolly attend his housewarming that summer.	
1898	-1900	Fannie Rowlee attends Chico State Normal School for 2 years	W^m McKinley Pres 1897-1901
1899	5-30	Julietta Cooley (b. abt 10-13-1826), wife of Charles Wellington Canfield, dies at the age of 72 yrs, 8 mos, 17 days. (Elsewhere: 5/31)	
1900	1-20	Pioneer Society honors Ferdinand & Ellen Tracy on their 25th wedding anniversary. Ellen's grandsons Tracy & Roy Baker & Edgar Cowgill attend.	
1900		William Tracy completes the first phase of a ranch house on the site where, in 1984, JB Selvidge would build his home. He and Fannie Rowlee meet again at the housewarming.	1901 Queen Victoria dies, succeeded by her son, Edward VII
1900	- 1902	Fannie Rowlee transfers to San Diego State Normal School	
1902	6-12	Fannie Rowlee graduates from San Diego State Normal School at age 25	Teddy Roosevelt Pres 1901-09
1902	-1903	Fannie Rowlee teaches at the Wildwood School.	
1903		Nellie Rowlee Tilton dies; Fannie Rowlee Tracy, Dolly Rowlee Taylor and their mother Martha Martin Rowlee at some point take in her children	
1903	-1904	Fannie Rowlee, by now engaged to William Tracy, teaches at the Wicker School	
1904	4-3	William Tracy weds Fannie Rowlee	
1905	1-14	Cecil Foster Tracy born to Fannie Rowlee and William Tracy	
1905	6-13	William Tracy's sister Anna (Tracy) Clark weds James Arp	
1905	11-02	Marie Fine (later to wed Darrel Tracy) born in Oklahoma	
1906		Kern River floods & wipes out Charles Rowlee's crops; had erected a levee by 1908	

1906		Fannie & William import the ostriches "Phoenix" & "Tempe" to Ca. from Arizona. The feather business is soon thriving.	
1906	9/29	Elden J. "Skip" Selvidge born to James Wesley and Fannie Selvidge in Manhattan, Kansas	
1906	12-03	Darrel Tracy is born to Fannie & William Tracy	
1907	9-07	William Tracy buys another ostrich	
1907	or early '08	Charles Rowlee erects a levee to protect his crops from Kern River floods	
1908	1-10	Ferdinand Tracy dies in Bakersfield. He and Ellen Baker had two children, one of which died at birth, the other at age two. They are buried with Ellen in the Baker plot. Obituary "… born Wilkes Barre PA Oct 21, 1829. Came first to Carissa. A staunch Republican, was once county surveyor. A man of the highest character. 11th: pall bearers: Lightner, McKelvey, Blodget, Said, Maul, Roberts: all old pioneers. The hearses reached Union before the last turned onto Truxtun."	
1908	1-18	Estate of F. A. Tracy: nephews William & Theodore trade town lots for ranch property	
1908	5/04	Charles Wellington Canfield dies at age 81; he spent the last two years of his life living with Fannie and Will Tracy. (Elsewhere: death on 6/05.) Funeral May 7, pallbearers J. Myers, Lewis McKelvy, Mattley, W H Thomas, Said, Lightner.	
1909	6-27	Frances Fay Tracy born to Fannie Rowlee & William Tracy	William Howard Taft Pres 1909-13
1911		Lydia Moore Martin Bullock dies at age 89 in Marion, Iowa. Family was from England; were Quakers who settled in Philadelphia soon after Wm. Penn. Benjamin L. probably named after Benjamin Franklin.	George V, second son of Edward VII, reigns 1910-36, changes family name from Saxe-Coburg-Gotha to Windsor
1911	6-	Lawrence John Frey weds Irene Conkey	
1912		Charles Wellington "Teddy" Tracy born to Fannie & William Tracy	*Titanic* sinks on April 15
1912	12-10	Effie Laura Frey [Parker] born on Ingram Island Oregon (near Junction City) to Irene Conkey and Lawrence Frey	
1913		First batch of ostrich chicks hatch in Kern Co. at the Tracy farm	Woodrow Wilson Pres 1913-21

1913		Charles Rowlee builds house on Chester Lane	
1913		Edgar Vernet (Jr.) Tracy dies in Northern California	
1914	6-21	Jack Myron Frey born to Irene Conkey & Lawrence Frey on Ingram Island near Junction City, Oregon	World War I begins in Europe
1915		Will builds a larger barn (the "white barn") 300 feet south of the original	America tries to remain neutral in war
1914		Morgan's History of Kern Co. published.	
1915	5-07	Martha May Tracy (Frey) born to Fannie & William Tracy in Bakersfield in Dolly Rowlee Taylor's front bedroom	German U-boat sinks British liner *Lusitanian*, killing Americans
1915		Fannie Tracy exhibits her ostriches at the S.F. Exposition	
1917	-1918		US joins WWI in Europe
1919	7-16	Albert Tilton Tracy born to Fannie Rowlee & William Tracy at Homeplace	Prohibition enacted by 18the Amendment
1920		Charles Wellington "Teddy" Tracy dies at 8 years, complications from scarlet fever	
1920	-1921	Will Tracy, persuaded by slick promoters and against Fannie's advice, invests heavily in the Imperial Livestock and Mortgage Co. and in the Verdun Packing Company. Both fail, setting off a 20-year struggle for survival.	Women's vote guaranteed by 19th amendment, Republicans become dominate party
1920-		Cecil Tracy strikes out on his own to become a range cowboy for giant cattle barons, Miller & Lux; Cecil would later become a professional rodeo bronco and bull rider	
1921	8-1	Kathryn Nelson Pollard born to Willa Caperton and Robert E. Pollard in San Diego, Ca	
1921	10-12	Helena Hazel Rowley marries Glen I. Bunton in Strathmore, Tulare Co.	Warren Harding Pres 1921-23
1926	1-28	William Tracy sells property, including Section 5 (SW corner of 7th Standard Road and Tracy Avenue). Section 5 brings \$16/acre, and the (duck hunter) investors name it the "Buttonwillow Land & Cattle Co., Inc."	Warner Brothers Studio experiments with sound-on-disc talking pictures
1928		Fannie purchases Huntington Lake cabin	
1930		Fannie and William Tracy sell the	Movie houses begin

		homeplace on Wildwood Ave. to a Mr. Napoleon Merritt of LA, whose son becomes caretaker and neglects facilities & livestock	to be wired for sound
1929	Oct 29	Stock Market Crash	Herbert Hoover Pres 1929-33
1930	-1933	Fannie Tracy owns Buttonwillow grocery store & post office	
1931	9-	William Tracy suffers a stoke at Huntington Lake	
1933		Charles Wesley Rowlee dies; is interred in Union Cemetery, Bakersfield	Franklin Roosevelt Pres 1933-45
1934		Darrel Tracy marries Marie Fine	
1936			George VI follows brother Edward VIII as King of England
1937	6-11	Frances Tracy marries Elden J. Selvidge (b. 9-27-1906, D. 7-28-1982)	Edward VIII marries Wallis Simpson. As Duke of Windsor after abdicating throne
1938		Fannie leases out much of Tracy Ranch for oil & gas exploration. Cecil remodels the ranch house, buys a 1930's Art Deco Wurlitzer jukebox. William Tracy has become an invalid; Cecil helps care for him & tends the ranch.	
1938	12-23	Martha Tracy and Jack Myron Frey wed in Bakersfield Baptist Church	
1939	10-17	Kathryn Darrelen born to Marie Fine & Darrel Tracy	Hitler invades Poland; US tries again to remain neutral
1940	2-11	Nancy Jane born to Martha Tracy & Jack Frey, Coalinga, Ca.	Battle of Britain fails to conquer Britain
1941	1-27	John Bentley "JB" born to Frances Tracy & E.J. Selvidge	
1941	3-16	Albert Tilton Tracy weds Kathryn "Kit" Nelson Pollard in Bakersfield	
1941	7-04	William Tracy dies at Camarillo State Hospital, is interred in Bakersfield Union Cemetery	
1941	11-07	Margaret Marilyn born to Marie Fine and Darrel Tracy	
1941	11	With WWII threatening, Fannie calls a family council. To assure at least one male family member will be left to operate the	

		ranch, all children and their spouses agree to come back and farm.	
1941	12-07		Japan bombs Pearl Harbor, 12/7; US at war
1942	10-16	Robert William "Bill" born to Kathryn Pollard & Tilton Tracy	
1942	2-03	Lawrence William "Larry" born to Martha Tracy & Jack Frey	
1942	8-08	James Wesley born to Frances Tracy & E.J. "Skip" Selvidge	
1942	8-?	Martha and Jack Frey return to Tracy Ranch	
1943	1-01	The Tracy Ranch family partnership is formed. The Federal Land Bank of Berkeley turns down a $23,000 loan application. Jack Frey personally borrows $10,000 from friend Mrs. Baker of Baker's Transfer & Storage. The ranch's first cotton crop is planted.	
1945	10-24	Tracy Ann born to Frances Tracy & E.J. "Skip" Selvidge in Bakersfield, Ca.	
1945	6-16	James Tilton Tracy is born to Kathryn Pollard and Tilton Tracy	Harry S. Truman pres 1945-53. Germany surrenders May 7, Japan Aug 14, after 2 atomic bombs
1945	8-18	Randall Ernest Trahan born	
1946		The Federal Land Bank turns down a $25,000 loan application, implying that the ranch is a worse risk than in 1943. Ranch buys its first new tractor, a model A Case wheel tractor of 55 DBHP & a 10' offset disc	Benjamin Spock publishes *Baby and Child Care*, becomes a runaway best seller
1947	10-13	John Richard born to Kathryn Pollard and Tilton Tracy	Harry Truman gives aid to French in Vietnam
1948	7-17	Martha Rowlee dies at age 98; is interred at Union Cemetery, Bakersfield	
1948		Gilbert Rowlee dies; is interred at Union Cemetery, Bakersfield	
1948	12-21	John C. (Crummy?) Martin born (to ?)	
1949		Formation of Tracy Ranch, Inc from the original partnership	
1949	9-07	Fannie Tracy joins a post-war Farmer's Friendship Tour of Europe. She left the tour in Germany and lived there for several	

		months, returning to the ranch at Christmas.	
1949	6	Cecil Tracy and his horse 'Captain," both gilded in gold, win the Santa Barbara Fiesta Parade Grand Sweepstakes Award	
1949	8	Mechanical cotton pickers introduced on Tracy Ranch	
1949	9-13	David Grant born to Kathryn Pollard and Tilton Tracy	
1950	11-20	Heavy rains cause the Kern River to flood; the men are able to clear the Jerry Slough Channel just in time to save most of the crops	US enters Korean War; armistice comes in 1953
1952		George Taylor dies in Bakersfield; is interred in Union Cemetery, Bakersfield	Queen Elizabeth II succeeds father George VI
1952		Tracy Ranch begins development of Kern Co. Land Co. Brushfield on a crop-share basis. Eventually purchased, the field becomes the ranch's most bountiful producer.	
1952	5-04	Irene Conkey Frey killed in auto accident near Lancaster, CA; is interred in Union Cemetery, Bakersfield	
1952	5	Kern River floods for the second time in the decade. A 40' break in a levee north of Lerdo Highway floods acres of cropland. Flood danger greatly diminished after completion of Isabella Dam in 5-19-54.	
1955	8-31	Lawrence John Frey dies of pancreatic cancer in Bakersfield	Dwight Eisenhower Pres 1953-61
1960	1-20	Kathryn Darrelen Tracy weds Kenneth George Holden	Eisenhower increases aid to France in Viet Nam
1960		Dolly Taylor dies, is interred in Union Cemetery in Bakersfield	
1961	2-03	Carla Anne [Seidel] born to Nancy Frey and Covell Brown, Mt. Auburn Hospital, Cambridge, MA	John F. Kennedy Pres 1961; Bay of Pigs invasion April 17
1961	9-09	Nancy Jane Frey and Covell Brown wed in Bakersfield (div. 1967)	Kennedy triples aid to South Vietnam
1961	10-26	Hazel Rowlee Cooper dies; is interred in Union Cemetery, Bakersfield	
1962	12-01	Fannie Tracy publishes *The Tracy Saga*	10.62 Cuban missile crisis
1962	3-15	Tracy Annette born to Nancy Frey and Covell Brown, Boston Lying-In Hospital	

		(MA)	
1963		Margaret Marilyn Tracy weds Terry Lee McDonough	Kennedy assassinated; Lyndon Johnson Pres 1963-69
1963	6-20	Janet Louise Funk & James Wesley Selvidge wed in Wasco	
1964	7-24	Darrel Tracy dies the year after retiring from 33 years as Postmaster of Buttonwillow; is interred in Union Cemetery, Bakersfield	
1965	2-28	Lawrence William "Larry" Frey weds Claudia Irene Lee in Bakersfield (div. 1996)	
1965	7-11	James Tilton "Jimmy" Tracy killed in plane crash; is interred in Wasco cemetery	Johnson escalates war in Vietnam, but avoids all-out effort
1965	7-13	Ed De Fraine born in Detroit, Michigan to Trudy Heinzelman and James De Fraine	
1965	8-21	Robert William Tracy weds Antoinette "Toni" Leigh Torigiani in Buttonwillow (div. 1973)	
1965	9-21	Michael Tilton born to Claudia Irene Lee & Larry Frey in Bakersfield	
1966	3-13	Kenneth Charles born to Kathryn Darrelen Tracy and Kenneth George Holden in PA.	
1966	4-16	Fannie California Rowlee Tracy dies; is interred in Union Cemetery, Bakersfield on 4-21-66	
1966	12-17	Mark James born to Antoinette & Robert William "Bill" Tracy in Bakersfield	
1968	8-07	Jack Randolph born to Claudia Irene Lee & Larry Frey in Bakersfield	Martin Luther King, Jr. assassinated
1968	9-07	Todd Blake Tracy born to Susie Jackson	
1969	11-19	Robert Lee born to Antoinette Torigiani & Robert William "Bill" Tracy in Bakersfield	Richard Nixon Pres 1969-74
1970	1-10	John Richard Tracy weds Donna Kay Allen in Redondo Beach, CA	
1970	5-04	Albert Tilton Tracy dies in accident at Jacolitos Canyon Ranch	
1970	Jan?	Formation of Buttonwillow Land & Cattle Co (BLACCO)	
1971	1-20	Andrea Susanne born to Janet Funk & James Wesley Selvidge in Bakersfield	
1971	5-21	Stephanie Ann Tracy born to Susie Jackson	
1971		Tracy Ranch purchases Brushfield from Tenneco (formerly Kern Co. Land Co.)	
1972	11-08	Jason Wesley born to Janet Funk & James	

		Wesley Selvidge in Bakersfield	
1972	8-04	Stephen Lawrence born to Claudia Lee & Lawrence William "Larry" Frey in Bakersfield	
1974	8-09	Elizabeth A. "Susie" Jackson (b.9-14-47), w/ children Todd Blake (b.9-7-68) & Stephanie Ann (b.5-21-71), weds Robert William "Bill" Tracy, w/ children Mark James and Robert Lee, at home place	Gerald Ford Pres 1974-77
1974		BLACCO purchases Gooselake from Tenneco	
1974	11-04	Cecil's house (the "old home place") burns to the ground (Wednesday)	1975—US withdraws from Vietnam
1977	5-12	James Tilton "J.T." born to Donna Allen & John Richard Tracy in Bakersfield	Jimmy Carter Pres 1977-81. Elizabeth II Silver Jubilee
1978	8-03	Cecil Foster Tracy dies; is interred in Union Cemetery, Bakersfield	
1979	9-08	John Bentley "JB" Selvidge weds Diane Marie Davis (div. 1988)	
1980	2-16	Allen Richard Tracy born to Donna Allen & John Richard Tracy	
1980	?	Roland L. Tilton (son of Nellie) dies; is interred in Union Cemetery, Bakersfield	
1981	5-14	James Tilton Tracy "J.T." dies at age 4 of leukemia; is interred in Wasco Cemetery	Ronald Reagan Pres 1981-89
1982	7-28	Elden J. Selvidge dies; is interred in Tracy Ranch memorial park	
1982	10-25	Jessica Christine born to Donna Allen & John Richard Tracy	
1984	9-15	Tracy Ann Selvidge marries Gary Gill (b. 10-24-45)	
1985	12-30	Casey Garrett Gill born to Tracy Ann Selvidge and Gary Gill	
1987	5-30	Michael Tilton Frey weds Jennifer Ann Goins (b. 9-13-65) in Fresno	
1987	9-12	Tracy Family celebrates its 125th anniversary of the ranch with a barbecue & party for 2500 of their friends. The grounds and some buildings around the original home place are restored for the event. Junk pile becomes a museum.	
1989	10-30	Marie Fine Tracy dies in Bakersfield, is interred in Union Cemetery, Bakersfield	George H. Bush Pres 1989-93
1990	5-25	Justin Michael born to Jennifer Goins & Michael Tilton Frey in Bakersfield	Desert Storm conflict, Dec '90—

			Jan '91
1991	5-04	Tracy Annette Brown marries Paul Michael Koll (b. 3-23-58) at home place	
1992	9-29	Robert Lee Tracy weds Kathy Lynn Twisselman (b. 1-12-70) at home place	
1992	12-18	Ian Michael born to Tracy Brown and Paul Michael Koll in Westlake Village, CA	
1993	1-12	Nicholas Daniel born to Jennifer Goins & Michael Tilton Frey in Bakersfield	Bill Clinton Pres 1993-2001
1993	10-9	Carla Anne Seidel weds Edward De Fraine in Nashville, TN	
1994	6-18	Surprise party for Jack Myron Frey's 80th birthday. He receives yet another yellow pick-up.	
1994	8-27	Jack Randolph Frey weds Kelie Marie Wilson (b. 4-10-68) at home place	
1995	9-09	Nancy Frey Stacy weds Randall "Randy" Earnest Trahan at Martha & Jack's on the ranch	
1995	9-26	Stephanie Tracy and Kevin O'Callaghan (b. 2-18-71) wed in Bakersfield; reception follows at home place	
1996	2-01	Evan Anders born to Carla & Edward De Fraine in Atlanta, GA	
1996		Claudia Lee and Larry Frey end their marriage	
1996	9-11	Haylee Marie born to Kathy Twisselman & Robert Lee Tracy in Bakersfield	
1996	9-17	Mark James Tracy weds Allison Gabriella Sobocinski (b.4-30-73) in Bakersfield	
1997	7-18	Kathryn Pollard Tracy passes away in Bakersfield, CA; ashes at Tracy Memorial Garden	
1998	1-28	Wade Wilson born to Kelie Wilson and Jack Randolph Frey	
1998		Lawrence Parker dies; interred at Union Cemetery, Bakersfield	
1998	9-23	Kathryn Alyse born to Kathy Twisselman & Robert Lee Tracy in Bakersfield, CA	
1999	2-24	Madeline Kinsey born to Kelie Wilson and Jack Randolph Frey in Bakersfield	
1999	3-31	Grace Ann born to Stephanie Tracy and Kevin O'Callaghan in Placerville, CA	
2000	2-14	Kenneth Charles Holden weds Maria Skytta in Oulu, Finland	
2000	2-18	Trent Jackson born to Kathy Lynn	

		Twisselman & Robert Lee Tracy in Bakersfield, CA	
2000	10-01	Steve Frey weds Melissa Midzor in Orange Co, CA	
2001	1-17	Magdelyn Kae born to Stephanie Tracy & Kevin O'Callaghan in Placerville, CA	George W. Bush Pres 2001-
2001	2-4	Isabel Celeste born to Allison Sobocinski & Mark Tracy in Mt. Laurel, NJ	
2001	5-7	Aislin Isabella De Fraine born in Guatemala City; parented by Carla & Ed De Fraine.	
2001	9-11	America enters a new war—and a new world	Terrorists destroy the World Trade Center in New York
2001	9-15	Todd Tracy weds Annette Clarisse Easton (b. 9-17-71) in Bakersfield, CA. Reception on Tracy ranch	
2001	11-17	Rebecca Lake (b. 9-7-52) marries Larry Frey in Las Vegas, NV	
2002	1-21	Jack M. Frey dies at home in Buttonwillow, age 87; ashes at Tracy Memorial Garden	US engaged in Afghanistan in War on Terror
2002	4-9	Martha Frey at last realizes her dream of living "in town" and moves to the Carriage House residence in Bakersfield. Jack & Kelie Frey purchase her home on the ranch.	
2002	2-2	Effie Frey Parker dies in Paso Robles, CA, 12 days after her younger brother Jack. She is survived by two sons, John and Robert.	
2002	4-18	After 5½ years of work collaboration, Martha Frey and daughter Nancy begin transmission of ***Portrait of a Western Family*** to the publisher	
2003	1-15	Clarisse Ann born to Annette Clarisse Easton and Todd Blake Tracy